Advances in Understanding Strat

Advances in Understanding Strategic Behaviour

Game Theory, Experiments and Bounded Rationality

Essays in Honour of Werner Güth

Edited by Steffen Huck

palgrave
macmillan

First published 2004 by
PALGRAVE MACMILLAN
Houndmills, Basingstoke, Hampshire RG21 6XS and
175 Fifth Avenue, New York, N.Y. 10010
Companies and representatives throughout the world

PALGRAVE MACMILLAN is the global academic imprint of the Palgrave
Macmillan division of St. Martin's Press, LLC and of Palgrave Macmillan Ltd.
Macmillan® is a registered trademark in the United States, United Kingdom
and other countries. Palgrave is a registered trademark in the European
Union and other countries.

ISBN 1–4039–4152–1 hardback
ISBN 1–4039–4167–X paperback

This book is printed on paper suitable for recycling and made from fully
managed and sustained forest sources.

A catalogue record for this book is available from the British Library.

Library of Congress Cataloging-in-Publication Data
Advances in understanding strategic behaviour:game theory, experiments,
    and bounded rationality/edited by Steffen Huck.
        p. cm.
Includes bibliographical references and index.
ISBN 1–4039–4152–1 — ISBN 1–4039–4167–X (pbk.)
1. Game theory.   2. Decision making.   3. Econometric models.
I. Güth, Werner, 1944–   II. Huck, Steffen, 1968–

HB144.A38 2004
519.3—dc22

                                                          2004056082

10   9   8   7   6   5   4   3   2   1
13 12 11 10 09 08 07 06 05 04

Printed and bound in Great Britain by
Antony Rowe Ltd, Chippenham and Eastbourne

Contents

v

List of Figures

List of Tables

Preface

Between Rationality and a Hard Place: In Honour of Werner Güth

The first ultimatum game, conducted by Werner Güth and collaborators in the late 1970s in Cologne, marks a crucial point in the history of modern economics – the point at which game theory lost its innocence. Suddenly, there was a chasm – between the beauty and elegance of the theory of games on the one hand, and the dour facts of behaviour observed in games on the other. In between: a void.

In the decades that followed, the economics literature slowly started filling this gap. Evolutionary game theory, preference evolution, learning models, models of bounded rationality with and without optimization, and, of course, more and more systematic experimental evidence – all these approaches flourished, competing with as well as complementing each other. Yet, compared to other fields, progress has not been particularly fast and the task before us is still immense. The task is, of course, to *understand* behaviour, and understanding is not easy. Life between hyperrationality on the one hand and simple facts on the other is not easy. There are no fixed axioms, and there is no fast-track methodology. There are many ways that can and need to be tried. Unavoidably, most of these are arduous, and some turn out to be dead ends. Looking for new routes that fill the void needs an adventurous spirit, fearlessness, creativity and stubbornness.

Werner Güth is among the few who have been endowed with plenty of all these qualities, and he used them to great effect. His journey since the 1970s has been nothing but amazing. To stick to the metaphor, he has not only discovered many new ways and broadened others but has also invented new techniques for finding and building new routes. In particular, his work on preference evolution, the 'indirect evolutionary approach' he first studied with Menahem Yaari, appears to be to Economics what the Golden Gate Bridge is to San Francisco: beautiful and useful in equal measure.

This volume collects sixteen articles written in honour of Werner Güth. In their variety they reflect the spectrum of approaches towards understanding strategic behaviour that, at least for now, necessarily coexist. Each is inspiring on its own, and together they illustrate what life is like between rationality and a hard place – there are plenty of new insights but also contradictions. And this is what makes this volume exciting.

The volume begins with Reinhard Selten's model of qualitative reasoning on comparitive statics, a bounded-rationality approach that helps us to understand and model chains of economic arguments. Chapter 2, by Tilman Börgers and Antonio J. Morales, also presents a model of bounded rationality, though one that is build on optimization: agents use a learning algorithm that, subject to a complexity constraint, maximizes expected utility. Chapter 3, by Daniel Friedman and Nirvikar Singh, employs Güth's indirect evolutionary approach (which mixes elements of 'orthodox' game theory and 'blind' evolutionary dynamics) to study how vengefulness may have evolved in small groups. Chapters 4 and 5 move even further towards orthodox theory. In Chapter 4, Siegfried K. Berninghaus and Bodo Vogt study endogenous network formation in co-ordination problems. It is shown how different social structures emerge, depending on intricate aspects of the payoff structure of the underlying game. In Chapter 5, Helmut Bester uses the theory of signalling in games (which requires comparatively sophisticated Bayesian updating) to study why job rationing may occur when firms provide specific training. This sequence of chapters – moving towards ever stronger rationality assumptions – then leads to Jörgen Weibull's devilishly provocative Chapter 6, in which he questions whether conventional economic experiments have ever really tested game theory. This is a contribution with a pronounced methodological flavour, and so are the following two chapters. In Chapter 7, Hartmut Kliemt and Axel Ockenfels present a dialogue between an orthodox economist, an economic psychologist, and a second economist who is an adherent of bounded rationality. The reader might guess how much agreement prevails between the three. Chapter 8, by Colin F. Camerer, Teck-Hua Ho and Juin Kuan Chong, is the longest and broadest in this volume. It presents an up-to-date overview of 'behavioural game theory', coupling an index of players' bounded rationality (their steps of thinking) with a learning model. It provides many applications within a very rigorous framework.

The second half of the volume is devoted to new experimental evidence. In Chapter 9, John H. Kagel studies double auctions with stochastic supply and demand schedules. Interestingly, he finds that (hyperrational) Bayesian Nash equilibrium models can organize his data better than non-strategic models. In Chapter 10, James E. Parco, Amnon Rapoport, Darryl A. Seale, William E. Stein and Rami Zwick study bilateral bargaining in multistage games. Again, standard equilibrium predictions work well, in particular for experienced players. Bargaining in the presence of an arbitration mechanism is analysed in Chapter 11 by Gary E. Bolton and Elena Katok. Studying a more complex bargaining problem than earlier papers, they show how arbitration can help to reduce conflicts in the long run by way of simplyfying the complexity of the task (which highlights a dimension of mechanism design that is neglected in orthodox approaches). In Chapter 12, Juan-Camilo Cardenas, T. K. Ahn and Elinor Ostrom study the effects of communication in common-pool resource dilemmas. Rather than being carried out in the laboratory, their experiments were conducted with villagers in rural Colombia, which offers many interesting insights that could not otherwise have been gained. Chapter 13, by Martin Dufwenberg, Uri Gneezy and Aldo Rustichini, analyses an important dimension of human life that is typically defined away in economic theory: the role of gender. Specifically, they study gender differences in prize competitions (where they appear negligible) and in other forms of competition, where success can reveal 'talent' (and where they are significant). Chapter 14 contains, somewhat unusually, a contribution by the man this volume honours. Werner Güth and Menahem E. Yaari started a joint project on parity, sympathy and reciprocity more than a decade ago; the second author has now finally completed it (surprising the first author). It is a startling illustration of both authors' creativity and far-sightedness that the chapter reads extremely freshly despite the long publication lag. Chapter 15 brings together (for the first time) three Güth pupils, Manfred Königstein and Wieland Müller as well as myself. Studying Stackelberg games, they show that deviations from orthodox predictions are caused not only by subjects who dislike unfair outcomes but also by others who actually enjoy inequality. Finally, Chapter 16 contains an interview with Alvin E. Roth, going back to the days just after the first ultimatum game experiment and sketching a history of continual learning from it until the present day.

STEFFEN HUCK

Acknowledgements

The editor wishes to thank Gian Luigi Albano, Iris Bohnet, Dirk Engelmann, Armin Falk, Antonio Guarino, Ed Hopkins, Philippe Jehiel, Michael Kosfeld, Dorthea Kübler, Hans-Theo Normann, Jörg Oechssler, Ylva Sovik, and Georg Weizsäcker, who all helped to improve this volume. Further thanks are due to Heike Harmgart and Rosie Mortimer, for patient help whenever it was needed.

S. H.

List of Contributors

T. K. Ahn is Assistant Professor of Political Science, Florida State University.

Helmut Bester is Professor of Economics, Free University of Berlin.

Siegfried K. Berninghaus is Professor of Economics, University of Karlsruhe.

Gary E. Bolton is Professor of Managment Science, Smeal College of Business Administration, Penn State University.

Tilman Börgers is Professor of Economics and Director of the Centre for Economic Learning and Social Evolution (ELSE), University College London.

Colin F. Camerer is Rea A. and Lela G. Axline Professor of Business Economics, California Institute of Technology.

Juan-Camilo Cardenas is Professor of Economics, Universidad de Los Andes, Bogota.

Martin Dufwenberg is Eller Professor of Economics, University of Arizona.

Daniel Friedman is Professor of Economics, University of California Santa Cruz.

Uri Gneezy is Associate Professor of Behavioral Sciences, Graduate School of Business, University of Chicago.

Werner Güth is Professor of Economics and Director of the Strategic Interaction Group, Max Planck Institute for Research into Economic Systems.

Teck-Hua Ho is William Halford Jr. Family Professor of Marketing, Haas School of Business, University of California, Berkeley.

Steffen Huck is Professor of Economics, University College London.

John H. Kagel is University Chaired Professor of Economics, Ohio State University.

Elena Katok is Assistant Professor of Management Science, Smeal College of Business Administration, Penn State University.

Hartmut Kliemt is Professor of Philosophy, Gerhard Mercator University, Duisburg.

Manfred Königstein is Professor of Economics, University of Erfurt.

Juin Kuan Chong is Assistant Professor of Management, NUS Business School, National University of Singapore.

Antonio J. Morales is Associate Professor of Economics, University of Malaga.

Wieland Müller is Assistant Professor of Economics, Tilburg University.

Axel Ockenfels is Professor of Economics and Director of the Institute for Energy Economics (EWI), University of Cologne.

Elinor Ostrom is Arthur F. Bentley Professor of Political Science and Co-Director of the Workshop in Political Theory and Policy Analysis and of the Center for the Study of Institutions, Population, and Environmental Change, Indiana University.

James E. Parco is Assistant Professor of Management and Major of the US Air Force, United States Air Force Academy.

Amnon Rapoport is Karl Eller Professor of Management and Policy, University of Arizona.

Alvin E. Roth is George Gund Professor of Economics and Business Administration, Harvard University.

Aldo Rustichini is Professor of Economics, University of Minnesota.

Darryl A. Seale is Associate Professor of Management, University of Nevada.

Reinhard Selten is Nobel Laureate and Professor emeritus, University of Bonn.

Nirvikar Singh is Professor of Economics, University of California, Santa Cruz.

William E. Stein is Associate Professor of Operations Management, Texas A&M University.

Bodo Vogt is Privatdozent, University of Magdeburg.

Jörgen W. Weibull is Professor of Economics, Boston University.

Menahem E. Yaari is Professor of Mathematical Economics, Hebrew University, Jerusalem.

Rami Zwick is Professor of Marketing and Director of the Center for Experimental Business Research, Hong Kong University of Science and Technology.

1

Boundedly Rational Qualitative Reasoning on Comparative Statics

Reinhard Selten

Introduction

The purpose of this chapter is to throw light on the structure of verbal economic reasoning as we find it in newspaper articles on questions of economic policy, or in the non-mathematical economic literature. Presumably, the thinking of practical decision-makers in business and public administration has a similar structure. A better understanding of this structure seems to be important for the development of a realistic theory of boundedly rational economic behaviour.

In verbal economic reasoning one often finds statements of the following kind: 'An increase of X causes an increase of Y' or 'An increase of X causes a decrease of Y'. Such descriptions of causal relationships are *qualitative* in the sense that they concentrate on directions of change; nothing is said on the quantitative strength of the effect.

Qualitative reasoning proceeds directly from qualitative assumptions to qualitative conclusions. In this connection, it is necessary to emphasize the word 'directly'. Mathematical modelling may also pursue the goal to reach qualitative conclusions on the basis of quantitative assumptions. Sometimes theorists do not want to determine more than the sign of partial derivatives with respect to parameters. Qualitative reasoning avoids the intermediate step of formulating a quantitative model.

A Bayesian decision-maker has no use for qualitative reasoning. In order to find the decision parameters that maximize expected utility, one needs a quantitative model and a subjective probability distribution over its unknown structural parameters. Purely qualitative information is usually insufficient. Nevertheless, even mathematical theorists aim at qualitative conclusions. This suggests that such conclusions are valuable for decision-makers in view of their boundedly rational decision procedures. A boundedly rational decision-maker may be interested primarily in which direction a decision parameter should be adjusted. After the answer to this qualitative

1

question has been obtained, the parameter can be adjusted cautiously in the indicated direction.

Qualitative reasoning does not mean that quantitative information is ignored completely. The selection of those causal relationships that are taken seriously may be guided by quantitative information. Thus the influence of some variables may be neglected because of their quantitative insignificance. In fact, statistical figures are often mentioned in verbal discussions but mainly in order to argue that something is important or unimportant. Only rarely are such numbers combined and manipulated by arithmetical operations.

In this chapter, attention is concentrated on qualitative reasoning about comparative statics. Qualitative reasoning can also be applied to dynamic problems, but not much will be said about this here.

In the psychological literature, the idea has been proposed that human reasoning is based on mental models (Gentner, 1983; Johnson-Laird, 1983). In the next section, the concept of the causal diagram will be introduced. Causal diagrams can be looked upon as mental models underlying qualitative reasoning on comparative statics. They can also be described as belief systems composed of simple qualitative statements.

The view of qualitative reasoning proposed here will be exemplified by the *Newsweek* article 'Saving, Not the American Way' by Rich Thomas (*Newsweek*, 8 January 1990, pp. 42–3).

The concept of a causal diagram

Consider a policy question such as this: what happens if the income tax rate is increased? In order to answer such questions by qualitative reasoning, one needs a qualitative belief system. In this chapter, we shall restrict our attention to qualitative belief systems formed by a finite set of simple statements of the form (a) or (b):

(a) *Ceteris paribus* a change of x causes a change of y in the same direction; and
(b) *Ceteris paribus* a change of x causes a change of y in the opposite direction.

In the case of (a) we speak of a *positive influence* of x on y, and in the case of (b) of a *negative influence* of x on y.

A qualitative belief system composed of a finite set of statements of the form (a) or (b) can be described by a *causal diagram*. Formally, a *causal diagram* is a finite signed and directed graph with some additional properties that are specified later. The full definition will be given after the introduction of some auxiliary notions.

The nodes $x, y, \ldots$ of a causal diagram correspond to variables, and a directed link from x to y represents a causal influence. The influence is positive if the sign is '+' and negative if it is '−'.

A *causal chain* from x_1 to x_n is a sequence of at least two nodes $x_1, \ldots, x_n$ together with $n-1$ links from x_i to x_{i+1} for $i = 1, \ldots, n-1$. A causal chain is called *positive* if the number of its links with a negative sign is even, and it is called *negative* if this number is odd. A *loop* is a causal chain with the property that $x_1 = x_n$ holds for its nodes $x_1, \ldots, x_n$.

Qualitative reasoning on the basis of a causal diagram is little more than the evaluation of signs of causal chains. We say that x exerts an *indirect causal influence* on y if there is at least one causal chain from x to y. This influence is *positive* if all causal chains from x to y are positive, and *negative* if all these chains are negative. It may also happen that the diagram exhibits positive as well as negative causal chains from x to y. In this case, we say that x has an *indefinite* influence on y. An indefinite influence does not justify a conclusion of the form (a) or (b). Positive and negative causal influences are *definite*. A causal diagram is called *balanced* if all indirect causal influences are definite.

A causal diagram need not be balanced. However, indefinite indirect influences do not provide answers to qualitative policy questions. Therefore, one can expect that belief structures described by causal diagrams are formed in a way that avoids indefiniteness as far as possible, at least where it matters for decision-makers. This works in the direction towards balance.

The graph structure of a causal diagram has to be complemented by additional information about the strategic possibilities and the motivation of the decision-maker. In order to describe the strategic possibilities, a subset of the nodes of the causal diagram is specified as a *set of instruments*. An instrument represents a variable controlled by the decision-maker. An instrument x must have the property that no direct causal influence is exerted on x.

In order to describe the motivation, a subset of nodes must be specified as a *set of goals*. Goals are interpreted as variables whose increase is valued by the decision-maker. In this chapter, attention will be concentrated on the case of only one goal. In the presence of several goals, one would have to answer the question of how goal conflicts are solved in qualitative decision-making. One could try to do this with the help of aspiration adaptation theory (see Selten, 1998). No attempt to do this will be made here, since our subject matter is qualitative reasoning rather than decision-making.

We now present a formal definition of a causal diagram: a *causal diagram* is a finite signed and directed graph together with two non-empty subsets of nodes, a set of *instruments* and a set of *goals*. The following conditions must be satisfied:

(a) There is at most one link from one node to another;
(b) There are no loops;
(c) An instrument is a node x such that there is no link from a node y to x (however, not all such nodes must be instruments); and
(d) For every instrument x there is at least one causal chain to a goal y.

Condition (a) means that the direct *ceteris paribus* influence of one variable on another is specified unambiguously. The absence of loops required by (b) is a property that has been observed by Axelrod (1976) in his studies on qualitative reasoning in *Structure of Decision*. Axelrod does not look at causal diagrams but rather at undirected signed graphs which are used to describe opinions expressed by speeches of politicians. The use of such graph structures for the purpose of representing opinion structures goes back to the paper entitled 'Symbolic Psycho-Logic' by Abelson and Rosenberg (1958). Perhaps the most striking result of Axelrod was the absence of loops. The causal diagram does not just represent positive and negative connections, but also expresses causal relationships through the directions of links. An instrument is thought of as a variable controlled by the decision-maker. It is therefore natural to require that no causal influences other than the will of the decision-maker are exerted on an instrument.

In my paper entitled 'Investitionsverhalten im Oligopolexperiment' (Selten, 1967), the definition of a causal diagram was further restricted by the condition that there should be no influences on other variables exerted by goals. This condition seems to be unnecessarily restrictive and is therefore not a part of the definition proposed here.

An instrument without any indirect influence on any goal would be irrelevant for decisions on policy questions. Therefore, condition (d), which was absent from Selten (1967) is included in the definition of the causal diagram. One might want to require the stronger property that every instrument indirectly influences every goal, but this would be unnecessarily restrictive. In fact, a study by Williamson and Wearing (1996) indicates that economic opinions of lay people are often described by diagrams with several components without any connection to one other.

Williamson and Wearing investigated the economic opinions of many lay people in Australia, with very interesting results. In their highly remarkable paper they describe 'cognitive models' similar to causal diagrams, but also different in important respects. Thus, not only variables but also 'needs', such as for example, 'government should encourage Australians to work' are represented as nodes of the graph. Links may express relatedness or unrelatedness as well as positive or negative causal connections. In this way, Williamson and Wearing obtain a close agreement with the verbal statements of their subjects. The concept of a causal diagram restricts itself to positive and negative causal links between variables. In this way, one obtains clear qualitative descriptions of the perceived causal structure. The causal diagram is a mental model about the underlying reality rather than a detailed elaboration of all opinions, including evaluations and action recommendations.

The basic idea of a causal diagram is very simple and I would not be surprised to be told that it can be found in the literature before 1967. In the newer literature, sometimes explicit use of a causal diagram is made – for example, in the book by Frederic Vester (1990) *Ausfahrt Zukunft*. However,

Vester adds an additional feature to the causal diagram that is not really qualitative. Each link has a degree of strength. There are finitely many levels of this degree of strength.

The fact that there seems to be a widespread use of causal diagrams and similar graph structures suggests that it is a natural tool for the description of qualitative causal reasoning. It is maybe necessary to pay more attention to this formal structure even if it is a very simple one.

The causal diagram of a *Newsweek* article

Figure 1.1 shows a causal diagram constructed on the basis of a *Newsweek* article by Rich Thomas (8 January 1990) with the title 'Saving, Not the American Way'. Before this diagram can be explained in detail it is necessary to say something about the structure of this article. First, it is shown with the help of statistical figures that the US savings rate is very low in comparison to that in Germany and Japan. In view of this situation, it had been proposed to create new tax incentives for saving.

In the diagram, the only goal variable considered is growth, even though the word 'growth' is not explicitly mentioned in the text. However, after the discussion of the high savings rate in Germany and Japan, we find the following remark: 'Has this saving led to better living standards? In Germany, incomes have increased over the past 20 years, inflation is negligible and the Deutsche Mark has become Europe's most powerful currency. Japan, too, has emerged as an economic superpower.' Since living standards are connected to growth, it seems to be justified to look at growth as the goal variable. The text focuses on investment rather than growth. But it seems to be assumed that an increase of investment leads to a higher growth rate. Therefore, in the diagram, the influence of investment on growth is positive.

In the diagram, some of the influences are indicated by lines with solid black arrowheads. The variables tax incentives for saving, savings, funds available for investment, interest rate, investment and growth, as well as the lines with solid black arrowheads reflect the main argument for a positive indirect influence of tax incentives for savings on growth. We refer to this part of the diagram as the *main diagram*.

The lines with white arrowheads represent other influences and critical objections to the main argument. In most cases these influences involve variables outside the main diagram.

A '+' or '−' on a line represents a direct influence and indicates whether this influence is positive or negative. Numbers indicate quotations that support the relevant influence. These quotations are listed below the diagram in Figure 1.1.

The main diagram has two positive causal chains from the instrument variable 'tax incentives for saving' to the goal variable 'growth'. Obviously, the main diagram is balanced. If the causal chain from savings over

6

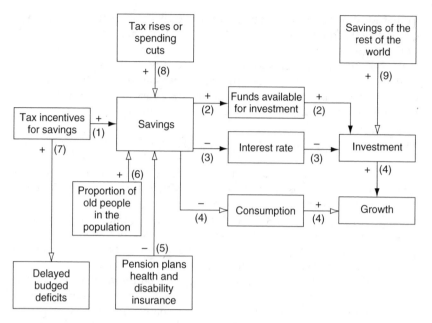

Quotations from the text of the article

(1) Tax law which generally benefits borrowing far more than saving and is in part responsible (for the decline in savings) . . . Most of the coming proposals assume that tax policy can motivate people to save more.

(2) Savings regardless where deposited act as a pool of money that businesses can tap for new plants, equipment and employees.

(3) Greater savings keep interest rates lower, making it cheaper to borrow in the United States.

(4) Because of these advantages [(2) and (3)] most economists dismiss the commonly held view that curbing consumption would drag the economy into a recession.

(5) In addition, the slide in savings is 'a byproduct of our social progress' says Harvard economics Professor Lawrence Summers. The spread of pension plans, health and disability insurance and large increases in social security benefits have vastly reduced the need of many Americans to save.

(6) Edward Yardeni of Prudential-Bache Securities Inc. says the aging population will push the savings rate above 10 percent.

(7) Critics say the plan would eventually cost the treasury billions of dollars. 'What you're really doing is planting a lot of delayed budget deficits . . . says Henry Aaron of the Brookings Institution'.

(8) Critics like Aaron believe that tax incentives are a poor way to attack the low savings rate in any case. They argue that balancing the budget – a tax rise or spending cuts – would accomplish the same aims.

(9) John Makin of the American Enterprise Institute says that the rest of the world is saving so much and investing it in this country that the U.S. savings rate isn't as relevant as before.

Figure 1.1 Causal diagram of the *Newsweek* article and quotations from it

consumption on growth were not to be dismissed by quotation (4), the balance of the main diagram would be destroyed, since tax incentives would then have a negative influence on growth via the causal chain involving consumption.

We now discuss the influences of the causal diagram following the numbering of the quotations. Tax incentives for savings supposedly have a positive influence on savings. This is clearly indicated by quotation (1). The causal chain from 'savings' over 'funds available for investment' to 'investment' reflects quotation (2). The causal chain 'savings' over the 'interest rate' to 'investment' is suggested by quotation (3), even if the second influence from the interest rate to investment is not mentioned explicitly. However, the words 'making it cheaper to borrow in the United States' suggest that more would be borrowed and invested if interest rates were lower. We have already discussed the influence of investment on growth, which is implied by the text but not expressed explicitly. Quotation (4) also gives a hint about growth as a goal because it is seen as a bad side effect that curbing consumption would drag the economy into a recession.

After having discussed the main diagram and the dismissed causal chain from 'savings' over 'consumption' to 'growth', we now turn our attention to the remaining causal influences.

The negative influence of 'pension plans etc.' on 'savings' is offered as an explanation for a lower savings rate compared to a time in the past when there was less social security. Of course, this explanation is dubious in view of the fact that the social security system of Germany is by no means less well developed than that of the United States.

One of the critical remarks reported in the article was the objection that tax incentives for savings create delayed budget deficits. This means that attention is directed to future budget balances as a second goal neglected by the main argument. Another objection is based on the opinion that the growing proportion of older people will lead to an increased savings rate, and that therefore tax incentives are not necessary. A third objection recommends a tax rise and a spending cut instead of tax incentives in order to raise savings, and is understood as including not only private but also public savings or dissavings. A fourth criticism points to the importance of foreign savings for investment and denies the significant influence of domestic savings. The positive influence of consumption on growth is also mentioned as a possible counter-argument. The balance of the main diagram would be destroyed by the inclusion of this influence.

Concluding remarks

The article by Rich Thomas discussed above is very well reasoned and therefore naturally lends itself to description by a causal diagram. Unfortunately, not every journalist writes that well. In many newspaper articles it is hard to find any reasoning to work with. However, the author has found it useful to translate discussions of policy questions into causal diagrams, and has taught this technique in courses on bounded rationality. The causal diagram formalizes the structure of qualitative reasoning about comparative statics.

Theories of qualitative reasoning about dynamical systems have been proposed in the artificial intelligence literature (Bobrow, 1985). The articles collected in the book by Bobrow concern physical systems, but the method can also be applied to economic systems. In fact, the old business cycle theories in vogue before mathematical models became popular were based almost exclusively on qualitative arguments. The methods described in Bobrow's book throw light on these theories. It would be interesting to say more about the effect of qualitative reasoning on dynamic economic systems, but this cannot be discussed here: the author hopes that he will be able to do this elsewhere.

References

Abelson, R. P. and M. J. Rosenberg (1958) 'Symbolic Psycho-Logic: A Model of Attitudinal Cognition', *Behavioral Science*, 3, 1–13.

Axelrod, Robert (1976) 'The Analysis of Cognitive Maps', in: Robert Axelrod (ed.), *Structure of Decision, The Cognitive Maps of Political Elites*, Princeton, NJ, Princeton University Press.

Bobrow, Daniel G. (ed.) (1985) *Qualitative Reasoning about Physical Systems*, Cambridge, Mass., MIT Press.

Gentner, D. and Gentner D. R. (1983) 'Flowing Waters or Teeming Crowds: Mental Models of Electricity', in D. Gentner and A. L. Stevens (eds), *Mental Models*, Hillsdale, NJ, Erlbaum, 99–129.

Johnson-Laird, P. (1983) *Mental Models*, Cambridge University Press.

Selten, Reinhard (1967) 'Investitionsverhalten im Oligopolexperiment', in H. Sauermann (ed.), *Beiträge zur experimentellen Wirtschaftsforschung*, Tübingen: J. C. B. Mohr (Paul Siebeck), 60–102.

Selten, Reinhard (1998) 'Aspiration Adaptation Theory', *Journal of Mathematical Psychology*, 42, 191–214.

Thomas, Rich (1990) 'Saving, Not the American Way', *Newsweek*, 8 January, 42–3.

Vester, Frederic (1990) '*Ausfahrt Zukunft – Strategien für den Verkehr von morgen. Eine Systemuntersuchung*', Munich, Wilhelm Heyne Verlag.

Williamson, Maureen R. and Wearing, Alexander (1996) 'Lay People's Cognitive Models of the Economy', *Journal of Economic Psychology*, 17, 3–38.

2
Complexity Constraints and Adaptive Learning: An Example

*Tilman Börgers and Antonio J. Morales**

Introduction

Experimental research suggests that it is very difficult for most people to make optimal decisions if their payoffs are affected by a sequence of random shocks that are independent and identically distributed (i.i.d.). Many subjects never learn to make optimal choices; others do, but take a very long time.

A simple experimental set-up, in which results of the type outlined in the previous paragraph have been obtained, is as follows: subjects repeatedly choose between two actions, say 'Left' or 'Right'. Payoffs can have only two values: one ('success'), or zero ('failure'). In each trial, one of the two choices is successful, and the other one fails. After each trial, subjects observe which choice was successful in that trial, and which choice was not successful. The probability that one of the two actions, say 'Left', is successful is $\pi < 0.5$, whereas the other action, say 'Right', has probability $1 - \pi > 0.5$ of being successful. Success in one trial is stochastically independent of success in any other trial. Subjects are not informed about the stochastic process that determines the success of different choices. Note that if this process were known, then expected payoff maximization would mean that the action with the higher likelihood of success would be chosen in each trial.

This binary choice experiment has been conducted by experimental psychologists for a very long time (see Myers, 1976, and Winter, 1982, for surveys). One of the most prominent findings in the older literature was that

* This chapter is offered as a tribute to Werner Güth on the occasion of his sixtieth birthday. The first author was Werner Güth's student as an undergraduate at the University of Cologne, and remains immensely grateful for the advice and guidance offered by Werner Güth at that time. We are grateful to Steffen Huck and an anonymous referee for their comments. Tilman Börgers' research was supported financially by the ESRC through the grant awarded to the Centre for Economic Learning and Social Evolution (ELSE) at University College London. Antonio J. Morales acknowledges financial support from MCYT and FEDER grant number BEC 2002-02852.

subjects *probability match* instead of optimize (Grant *et al.*, 1951). By this, the literature meant that, in the long run, the fraction of trials in which subjects chose the more successful action equalled the probability of success of that action, that is, $1 - \pi$ (using the above notation).

Whether probability matching is really a robust finding has been questioned regularly (see, for example, Edwards, 1961; Peterson and Ulehla, 1965; and Shanks *et al.*, 2002). These authors have found that subjects who are given enough time to learn, who receive sufficiently informative feedback, and who are incentivized with sufficiently large monetary rewards come closer to optimization than the theory of probability matching suggests.

The most recent experiments we know of in this context are those of Shanks *et al.* (2002). In one treatment, they gave twelve subjects the opportunity to make choices in the binary choice problem described above over 1,500 periods. They found that out of those twelve subjects, seven eventually optimized. These subjects took, however, a very long time to optimize. Typically, the subjects needed about 1,000 trials before settling for the best choice.[1]

While probability matching thus does not seem to be a robust experimental finding, Shanks *et al.*'s data also show that a more general point remains valid: subjects find it hard to make optimal choices in i.i.d. environments. Some subjects never do (in experimentally observed time frames); and others take a very long time.

In this chapter we are concerned with the question of what explains subjects' poor performance in the binary choice task with i.i.d. shocks. The theory we propose rests on the assumption that, when facing any learning task, people use a learning algorithm that is optimal in a suitable sense in some environments. This algorithm will be embedded in people's brains. Therefore, it will have been shaped by evolution. This makes our assumption of optimality of the learning algorithm not entirely implausible.

The most prominent model of optimal learning in economics is that of Bayesian learning. However, it is very hard to match observed learning behaviour with a Bayesian model. If people hold subjective prior beliefs about the nature of the stochastic process they are confronting, which attaches at least some positive probability to the possibility that this process is i.i.d., and if their prior beliefs are not biased in favour of either of the two alternatives, then Bayesian learning typically implies very fast convergence to the payoff maximizing action.

We therefore explore a new hypothesis. It is that subjects are using a learning algorithm that maximizes expected utility subject to a complexity constraint. We model the complexity constraint as the requirement that subjects need to use a learning algorithm that can be implemented by a stochastic automaton with a very small number of states.

Intuitively, we interpret the number of states as a measure of the amount of memory, or of attention, that the decision-maker allocates to the learning problem.

The formal setting we study is the learning task described above. As a first and crude formalization of our idea, we restrict attention to the extreme case that the decision-maker's prior places probability one on the event that the environment is i.i.d., and that all that is needed to learn is which of the two choices has the higher chance of success. Another extreme assumption we make is that the decision-maker wants to use an automaton with two states only. However, we do give the decision-maker some flexibility by allowing the use of an automaton with random transitions.

We calculate the optimal strategy of the decision-maker in this set-up, and find that it is very simple: in each period, the decision-maker acts as if what was successful yesterday will also be successful today. Thus, if 'Left' was successful in the last trial, then it will be chosen again in the next trial. If 'Right' was successful in the last trial, then the decision-maker will choose 'Right' in the next trial.

Notice that an implication of this algorithm is that, from period 2 onwards, the probability that the decision-maker will choose any particular action is equal to the probability that this action is successful. Thus our model predicts probability matching.

At this stage, however, our theory is not meant to fit the data in a quantitative sense. Although the particular model we have investigated predicts probability matching, we are aware that this is not what most subjects do in the long run in the experiments. A more general analysis of the role of complexity constraints, and a closer look at the data, are needed before we can draw any conclusions about the relevance of complexity constraints for experimentally observed learning behaviour in the binary choice experiment. Such an analysis needs to consider more general priors, as well as complexity constraints that are less tight than the ones assumed here. This chapter is meant as a tentative exploration.

The paper that is most closely related to our work is Winter (1982). Winter pointed out that limited memory decision-making may lead to probability matching or similar choice behaviour. However, Winter did not study the optimal use of a given amount of limited memory. He made exogenous assumptions about how a given limited memory would be used. By contrast, our work considers the optimal use of a given amount of memory. We show that optimal use of a very small amount of memory leads to probability matching.

Kalai and Solan (2003) have presented a general study of optimal finite stochastic automata for Markov decision problems. What we present here is close to being an example of Kalai and Solan's general framework, and we explain later in more detail the connection between our work and theirs. Schlag (2002) has also studied several desirable properties of simple learning algorithms in i.i.d. environments. However, he focuses on non-Bayesian criteria, whereas we use entirely orthodox Bayesian criteria to evaluate different algorithms.

This chapter is structured as follows. In the next section we describe the set-up for our analysis. In the third section we present our main results. The fourth section comments on the results, and the final two sections contain proofs.

Set-up

There is a single decision-maker. Time is discrete, and the time horizon is infinite, so that the time periods are: $t = 1, 2, 3, \ldots$. In every period t, the decision-maker chooses one of two actions which, for concreteness, we now label as: 'take an umbrella', or 'don't take an umbrella'. After the decision-maker has taken the decision, a random event occurs: it either rains, or it doesn't. The probability that it will rain is independent of the period, and the occurrence of rain in any one period is stochastically independent from the occurrence of rain in any other period. In other words, rain is an i.i.d event.

The decision-maker knows that rain is i.i.d., but does not know the probability of rain. This probability can be either 'low' or 'high'. If it is low, then it is equal to some probability $\pi \in (0, 0.5)$. If it is high, then it is equal to the complementary probability, that is, $1 - \pi$. This particular specification of the probabilities makes our calculations simple (see below). The prior probability the decision-maker attaches to the possibility that the probability of rain is low is 0.5, and the same probability is attached to the possibility that the probability of rain is high.

In each period, the decision-maker first chooses the action, and then observes whether it rains. S/he can thus learn about the probability of rain. Observe that this is not a bandit problem. The decision-maker's observation in each period is independent of his/her action, whereas in a bandit problem the decision-maker's observation would depend on his/her action.

The decision-maker's payoff in each period depends on his/her action and on the state of the world. Figure 2.1 shows actions in rows, states in columns, and each entry indicates the decision-maker's von Neumann Morgenstern utility if the row action is taken, and the column state occurs. The decision-maker's objective is to maximize the expected discounted value of his/her payoffs. S/he uses a discount factor $\delta \in (0, 1)$.

	Rain	No rain
Umbrella	1	0
No umbrella	0	1

Figure 2.1 The payoff matrix

Optimal strategies

We begin by describing the optimal strategy for the decision-maker if the complexity of the strategy is of no concern to that person. The following result follows from the simple calculations of conditional probabilities which we shall carry out in the proof of Proposition 1 below. The result is a consequence of the symmetry of our model with respect to actions.

PROPOSITION 1: *If there is no complexity constraint, then the optimal strategy of the decision-maker is as follows:*

(i) *In period 1 choose some arbitrary action.*
(ii) *In periods $t \geq 2$ count the number t_R of earlier periods in which it rained, and the number t_N of earlier periods in which it did not rain. If $t_R > t_N$, then take your umbrella. If $t_N > t_R$, then do not take your umbrella. If $t_R = t_N$, then take some arbitrary action.*

In other words, this strategy simply says that the decision-maker's updated probability of rain in any period is equal to the relative frequency of rain in the past. The decision-maker takes optimal actions with respect to these beliefs.

REMARK 1: What will this decision-maker's behaviour look like? Let us denote by e_t the *ex ante* probability that the decision-maker makes the objectively 'wrong' decision in period t – that is, that s/he chooses to take his/her umbrella even though the probability of rain is π, or that s/he leaves his/her umbrella even though the probability of rain is $1 - \pi$. In period 1, independent of the decision-maker's choice, $e_1 = 0.5$. In period 2, the decision-maker makes a wrong choice if and only if the weather in period 1 was of the type that is objectively less frequent. Hence, the decision-maker makes the wrong choice with probability $e_2 = \pi$. Clearly, as t increases, e_t decreases, because the decision-maker has more information. The weak law of large numbers, moreover, implies: $\lim_{t \to \infty} e_t = 0$. Finally, by the strong law of large numbers, with probability 1 there will be some random time $\tilde{t}$ such that, in all periods $t > \tilde{t}$, the decision-maker will make the 'right' decision.

REMARK 2: What is the decision-maker's expected utility? Denote by u_t his/her expected utility in period t, where expected values are taken from the *ex ante* perspective. Then:

$$u_t = (1 - e_t)(1 - \pi) + e_t \pi$$

Clearly, in period 1 we have: $u_t = 0.5$, independent of which action the decision-maker chooses. In period 2, expected utility is: $u_2 = (1 - \pi)^2 + \pi^2$. As e_t is decreasing in t, it follows that u_t is increasing in t. The weak law of large numbers implies that in the long run we have: $\lim_{t \to \infty} u_t = 1 - \pi$.

The present value of the decision-maker's expected utility will depend on the discount factor δ. But as $\delta \to 1$, the present value, if normalized in the standard way through multiplication by the factor $(1 - \delta)$, will converge to $1 - \pi$. On the other hand, as $\delta \to 0$, the present value will converge to 0.5.

A decision-maker who adopts the above, fully rational strategy has to keep track of the difference between the number of rainy days and the number of dry days. This difference can be any integer number. In this sense, it is required that there is an infinite number of possible states in which the memory of this decision-maker can be. For some decision-makers, this might demand more attention and memory space than they want to allocate to this problem. Such decision-makers might look for a simpler strategy. Motivated by this consideration, we shall next study a decision-maker who is only willing to choose strategies that can be implemented using a finite automaton.

We shall make the extreme assumption that the number of states of the automaton that implements the decision-maker's strategy has to be two. Note that, for a non-trivial analysis, the number of states must not be smaller than two, because each state will have to be assigned some (possibly mixed) action, and if there is only one state, then the decision-maker's behaviour has to be constant for ever. Thus we study the case in which the strategy has to be implemented by an automaton with the minimal number of states that allows a non-trivial learning process.

The two states of the automaton are 'umbrella' (U) and 'no umbrella' (N). If the decision-maker is in the state 'umbrella', then s/he takes his/her umbrella. If s/he is in the state 'no umbrella', then the umbrella is left at home.

One of the two states will be chosen as the initial state. We shall study below the optimal choice of the initial state. We shall also study the optimal choice of the transition rule. The transition rule determines, as a function of the current state (and of the weather that the decision maker experiences) what his/her state in the next period is going to be. We shall allow for stochastic transition rules – that is, rules where the next state is a random variable that depends only on the current state, and on the weather experienced in the current period.

Kalai and Solan (2003) have demonstrated that a finite automaton with *stochastic* transitions can sometimes achieve a larger expected utility than a finite automaton with *deterministic* transitions. They also demonstrate that nothing is gained if actions, conditional on the state, are allowed to be stochastic. Their set-up is different from ours in that they do not allow for discounting, but their observation that stochastic transitions may be superior, and that randomization conditional on the state is redundant, are also true in our context. With regard to the first point, we shall provide an example in the fourth section, below. As for the second point, Kalai and Solan's proof can easily be adapted to our context.

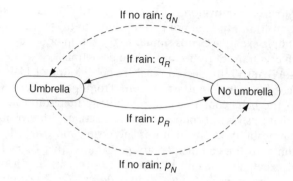

Figure 2.2 The two-state automaton

We now provide some formal notation for our two-state automaton. We shall denote the probability that the decision-maker changes his/her state, if the current state is 'umbrella', and if it has rained, by p_R. If the current state is 'umbrella', and it does not rain, then the probability that the decision-maker changes his/her state is p_N. In both cases, the decision-maker stays in the state 'umbrella' with the complementary probability. The exit probabilities from state 'no umbrella' are q_R (if it rains) and q_N (if it doesn't rain). Again, the state is left unchanged with the complementary probability. Figure 2.2 gives a simple graphical representation of the finite automata we are studying.

For a given finite automaton it is straightforward to calculate the decision-maker's expected utility. The next Proposition, which we shall prove in the final section of the chapter, describes the finite automaton for which this expected utility is maximized.

PROPOSITION 2: *The optimal transition probabilities are*

$$p_R = 0; \quad p_N = 1; \quad q_R = 1; \quad p_N = 0.$$

Any probability distribution over initial states is optimal.

This automaton implements the strategy according to which the decision-maker assumes that tomorrow's weather will be the same as yesterday's, and acts accordingly. Thus, the decision-maker has *adaptive expectations*.

REMARK 3: We emphasized above the importance of allowing for stochastic transitions. Proposition 2, however, says that *in our example*, it is optimal not to choose stochastic transitions. It is not at all trivial that in our example stochastic transitions turn out not to be optimal. In the fourth section below we give a simple example that is similar to the set-up of our chapter, but in which stochastic transitions do turn out to be optimal.

REMARK 4: What will this decision-maker's behaviour look like? In period 1 s/he will choose some arbitrary action. From period 2 onwards, s/he will exhibit the behaviour that the psychology literature calls *probability matching* – that is, s/he will choose the correct action with probability $1 - \pi$, that is, the probability with which it is successful, and s/he will choose the wrong action with the complementary probability, that is, π. To see this, suppose for example, that rain is infrequent – that is, that it occurs with probability π, and suppose that we are in period t. Then the probability that the decision-maker observed no rain in period $t-1$ is $1 - \pi$. Thus, in period t, s/he will take the correct action, with probability $1 - \pi$.

If we denote again by e_t the *ex ante* probability that the decision-maker makes the objectively 'wrong' decision, then we have $e_t = 0.5$, if $t = 1$; and $e_t = \pi$ if $t \geqslant 2$. Compare this with the performance of a decision-maker who adopts the strategy described in Proposition 1. In periods 1 and 2, this decision-maker will have the same error rates. But from period 3 onwards the fully rational decision-maker has a further declining error rate, which converges to zero as t tends to infinity, whereas the decision-maker in Proposition 2 has a constant error rate. He does not 'learn' any further.

REMARK 5: What is the decision-maker's expected utility? As before, denote expected utility in period t by u_t. Then: $u_t = 0.5$ if $t = 1$, and $u_t = (1 - \pi)^2 + \pi^2$ if $t \geqslant 2$. The present value of the decision-maker's expected utility will depend on the discount factor δ. But as $\delta \to 1$, the present value, if normalized in the standard way through multiplication by the factor $(1 - \delta)$, will converge to $(1 - \pi)^2 + \pi^2$. On the other hand, as $\delta \to 0$, the normalized present value will converge to 0.5.

In Figure 2.3 we plot the loss in expected utility that the decision-maker suffers if s/he uses a two-state automaton rather than the unconstrained optimal strategy. We pick three values of the discount factor δ, and then illustrate for these how the expected utility loss depends on the probability π.

Figure 2.3 shows that, for all three values of δ, the expected utility loss is small if π is close to 0. The intuitive reason is that, in this case it is easy to learn, and even the two-state automaton will almost always pick out the optimal action. Expected utility loss is also small if π is close to 0.5. In this case, learning is difficult, but it does not matter much, because the two actions yield almost the same expected utility. Correspondingly, the expected utility loss is largest for some interior value of π. Figure 2.3 also shows that the expected utility loss is increasing in δ. The intuition for this is that as δ increases, the future becomes more important, and therefore the incentive for accurate learning increases.

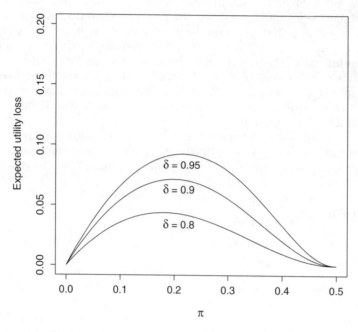

Figure 2.3   Expected utility loss because of the complexity constraint

Comments

Two important aspects of the automaton in Proposition 2 are, first, that the initial state does not matter, and second, that transitions are deterministic, rather than stochastic. We now give a very simple example of a setting in which the optimal automaton has neither of these properties. The purpose of this section is twofold. First, we seek to illustrate that the result of the previous section is somewhat surprising. Second, we wish to show that a fully developed theory of learning by boundedly rational agents will have to be more complicated than the example constructed in this chapter. The example we give below is generalized and analysed in more detail, in Börgers and Morales (2004).

EXAMPLE: There are two actions, *A* and *B*. The payoffs for each action are deterministic, but unknown. Each action's deterministic payoff can be either 0, 1 or 2. The decision-maker assigns probability 1/3 to each of these events. The payoffs of the two actions are stochastically independent.

Note that this example is a proper bandit problem. The decision-maker's observations depend on his/her actions. Suppose that the decision-maker uses an automaton with only two states, one for each action. Then the optimal automaton can easily be calculated to be the following (see Börgers and Morales, 2004): the automaton will select one of the two states as the initial state, say A. If the payoff is 0, the automaton will always switch states. If the payoff is 2, then the automaton will always stay in the current state. If the payoff is 1, and the current state is B, then the automaton will always stay in that state. The reason is that the automaton can be in state B only if payoff in state A was either 0 or 1. Thus, if in state B the payoff 1 is received, it is not worth switching back to A. Note how the automaton 'remembers' that it chose A first. It does not matter what the initial state is, but it does matter that it is not random.

The most interesting feature of the optimal automaton arises if payoff 1 is received in state A. Then the optimal transition to B is potentially stochastic. A simple calculation shows that the optimality probability with which the decision-maker switches to B is then:

$$p = \min\left\{1, \sqrt{\frac{1-\delta}{\delta^2}}\right\}$$

For δ close to 1 the decision-maker will thus experiment when s/he is in state A and receives payoff 1.

Intuitively, if the decision-maker receives payoff 1 when s/he first plays A, and if s/he is very patient, then s/he has an incentive to experiment with action B to find out whether B perhaps yields payoff 2. If B yields payoff 1 or 2, then the decision-maker would like to stay with B, but if it yields payoff 0 the decision-maker would wish to return to A. The problem is that state A is used both for the initial choice, and for the choice of A if the decision-maker has been disappointed by B. Thus, in state A, the decision-maker does not 'remember' whether s/he has already experimented with B or not. The randomization ensures that some experimentation occurs if the decision-maker plays A for the first time and receives 1, but that, on the other hand, A is played at least for some proportion of the time if B yields payoff 0.

Notice that the decision-maker, as described in the previous paragraph, suffers from 'absent-mindedness' in the sense of Piccione and Rubinstein (1997). Our result is therefore reminiscent of their finding that, with absent-mindedness randomization may be strictly optimal. In our example, in contrast to theirs, the absent-mindedness arises endogenously.

This example indicates that our results generalize only to a limited extent. One might conjecture that both of the important features of our optimal automaton – that is, that the initial state is irrelevant, and that transitions are non-stochastic, survive in settings in which there are only two possible payoff levels. We have, however, not proved this conjecture.

Proof of Proposition 1

Denote the decision-maker's subjective probability that the frequency of rain is low – that is, that it occurs with probability π, at the beginning of period t by ρ_t. Consider the evolution of the decision-maker's beliefs. If the decision-maker observes rain in period t, then his/her subjective probability that rain is infrequent is in the next period:

$$\rho_{t+1} = \frac{\rho_t \pi}{\rho_t \pi + (1-\rho_t)(1-\pi)}$$

If s/he observes no rain in period t, then his/her subjective probability that rain is infrequent is in the next period:

$$\rho_{t+1} = \frac{\rho_t(1-\pi)}{\rho_t(1-\pi)+(1-\rho_t)\pi}$$

Suppose the decision-maker observed rain in period t, and then no rain in period $t+1$. Using the above two equations, one can easily verify that the decision-maker's beliefs in period $t+2$ will be the same as they were in period t:

$$\rho_{t+2} = \rho_t$$

The same holds if s/he observes first no rain, and then rain.

Now suppose that the decision-maker starts out with beliefs $\rho_1 = 0.5$, and observes a sequence of periods of rain (R) and no rain (N), such as the sequence 'RRNNRNNNRN'. Note that any such sequence, as long as it does not contain only Rs or only Ns, must contain a pair of adjacent periods such that it rained in one of those periods, but not in the other. Our calculations above imply that the final beliefs of the decision-maker are the same, independent of whether these two adjacent periods are present in the sequence, or whether they are taken out. For example, in the above sequence we can eliminate the second entry, R, and the third entry, N, without affecting the final beliefs of the decision-maker: 'RNRNNNRN'.

A finite number of steps of the type described in the previous paragraph are enough to reduce any finite sequence of Rs and Ns which either consists only of Rs, or only of Ns, or which consists of no entries at all. If the final sequence consists only of Rs, then the original sequence had more periods of rain than of no rain, and the length of the final sequence is simply the difference between the number of rainy periods and the number of periods without rain in the original sequence. The analogous statement is true if the final sequence consists only of Ns. If the final sequence has no entries at all, then the original sequence had an equal number of rainy and dry periods.

If the reduced sequence has only Rs, then the subjective probability of rain, after Bayesian updating of the beliefs given the observations available, will be more than 0.5. If it has only Ns, then the subjective probability of rain will be less than 0.5. Finally, if it has no entries at all, then the subjective probability of rain will be the same as it was initially, that is, 0.5.

Because choices do not affect what the decision-maker learns, the decision-maker should choose in every period t whatever is optimal according to ρ_t.

Thus, any choice is optimal if $\rho_t = 0.5$. If $\rho_t < 0.5$, then it will be optimal to bring the umbrella, and if $\rho_t > 0.5$, it will be optimal to leave the umbrella at home. We have thus proved Proposition 1.

Proof of Proposition 2

Proof strategy

We shall prove that the transition probabilities indicated in Proposition 2 are optimal if it is assumed that the initial state is U. By symmetry, it then follows that these transition probabilities are also optimal if it is assumed that the initial state is N. Moreover, symmetry also implies that the maximal achievable utility is the same for both initial states. Thus, we can deduce that with optimally chosen transition probabilities, every probability distribution over initial states is optimal, and the proposition is proven.

Calculating the expected utility

Denote by V_U^π (resp. V_R^π) the decision-maker's expected utility if s/he is in state U (resp. R) and the true probability of rain is π. Define $V_U^{1-\pi}$ and $V_R^{1-\pi}$ analogously. If the true probability of rain is π, then the expected utilities satisfy:

$$V_U^\pi = \pi(1 + p_R\delta V_N^\pi + (1-p_R)\delta V_U^\pi) + (1-\pi)(0 + p_N\delta V_N^\pi + (1-p_N)\delta V_U^\pi)$$
$$V_N^\pi = \pi(0 + q_R\delta V_U^\pi + (1-q_R)\delta V_N^\pi) + (1-\pi)(1 + q_N\delta V_U^\pi + (1-q_N)\delta V_N^\pi)$$

The solution of these two equations is:

$$V_U^\pi = \frac{1}{1-\delta}\frac{(1-\delta)\pi + \delta(\pi(1-\pi)p_R + (1-\pi)^2 p_N + \pi^2 q_R + \pi(1-\pi)q_N)}{1-\delta+\delta(\pi p_R + (1-\pi)p_N + \pi q_R + (1-\pi)q_N)}$$

$$V_N^\pi = \frac{1}{1-\delta}\frac{(1-\delta)(1-\pi) + \delta(\pi(1-\pi)p_R + (1-\pi)^2 p_N + \pi^2 q_R + \pi(1-\pi)q_N)}{1-\delta+\delta(\pi p_R + (1-\pi)p_N + \pi q_R + (1-\pi)q_N)}$$

If the true probability of rain is $1 - \pi$, then the expected utilities satisfy:

$$V_U^{1-\pi} = (1-\pi)(1 + p_R\delta V_N^{1-\pi} + (1-p_R)\delta V_U^{1-\pi}) + \pi(0 + p_N\delta V_N^{1-\pi} + (1-p_N)\delta V_U^{1-\pi})$$

$$V_N^{1-\pi} = (1-\pi)\pi(0 + q_R\delta V_U^{1-\pi} + (1-q_R)\delta V_N^{1-\pi}) + \pi(1 + q_N\delta V_U^{1-\pi} + (1-q_N)\delta V_N^{1-\pi})$$

The solution of these two equations is:

$$V_U^{1-\pi} = \frac{1}{1-\delta} \frac{(1-\delta)(1-\pi) + \delta(\pi(1-\pi)p_R + \pi^2 p_N + (1-\pi)^2 q_R + \pi(1-\pi)q_N)}{1-\delta + \delta((1-\pi)p_R + \pi p_N + (1-\pi)q_R + \pi q_N)}$$

$$V_N^{1-\pi} = \frac{1}{1-\delta} \frac{(1-\delta)\pi + \delta(\pi(1-\pi)p_R + \pi^2 p_N + (1-\pi)^2 q_R + \pi(1-\pi)q_N)}{1-\delta + \delta((1-\pi)p_R + \pi p_N + (1-\pi)q_R + \pi q_N)}$$

As indicated above, under the heading 'Proof strategy', we focus on the case in which the decision-maker chooses as the initial state U. His/her expected utility is then:

$$U = \frac{1}{2}(V_U^\pi + V_U^{1-\pi})$$

Maximizing U is the same as maximizing

$$V = 2(1-\delta)U = A + B$$

where

$$A = \frac{(1-\delta)\pi + \delta(\pi(1-\pi)p_R + (1-\pi)^2 p_N + \pi^2 q_R + \pi(1-\pi)q_N)}{1-\delta + \delta(\pi p_R + (1-\pi)p_N + \pi q_R + (1-\pi)q_N)}$$

and

$$B = \frac{(1-\delta)(1-\pi) + \delta(\pi(1-\pi)p_R + \pi^2 p_N + (1-\pi)^2 q_R + \pi(1-\pi)q_N)}{1-\delta + \delta((1-\pi)p_R + \pi p_N + (1-\pi)q_R + \pi q_N)}.$$

In the following, we shall study how p_R, p_N, q_R and q_N should be chosen so as to maximize V.

Switch after success only if you always switch after failure

We begin by demonstrating the following claim:

CLAIM 1: *In the optimum, if the decision-maker switches with positive probability after a successful choice, then s/he switches after unsuccessful choices with Probability 1. Formally:*

$$p_R > 0 \Rightarrow p_N = 1$$

and

$$q_N > 0 \Rightarrow q_R = 1$$

We prove only the first implication. The second follows by an analogous argument. Observe that A depends on the value of $\pi p_R + (1 - \pi)p_N$, but different values of p_R and p_N which give rise to the same $\pi p_R + (1-\pi)p_N$ lead to the same value of A. Moreover, A is increasing in $\pi p_R + (1-\pi)p_N$. In the same way, B only depends on $(1-\pi)p_R + \pi p_N$. Moreover, B is decreasing in $(1-\pi)p_R + \pi p_N$. Thus, when choosing p_R and p_N, the decision-maker will prefer among all combinations that give rise to a given value of $\pi p_R + (1-\pi)p$ those combinations that minimize $(1-\pi)p_R + \pi p_N$. Thus, whenever it is possible to achieve the given value of $\pi p_R + (1-\pi)p_N$ by choosing $p_N > 0$ but $p_R = 0$, the decision-maker will do so. The decision-maker will choose a positive p_R only if even a choice of $p_N = 1$ will not give rise to the sought value of $\pi p_R + (1-\pi)p_N$. The first implication in Claim 1 thus follows.

Switch after failure only with probability zero or with probability 1

We next demonstrate the following claim:

CLAIM 2: *In the optimum, the probability that the decision-maker switches state after a successful choice is either zero or one*, that is:

$$p_R \in \{0, 1\}$$

and

$$q_N \in \{0, 1\}.$$

We prove only the first of the two claims. The second claim can be proved analogously. To prove the claim we show that the following implication holds:

$$\frac{\partial V}{\partial p_R} = 0 \Rightarrow \frac{\partial^2 V}{(\partial p_R)^2} > 0.$$

This implies that any stationary point of the objective function V is a local minimum. This immediately implies that the value of p_R which maximizes V must be one of the two boundary values.

We begin by evaluating the derivatives involved:

$$\frac{\partial A}{\partial p_R} = \frac{\delta\pi(1-2\pi)(1-\delta+\delta\pi q_R+\delta(1-\pi)q_N)}{(1-\delta+\delta(\pi p_R+(1-\pi)p_N+\pi q_R+(1-\pi)q_N))^2}$$

$$\frac{\partial^2 A}{(\partial p_R)^2} = -\frac{2\delta^2\pi^2(1-2\pi)(1-\delta+\delta\pi q_R+\delta(1-\pi)q_N)}{(1-\delta+\delta(\pi p_R+(1-\pi)p_N+\pi q_R+(1-\pi)q_N))^3}$$

$$\frac{\partial B}{\partial p_R} = -\frac{\delta(1-\pi)(1-2\pi)(1-\delta+\delta(1-\pi)q_R+\delta\pi q_N)}{(1-\delta+\delta((1-\pi)p_R+\pi p_N+(1-\pi)q_R+\pi q_N))^2}$$

$$\frac{\partial^2 B}{(\partial p_R)^2} = \frac{2\delta^2(1-\pi)^2(1-2\pi)(1-\delta+\delta(1-\pi)q_R+\delta\pi q_N)}{(1-\delta+\delta((1-\pi)p_R+\pi p_N+(1-\pi)q_R+\pi q_N))^3}$$

We now assume that $\dfrac{\partial V}{\partial p_R} = 0$, i.e. $\dfrac{\partial A}{\partial p_R} = -\dfrac{\partial B}{\partial p_R}$, and we seek to prove:

$$\frac{\partial^2 V}{(\partial p_R)^2} > 0$$

$$\Leftrightarrow \frac{2\delta^2(1-\pi)^2(1-2\pi)(1-\delta+\delta(1-\pi)q_R+\delta\pi q_N)}{(1-\delta+\delta((1-\pi)p_R+\pi p_N+(1-\pi)q_R+\pi q_N))^3}$$

$$> \frac{2\delta^2\pi^2(1-2\pi)(1-\delta+\delta\pi q_R+\delta(1-\pi)q_N)}{(1-\delta+\delta(\pi p_R+(1-\pi)p_N+\pi q_R+(1-\pi)q_N))^3}$$

$$\Leftrightarrow -\frac{2\delta(1-\pi)}{1-\delta+\delta((1-\pi)p_R+\pi p_N+(1-\pi)q_R+\pi q_N)}\frac{\partial B}{\partial p_R}$$

$$> \frac{2\delta\pi}{1-\delta+\delta(\pi p_R+(1-\pi)p_N+\pi q_R+(1-\pi)q_N)}\frac{\partial A}{\partial p_R}$$

Now we use the assumption $\dfrac{\partial A}{\partial p_R} = -\dfrac{\partial B}{\partial p_R}$, and divide the left-hand side by $-\dfrac{\partial B}{\partial p_R}$ and the right hand side by $\dfrac{\partial A}{\partial p_R}$. It is easily verified that these expressions are positive.

$$\frac{2\delta(1-\pi)}{1-\delta+\delta((1-\pi)p_R+\pi p_N+(1-\pi)q_R+\pi q_N)}$$

$$> \frac{2\delta\pi}{1-\delta+\delta(\pi p_R+(1-\pi)p_N+\pi q_R+(1-\pi)q_N)}$$

$$\Leftrightarrow 2\delta(1-\pi)(1-\delta+\delta(\pi p_R+(1-\pi)p_N+\pi q_R+(1-\pi)q_N))$$

$$-2\delta\pi(1-\delta+\delta((1-\pi)p_R+\pi p_N+(1-\pi)q_R+\pi q_N)) > 0$$

$$\Leftrightarrow (1-\delta+\delta p_N+\delta q_N)(1-2\pi) > 0$$

which is evidently true.

We now summarize what we have learned above. If the decision-maker is in state U, s/he either switches with some probability if there has been no rain: $p_R \in [0,1]$, and never switches if there has been rain: $p_R = 0$, or he switches in both cases with probability 1: $p_R = p_N = 1$. Similarly, if the decision-maker is in state N, s/he either switches with some probability if there has been rain: $q_N \in [0,1]$, and never switches if there has been no rain: $q_R = 0$, or switches in both cases with probability 1: $q_N = q_R = 1$.

Never switch after success

CLAIM 3: *In the optimum, the probability that the decision-maker switches state after a successful choice is zero, that is:*

$$p_R = q_N = 0$$

We begin by proving: $p_R = 0$. As the previous subsection showed, the only alternative candidate for an optimal value of p_R is: $p_R = 1$. As indicated at the end of the previous subsection, we know that $p_R = 1$ can be optimal only if $p_N = 1$. Thus we shall assume that $p_N = 1$. We shall show that for all constellations of q_R and q_N that haven't yet been ruled out, it will be better to choose $p_R = 0$ rather than $p_R = 1$.

As indicated at the end of the previous subsection, there are two types of possible constellations of q_R and q_N. The first is: $q_R \in [0, 1]$ and $q_N = 0$. The second is: $q_R = q_N = 1$. We focus on the first of these constellations, and calculate the difference between the value of V if $p_R = 1$ and the value of V if $p_R = 0$, as:

$$\delta(1 - 2\pi) \left(\frac{\pi(\delta \pi q_R + 1 - \delta)}{(1 + \delta \pi q_R)(1 - \delta \pi (1 - q_R))} \right.$$
$$\left. - \frac{(1 - \pi)(\delta q_R(1 - \pi) + 1 - \delta)}{(1 + \delta(1 - \pi)q_R)(1 - \delta(1 - \pi)(1 - q_R))} \right)$$

To prove that this is negative we need to show:

$$\frac{(1 - \pi)(\delta q_R(1 - \pi) + 1 - \delta)}{(1 + \delta(1 - \pi)q_R)(1 - \delta(1 - \pi)(1 - q_R))} > \frac{\pi(\delta \pi q_R + 1 - \delta)}{(1 + \delta \pi q_R)(1 - \delta \pi (1 - q_R))}$$

But note that the left-hand side is the same expression as the right-hand side, except that we have replaced π by $1 - \pi$. Thus, if we can show that the right-hand side is strictly increasing in π, it follows that the left-hand side is larger than the right-hand side. To show this, we calculate the partial derivative:

$$\partial \left(\frac{\pi(\delta \pi q_R + 1 - \delta)}{(1 + \delta \pi q_R)(1 - \delta \pi (1 - q_R))} \right) \Big/ \partial \pi$$
$$= \frac{1 - \delta + q_R \delta \pi (2 - \pi \delta^2) + q_R^2 \delta^2 \pi^2 (\delta + 1)}{(1 + \delta \pi q_R)^2 (1 - \delta \pi (1 - q_R))^2}$$

Clearly, this derivative is positive. Thus, the claim follows.

The second constellation of values of q_R and q_N to consider is: $q_R = q_N = 1$. We calculate again the difference between the value of V if $p_R = 1$, and the value of V if $p_R = 0$. It is:

$$\delta(1 - 2\pi) \left(\frac{\pi}{(1+\delta)(1+\delta(1-\pi))} - \frac{1-\pi}{(1+\delta)(1+\delta\pi)} \right)$$

$$= -\frac{\delta(1 - 2\pi)^2}{(1+\delta\pi)(1+\delta-\delta\pi)} < 0$$

which is clearly negative, and thus $p_R = 0$ is also in this case optimal. Thus we can conclude that in the optimum we shall have: $p_R = 0$.

We now show that the optimal choice of q_N is $q_N = 0$. As the previous subsection showed, the only alternative candidate for an optimal value of q_N is: $q_N = 1$. As indicated at the end of the previous subsection, we know that $q_N = 1$ can be optimal only if $q_R = 1$. Thus, we shall assume that $q_R = 1$. We shall show that for all constellations of p_R and p_N which have not yet been ruled out, it will be better to choose $q_N = 0$ rather than $q_N = 1$.

As indicated at the end of the previous subsection, there are two types of possible constellations of p_R and p_N. The first is: $p_N \in [0, 1]$ and $p_R = 0$. The second is: $p_N = p_R = 1$. But the argument we gave above has ruled out the second constellation. Thus we shall assume the first one. We calculate the difference between the value of V if $q_N = 1$ and the value of V if $q_N = 0$, as:

$$\delta^2 p_N (1 - 2\pi) \left(\frac{\pi^2}{(1+\delta(\pi p_N - (1-\pi)(1-q_R)))(1-\delta+\delta(\pi p_N + (1-\pi)q_R))} \right.$$

$$\left. - \frac{(1-\pi)^2}{(1+\delta((1-\pi)p_N - \pi(1-q_R)))(1-\delta+\delta((1-\pi)p_N + \pi q_R))} \right)$$

This is negative if:

$$\frac{(1-\pi)^2}{(1+\delta((1-\pi)p_N - \pi(1-q_R)))(1-\delta+\delta((1-\pi)p_N + \pi q_R))}$$

$$> \frac{\pi^2}{(1+\delta(\pi p_N - (1-\pi)(1-q_R)))(1-\delta+\delta(\pi p_N + (1-\pi)q_R))}$$

The term on the left-hand side of this inequality is the same as the term on the right-hand side, except that π has been replaced by $1 - \pi$. Thus it suffices to show that the term on the right-hand side is increasing in π. For this, we calculate the partial derivative:

$$\partial \frac{\pi^2}{(1+\delta(\pi p_N-(1-\pi)(1-q_R)))(1-\delta+\delta(\pi p_N+(1-\pi)q_R))} \Big/ \partial\pi$$

$$= \frac{\pi(1-\delta(1-q_R))(\pi\delta+2\delta q_R(1-\pi)+2\pi\delta p_N+2(1-\delta))}{(1+\delta(\pi p_N-(1-\pi)(1-q_R)))^2(1-\delta+\delta(\pi p_N+(1-\pi)q_R))^2}$$

which is positive. Thus we can conclude that the optimal choice is: $q_N = 0$.

Switching probabilities after failure are state-independent

It remains to consider the optimal values of p_N and q_R. We begin by showing:

CLAIM 4: *In the optimum, the probability that the decision-maker switches state after an unsuccessful choice is independent of the state,* that is:

$$p_N = q_R$$

We assume that $p_N = p - \varepsilon$ and $q_R = p + \varepsilon$. If we calculate the value of V for $\varepsilon \neq 0$, and subtract the value for $\varepsilon = 0$, we find:

$$-\delta\varepsilon\frac{(1-2\pi)^2(4\delta^2\pi p\varepsilon(1-\pi)+(1-\delta)^2+\delta(1-\delta)(p+\varepsilon))}{(1-\delta+\delta(p-\varepsilon)+2\delta\pi\varepsilon)(1-\delta(1-p)+(1-2\pi)\delta\varepsilon)(1-\delta(1-p))}$$

which is negative. Therefore, $\varepsilon = 0$ is optimal.

Switch after failure with Probability 1

We can now complete our proof by showing the following claim:

CLAIM 5: *In the optimum, the probability that the decision-maker switches state after an unsuccessful choice is 1:*

$$p_N = q_R = 1$$

Setting $p_R = q_N = 0$, and $p_N = q_R = p$, we obtain for V:

$$V = \frac{(1-\delta)+2\delta(1-2\pi(1-\pi))p}{1-\delta(1-p)}$$

Therefore:

$$\frac{\partial V}{\partial p} = \frac{\delta(1-\delta)(1-2\pi)^2}{(1-\delta(1-p))^2} > 0$$

It follows that $p = 1$ is the optimal value.

Note

1 We would like to thank David Shanks, Richard Tunney and John McCarthy for making their data available to us. They ran a number of different treatments. The one to which we refer here is the one in which subjects received monetary incentives, and in which they received no feedback information beyond the payoff received in each round: that is, their 'Experiment 3 (payoff and no feedback condition)'.

References

Börgers, T. and Morales, A. (2004) 'Complexity Constraints in Two-Armed Bandit Problems: An Example', Mimeo. University College London.

Edwards, W. (1961) 'Probability Learning in 1000 Trials', *Journal of Experimental Psychology*, 62, 385–94.

Grant, D. A., Hake, H. W. and Hornsety, J. P. (1951) 'Acquisition and Extinction of a Verbal Conditioned Response with Differing Percentages of Reinforcement', *Journal of Experimental Psychology*, 42, 1–5.

Kalai, E., and Solan, E. (2003) 'Randomization and Simplification in Dynamic Decision-making', *Journal of Economic Theory*, 111, 251–64.

Myers, J. L. (1976) 'Probability Learning and Sequence Learning', in W. K. Estes (ed.), *Handbook of Learning and Cognitive Processes: Approaches to Human Learning and Motivation*, Hillsdale, NJ, Erlbaum.

Peterson, C. R. and Z. J. Ulehla (1965) 'Sequential Patterns and Maximizing', *Journal of Experimental Psychology*, 69, 1–4.

Piccione, M. and Rubinstein, A. (1997) 'On the Interpretation of Decision Problems with Imperfect Recall', *Games and Economic Behavior*, 20, 3–24.

Schlag, K. (2002) 'How to Choose – A Boundedly Rational Approach to Repeated Decision Making', Mimeo, European University Institute, Florence.

Shanks, D. R., Tunney, R. J. and McCarthy, J. D. (2002) 'A Re-examination of Probability Matching and Rational Choice', *Journal of Behavioral Decision Making*, 15, 233–50.

Winter, S. (1982) 'Binary Choice and the Supply of Memory', *Journal of Economic Behavior and Organization*, 3, 277–321.

3
Vengefulness Evolves in Small Groups

*Daniel Friedman and Nirvikar Singh**

Introduction

After a century of neglect, economists since the 1980s have begun to write extensively about social preferences. The vast majority of the articles so far have focused on altruism or positive reciprocity. Only a few examine the dark side – negative reciprocity or vengefulness. When a person harms you (or your family or friends), you may choose to incur a substantial personal cost to harm that person in return. Vengeance deserves serious study because it has major economic and social consequences, both positive and negative. For example, workers' negative reciprocity at the Decatur plant threatened to bring down Firestone Tyres (Krueger and Mas, 2004); terrorists often explain their actions as revenge against the oppressor; and successful corporate cultures succeed in forestalling petty acts of vengeance and other sorts of dysfunctional office politics.

A taste for vengeance, the desire to 'get even', is so much a part of daily life that it is easy to miss the evolutionary puzzle. We shall argue that indulging a taste for vengeance in general reduces material payoff or fitness. Without countervailing forces, vengefulness would have died out long ago, or would never have appeared in the first place.

Why, then, does vengeance exist? Economists' natural response is to think of vengeance as the punishment phase of a repeated game strategy supporting altruism. The models supporting this view are now taught to all Economics Ph.D. students and many undergraduates, and for good reason. Yet they hardly capture the whole story. The standard models have no place for the powerful emotions surrounding vengeance, and their predictions

*   While the ideas took shape for this chapter and its companions, we benefited greatly from the conversations with Ted Bergstrom, Robert Boyd, Bryan Ellickson, Jack Hirshleifer, Peter Richerson, Donald Wittman, and participants at the UC Davis conference on Preferences and Social Settings, 18–19 May 2001. Steffen Huck and an anonymous referee offered valuable guidance in writing the chapter, and the work of Werner Güth provided inspiration.

do not match up especially well with everyday experience. One often sees vengeance when the discount factor is too small to support rational punishment (for example, in one-off encounters with strangers), and often the rational punishment fails to appear (for example, when a culprit apologizes sincerely).

This chapter explores a different class of models. We consider repeated interactions in the context of small groups that enforce social norms. The norms are modelled not as traits of individual group members, but rather as traits of the group itself. We show that such group traits naturally support efficient levels of the taste for vengeance when encounters outside the group are also important. However, the model discloses two further problems. The threshold problem asks how vengeance can evolve from low values where it has a negative fitness gradient. The mimicry problem asks why cheap imitators do not evolve who look like highly vengeful types but do not bear the costs of actually wreaking vengeance. We argue that small-group interactions can overcome both problems.

The next section sets the stage with a simple illustration of the 'fundamental social dilemma': evolution supports behaviour that is individually beneficial but socially costly. We mention the standard devices for resolving the dilemma – genetic relatedness and repeated interactions – but focus on the more recent device of social preferences under the indirect evolution approach, as pioneered by Güth and Yaari (1992). The third section lays out the issues in more detail. It presents a simple Trust game, very similar to that analysed by Güth and various co-authors, and uses that game to lay out the social dilemma, and the threshold and mimicry problems.

The fourth and fifth sections are the heart of our analysis. We explain the role of group traits, their relationship to individual fitness, the time-scales governing their evolution, and how they can overcome the threshold and mimicry problems. The fifth section presents a more formal argument that group traits adjust behaviour in small groups towards a socially optimal level. The sixth section offers an extended discussion of how our approach relates to existing literature, and the seventh concludes with remarks on remaining open issues.

Vengefulness as an evolutionary puzzle

Figure 3.1 illustrates the fundamental social dilemma in terms of net material benefit ($x > 0$) or cost ($x < 0$) to 'Self' and benefit or cost ($y > 0$ or < 0) to counter-parties, denoted 'Other'.[1] Social dilemmas arise from the fact the Self's fitness gradient is the x-axis while, in contrast, the social efficiency gradient is along the 45-degree line. Social creatures (such as humans) thrive on co-operation, by which we mean devices that support efficient altruistic outcomes in II + and that discourage inefficient opportunistic outcomes in IV −. Such co-operation arises from devices that somehow internalize Other's costs and benefits.

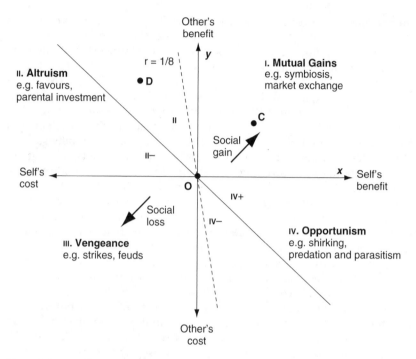

Figure 3.1   Payoffs to self and other

Quadrant III is anomalous; indeed, Cipolla (1976) refers to such behaviour as 'stupidity'. Behaviour producing quadrant III outcomes harms both Self and Other, contrary to efficiency as well as self-interest. How can it persist? We shall argue that the threat of visits to quadrant III (wreaking vengeance) helps to discipline opportunistic behaviour and encourages co-operation. But first we mention two other, better-known, devices that can serve the same purpose: genetic relatedness and repeated interactions.

Genetic relatedness

Biologists emphasize the device of genetic relatedness. If Other is related to Self to degree $r > 0$, then a positive fraction of Other's payoffs are internalized via 'inclusive fitness' (Hamilton, 1964) and iso-fitness lines take the form $[x + ry = k]$. For example, the unusual genetics of insect order *Hymenoptera* lead to $r = 3/4$ between full sisters, so it is no surprise that most social insects (including ants and bees) belong to this order, and that the workers are sisters. For humans and most other species, r is only 1/2 for full siblings and for parent and child, 1/8 for first cousins, and becomes zero exponentially for more distant relations. On average, r is rather small in human interactions, as in the steep dashed line in Figure 3.1, since we typically have only a few children, but work and live in groups with dozens of individuals. Clearly, non-genetic devices are needed to support human social behaviour.

Repeated interactions

Economists emphasize devices based on repeated interaction, as in the 'folk theorem' (Fudenberg and Maskin, 1986; Sethi and Somanathan, 2003). Suppose that Other returns the benefit ('positive reciprocity') with probability and delay summarized in discount factor $\delta \in [0, 1)$. Then that fraction of other's payoffs is internalized (Trivers, 1971) and evolution favours behaviour that produces outcomes on higher iso-fitness lines $[x + \delta y = k]$.[2] This device can support a large portion of socially efficient behaviour when δ is close to 1 – that is, when interactions between two individuals are symmetrical, predictable and frequent. But humans specialize in exploiting one-off opportunities with a variety of different partners, and here δ is small, as in the same steep dashed line. Other devices are needed to explain such behaviour.

Other regarding preferences and indirect evolution

Our focus is on other-regarding preferences. For example, suppose Self gets a utility increment of ry. Then Self partially internalizes the material externality, and will choose behaviour that attains higher indifference curves $[x + ry = k]$. Friendly preferences, $r \in [0, 1]$, can thus explain the same range of behaviour as genetic relatedness and repeated interaction.[3] However, by itself the friendly preference device is evolutionarily unstable: those with lower positive r will tend to make more personally advantageous choices, gain higher material payoff (or fitness), and displace the friendlier types. Friendly preferences therefore require the support of other devices.

Vengeful preferences rescue friendly preferences. Self's material incentive to reduce r disappears when others base their values of r on Self's previous behaviour and employ $r < 0$ if Self is insufficiently friendly. Such visits to quadrant III will reduce the fitness of less friendly behaviour and thus boost friendly behaviour. But visits to quadrant III are also costly to the avenger, so less vengeful preferences seem fitter. What then supports vengeful preferences: who guards the guardians?

In answering this question, our analysis must pass the following theoretical test: people with the hypothesized preferences receive at least as much material payoff (or fitness) as people with alternative preferences. Otherwise, the hypothesized preferences would disappear over time, or would never appear in the first place. In a seminal piece, Güth and Yaari (1992) described this test as indirect evolution, because evolution operates on preference parameters that determine behaviour rather than operating directly on behaviour. Precursors of this idea include Becker (1976), and Rubin and Paul (1979), but it was after Güth and Yaari's work that the literature expanded hugely, including papers such as those by Dekel *et al.* (1998), Huck and Oechssler (1999),

Kockesen *et al.* (2000), Ely and Yilankaya (2001), Samuelson and Swinkels (2001), and Possajennikov (2002a, 2002b). Many of these papers focus on positive reciprocity rather than negative reciprocity, or vengeance. For example, the key issue in Güth *et al.* (2001) is the cost of observing Other's true preferences for positive reciprocity (or altruism; in their game the two cannot be distinguished).

Modelling issues

We discuss the leading approaches to modelling social preferences, and then lay out a canonical Trust game. Using this game, we present the evolutionary problems of viability, threshold and mimicry.

Social preferences

Two main approaches can be distinguished in the recent literature. The distributional approach is exemplified in the Fehr and Schmidt (1999) inequality aversion model, the Bolton and Ockenfels (2000) mean-preferring model, and the Charness and Rabin (2001) social maximin model. These models begin with a standard selfish utility function and add additional terms capturing Self's response to how own payoff compares to other's payoffs. In Fehr–Schmidt, for example, my utility decreases (increases) linearly in your payoff when it is above (below) my own payoff. Put another way, I am altruistic when I am ahead and spiteful when I am behind you, irrespective of what you might have done to put me ahead or behind.

The other main approach is to model reciprocity in equilibrium. Building on the Geanakoplos *et al.* (1989) model of psychological games, Rabin (1993) constructs a model of reciprocity in two-player normal form games, extended by Dufwenberg and Kirchsteiger (1999), as well as Falk and Fischbacher (2001), to somewhat more general games. The basic idea is that my preferences regarding your payoff depend on my beliefs about your intentions – for example if I believe you tried to increase my payoff then I will want to increase yours. Such models are usually intractable. Levine (1998) improves tractability by replacing beliefs about others' intentions by estimates of others' type.

We favour a further simplification. Model reciprocal preferences as state-dependent: my attitude towards your payoffs depends on my state of mind – for example, friendly or vengeful, and your behaviour systematically alters my state of mind. This state-dependent Other-regarding approach is consistent with Sobel (2000) and is hinted at in some other papers including Charness and Rabin (2001). The approach is quite flexible and tractable, but in general requires a psychological theory of how states of mind change. Fortunately, a very simple rule will suffice for our present purposes: you become vengeful towards those who betray your trust, and otherwise have standard selfish preferences.

Again for simplicity (and perhaps realism), assume that, given the current distribution of v within the population, behaviour adjusts rapidly towards Nash equilibrium, but that there is at least a little bit of behavioural and observational noise.

Noise is present because equilibrium is not quite reached, or just because the world is uncertain. For example, Self may intend to choose N but may twist an ankle and find him/herself depending on Other's co-operative behaviour. Similarly, Other may intend to choose C but oversleeps or gets tied up in traffic. Such considerations can be summarized in a behavioural noise amplitude $e \geqslant 0$. Also, Other may imperfectly observe Self's true vengeance level v. Thus assume that Other's perception of v includes an observational error with amplitude $a \geqslant 0$.

The key task is to compute Self's (expected) fitness $W(v; a, e)$ for each value of v at the relevant short-run equilibrium, given the observational and behavioural noise. First consider the case $a = e = 0$, where v is perfectly observed and behaviour is noiseless. Recall from the previous section that in this case the short run equilibrium (N, D) with payoff $W = 0$ prevails for $v < c$, and (T, C) with $W = 1$ prevails for $v > c$. Thus $W(v; 0, 0)$ is the unit step function at $v = c$. One can show (Friedman and Singh, 2003b) that with a little behavioural noise (small $e > 0$) the step function slopes down, and with a little observational noise (small $a > 0$) the sharp corners are rounded off, as in Figure 3.3. In this case, a high level of vengefulness ($v > c + a$) brings high fitness and thus is viable.

The threshold problem

How will vengeful traits evolve in the Self population? It is inappropriate to assume standard replicator dynamics or monotone dynamics for a

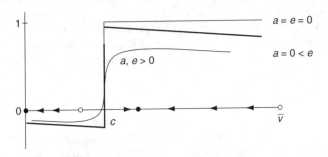

Figure 3.3   Fitness W as a function of vengefulness v

Note: For $a = e = 0$, the fitness function is a unit step function at $v = c$. Up to first order in behavioural noise amplitude e, the fitness function for $a = 0$ has slope $-e$ on the first segment and $-2e$ on the second segment. For signal noise amplitude $a > 0$, the fitness function is the convolution of the $a = 0$ fitness function with the signal noise density function. It has a local maximum at $v = 0$ and a global maximum near $v = c + a$ (solid dots) and a minimum near $v = c - a$ (open circle).

continuous trait like v.[5] Stochastic dynamics, such as noisy fictitious play, or the Kandori *et al.* (1993)/Young (1993) dynamic also apply to traits with discrete alternatives, such as eye colour, but have no natural application to traits with many ordered levels. Biological theorists from Wright (1949) through Eshel (1983) and Kaufman (1993) have routinely modelled continuous traits in terms of a fitness landscape in which evolution pushes the evolving trait v uphill. That is, selective pressure tends to increase (decrease) the level of the trait when higher (lower) levels are fitter. The underlying idea–that change is usually local and large jumps are rare–would seem to apply to a preference trait such as vengefulness as well as to standard biological traits such as height or foot speed (Friedman and Yellin, 1997).

Applying landscape dynamics to the fitness landscape in Figure 3.3, we see that evolution pushes v downwards towards 0 in the subpopulation initially below a level near $c - a$, and pushes v in the rest of the subpopulation to a level near $c + a$. Thus evolution in this case should lead to two types of individual. One type is just sufficiently vengeful to deter inefficient defection and has fitness $W \approx 1 - 2e$. The other type, recognizably different, is completely unvengeful and therefore unable to support co-operation. It has fitness $W \approx -e$.

There is a serious problem for the more vengeful type: how could it evolve from low values given the negative fitness gradient? How would a positive fraction of the subpopulation ever achieve levels above $c - a$ in the first place? We refer to this as the threshold problem, and will outline a solution in the next section.

The mimicry problem

Putting aside the threshold problem for the moment, assume that there are indeed two stable types – a vengeful type with v near $c + a$, and an unvengeful type with v near 0. The observational error amplitude, a, is small, so Other usually identifies Self's true type correctly. But the error amplitude itself is subject to evolutionary forces, creating what we shall call the mimicry or Viceroy problem.

An instructive example is a game played by butterflies and insect-eating birds. A butterfly can hide from birds (analogous to strategy N) or fly about freely (T), and the bird can prey on it (D) or leave it alone (C). Monarch butterflies (*Danaus plexippus*) feed on toxic milkweed and so are very unpalatable ($v > c$). Their striking Halloween markings (orange and black) make them easy for birds to avoid as in the efficient deterrence equilibrium (T, C). However, in Santa Cruz and many other areas where Monarchs are common, an unrelated species called the Viceroy (*Limenitis archippus*) has evolved markings that are almost identical to the Monarch's, a situation that biologists call Batesian mimicry. The Viceroys free-ride on the Monarch's high v reputation and are even fitter because they do not bear the dietary cost.

Note that we have not described evolutionary equilibrium in the butterfly–bird game. Although evolution favours population growth of Viceroys when scarce, it does not favour either species once the Viceroys become common. At that point it is worthwhile for hungry birds to sample the butterflies and spit out the unpalatable. An interior equilibrium with both Viceroys and Monarchs is possible if Monarchs can survive being spit out. If Monarchs cannot survive the experience, then two other evolutionary equilibria seem plausible: one where the Monarchs migrate ahead of Viceroys so the latter remains relatively scarce, and a second (called Müllerian mimicry), where Viceroys also evolve unpalatability. The field evidence for all three equilibria seems inconclusive.[6]

The mimicry or Viceroy problem surely arises in the extended Trust game. An individual with actual $v = 0$ who could convincingly mimic $v > c$ would gain a fitness increment of approximately $(1 + v)e$ over the object of his/her mimicry, and an increment of approximately $1 - e$ over his/her candid clone. Such increments are irresistible, evolutionarily speaking, so the assumption of near observability (small a) cannot be maintained in evolutionary equilibrium without some mechanism to suppress mimicry. We shall discuss possible mechanisms in the next section.

Group interactions and group traits

We do not know any way to overcome the threshold problem and the Viceroy problem within the context of unstructured interactions in a large population. Group interactions suggest an appealing solution to the threshold problem. Much of this section explicates the idea of group traits, which help solve the basic viability problem as well as the mimicry problem.

A solution to the threshold problem

Standard game theory shows how repeat interaction within a small group improves the adaptive value of sub-threshold v. Suppose Other expects that s/he and Self will switch roles from time to time, and that s/he can expect Self to reciprocate his current choice (C or D) into the indefinite future. Summarizing the probability and delay of reciprocation in the discount parameter δ, Other compares an immediate payoff $2 - v/c$ and continuation value 0 if s/he chooses D, to immediate payoff 1 and continuation value $\delta + \delta^2 + \delta^3 + \ldots = \delta/(1 - \delta)$ if s/he chooses C. Simple calculations reveal that it is advantageous to choose C if $\delta > \frac{1}{2}$ in the $v = 0$ case, and if $\delta > (c - v)/(2c - v)$ in case of positive v. The latter expression decreases towards 0 as v increases towards c. Thus small increments of $v < c$ increase the range of Other, who will find it in his/her interest to play C. This boosts Self's fitness and (depending on the distribution of δ within the group) can more than offset the increment's fitness cost (of order $-e$, as seen earlier).

Repeat interaction can also reduce the marginal cost of punishing culprits within the group. One does not have to retaliate immediately and directly, as assumed in Panels B and C of Figure 3.2. Instead, one can tell other group members about the culprit, and they can choose other partners for mutually productive activities at little or no cost to anyone except the culprit. If so, the effective value of c is quite small within the group. (Later we shall describe another group punishment technology with even lower cost.) Thus, within the group, the threshold is lower and moderate positive values of ν have positive incremental fitness, and the threshold problem is solved.

Interactions with outsiders and the mimicry problem

At first it seems that similar considerations also solve the mimicry or Viceroy problem. Given a lot of repeat interaction and communication among group members, and a small amount of behavioural noise, a player's true ν would soon be revealed to his/her group. Mimicry is not viable in this setting, but reputations are. Thus there are devices for overcoming first-order co-operation and second-order enforcement problems within the group.

The real problem arises from players' interactions outside the group. Assume, as might be reasonable, that a typical individual does not have significant repeat interaction with any particular person outside the group, but that the interactions with all people outside the group collectively do have a significant effect on his/her fitness.[7] Assume also that individuals can assess fairly reliably any individual's group affiliation and know the reputation of the group.[8] Then we have a free-rider problem with respect to group reputation. Each individual would benefit from using low ν in interactions outside the group, but the group's reputation, and hence its members' fitness, would suffer. The group must in some way regulate its members' behaviour or things would unravel. We hypothesize that groups themselves possess traits that evolve to solve such problems.

Note that social groups, unlike butterflies, use conscious mechanisms to control mimicry. Gangs may have secret handshakes and other codes of communication, but these are relevant only for identifying membership within the group. In Indian villages, one aspect of enforcing caste distinctions involves codes of dress and bodily decoration, so that lower castes cannot mimic upper castes, in general interactions, including with third parties. In that case, the higher caste is protecting its group reputation. In large, anonymous settings such as towns and cities, these codes are harder, if not impossible, to enforce, and mimicry is more common, with lower castes redefining their identities to claim higher caste status.[9]

Three different responses to the Viceroy problem now present themselves. The first, and the one most familiar to economists, would be the use of costly signalling. In the standard signalling model, one type (say, High) has a lower cost of signalling than another type (say, Low), and in a separating equilibrium, the Low type chooses not to mimic the High type. For example, 'toughness' may be signalled by acquiring tattoos, which would be too painful for those who are not 'tough'. Depending on the parameters of the situation, however, there may also be pooling equilibria, where the two types cannot be distinguished. In the kinds of situations we are interested in (across-group interactions where group reputations matter), signalling might be enforced by the group, when group benefits to signalling exceed individual benefits. As noted, certain kinds of dress codes and bodily decorations may be enforced within groups.[10]

A second possible response to the Viceroy problem is evasion, so that mimicry is avoided by physical separation. This is plausible in the context of migratory butterflies, but it is not clear how relevant it might be for human groups. One might also conceive of evasion and pursuit taking place in the space of characteristics, with the mimicked species or group evolving new traits as the old ones lose their distinctiveness. This would be akin to a dynamic signalling model, where multiple signals are possible: as the signalling characteristics of the Viceroy evolve towards those of the Monarch, the Monarch may evolve new distinguishing markers. Note once more that in the non-human species case, the evolution is necessarily through genetic mutation and selection, whereas in the case of human groups, conscious choices are involved, in choosing signal levels – evolution in the latter case would be cultural, and could be the result of learning.

The third possible response to the Viceroy problem is that of group enforcement. Here, we mean enforcement across groups, rather than within groups, which we discussed in the context of the signalling model. Thus high-caste groups may be willing to incur the costs of punishing low-caste groups that try to mimic them in encounters with third parties. The benefits are protection of reputation, and fitness gains associated with that protection. Note that this enforcement also requires overcoming free-rider problems within the group, but, as we have discussed, within-group interactions that are frequent allow repeated game mechanisms to come into play.

Group traits and individual fitness

We need to discuss the relevant traits before working out any of these responses in detail. A group trait is a characteristic of the group rather than an individual characteristic. Perhaps the sort of group trait most discussed in recent literature is a convention or norm: a Nash equilibrium of a co-ordination game in which it is in each member's interest to play a certain way, given that the other group members are doing so – for example, observe the Sabbath on Saturdays. But this is unnecessarily restrictive. Majority rule and primogeniture (or school mascots such as aggies or banana slugs) are group traits that need not be modelled as Nash equilibria of individual behaviour. Similarly for group traits such as use of a particular flag design, or language, or (closer to home) peer review protocols or the use of special jargon. Group traits are often discussed in the context of corporate culture and organizational routines (Nelson and Winter, 1982). Recent experiments by Weber and Camerer (2003) demonstrate that some facets of organizational culture are created by organization members but survive changes in individual membership of organizations.[11]

The relevant group traits for the present discussion are prescriptions for how individuals *should* behave in social dilemmas. Such prescriptions, when widely shared by group members, are group traits that are logically distinct from (but that co-evolve with) the individual traits that determine actual behaviour. For example, the group trait might be the shared belief that the appropriate level of the vengeance parameter is 3 and (as we shall see in the next section) that group trait might be in evolutionary equilibrium with actual behaviour being governed by the individual trait with a somewhat lower value, say $v = 2$.

One can imagine several different mechanisms by which group traits affect the fitness of an individual's traits. Perhaps the mechanism most familiar to game theorists is higher-order punishment strategies: deviations of actual behaviour from prescriptions are punished, as are failures to punish, failures to punish non-punishers, and so on, *ad infinitum*. We prefer to emphasize a different mechanism, mediated by status (for example, Catanzaro, 1992; Nisbett and Cohen, 1996). The mechanism has two parts: (a) the group's traits and the individual's behaviour affect status; and (b) status affects fitness.

To elaborate on (a), we recognize that status may depend on individual traits of all sorts, including age, sex, birth order and parental status. In all societies we know about, it also depends on contribution to local public goods. Local public goods include access to resources such as water supplies, sites for shelter and foraging, and military capabilities. Also included are intangibles such as the group's reputation among other groups, and its internal cohesiveness. Adherence to the group's prescribed level of vengefulness v^n contributes to that group's internal cohesiveness and external reputation. Thus it is reasonable to postulate that, other things being equal, an individual will have higher status when his/her behaviour reflects v closer to v^n. Such behaviour upholds the group's identity; see Akerlof and Kranton (2000).

Part (b) is straightforward. The group allocates many rival resources; depending on the context, these might include marriage partners, home sites, access to fishing holes and plots of land. Status is a device for selecting among the numerous co-ordination equilibria: the higher-status individuals get the first choice on available home sites and so on. The model in the next section uses a single parameter t to combine the sensitivity of fitness to status with the sensitivity of status to behaviour.

Evolution of group traits

Several authors recently have discussed the evolution of individual traits whose fitness depends on their prevalence in the group (for example, Sober and Wilson, 1998) and other authors have discussed the evolution of conventions (for example, Young, 1993), but our question is a bit different.

Unlike individual traits such as v, group traits cannot differ across individuals within a group: all know how they are supposed to behave in that group, and know the likely consequences of a deviation. Individuals of various sorts may enter or leave a group, and the group may grow or shrink, but these changes have no direct impact on group traits. Rather, over time, a particular group's trait may drift or occasionally change abruptly, as the members' common understanding reacts to experience.

A detailed micro-dynamic evolutionary model for a group trait would have to consider the joint time path of the traits across groups and group sizes. Such detail seems awkward and unnecessary. We need to know which group traits will displace others, but it does not much matter whether the displacement occurs through changes in group size, or through the numbers of groups. It seems sufficient to use aggregate dynamics that track the population shares for each group trait.

In specifying even aggregate dynamics one must consider a variety of transmission mechanisms for group traits, including imitation, proselytization, migration and conquest, as well as fertility and mortality. It is possible for horizontal transmission to increase the share of a group trait that reduces fitness (for example, encouraging tobacco consumption), but we do not believe that such considerations play a central role for the group traits of present interest. For simplicity we shall just hypothesize that the population shares respond positively to the average fitness of its members relative to the overall population average.

The relevant group traits here are prescriptions for responding to culprits and imitators from other groups, and for responding to deviations from the first-level prescriptions. Prescriptions for all permutations and combinations could be cumbersome, but are largely irrelevant for present purposes. Given the devices discussed earlier that ensure a high degree of co-operation within the group, the relevant group traits can be summarized in two parameters: the prescribed degree of vengefulness v^n towards culprits (or imitators) outside the group, and the tolerance parameter t for dealing with deviations from v^n by group members.

Recall from the previous section that deviations $x = v - v^n$ from prescribed behaviour are dealt with by reducing status, which leads to an adverse redistribution of resources and reduced fitness for the deviator. We assume simply that the fitness reduction $\rho(x)$ is smooth and convex (that is, the incremental fitness reduction increases with the magnitude of the deviation) and is minimized with value 0 at $x = 0$. The second order Taylor expansion approximation can therefore be written $\rho(x; t) = x^2/(2t)$, where deviations are treated less harshly the larger the tolerance parameter $t > 0$.

Evolutionary time-scales and equilibrium

A few remarks may be in order about fitness, monotone dynamics and time-scales. The analysis becomes very simple if there is a hierarchy of time-scales, so only one sort of trait is evolving significantly in any time-scale. One can assume that individual levels of v adjust rapidly within the genetically feasible range $[0, v^{max}]$; the idea is that people learn and accommodate themselves to the group's meme within a short period – say, weeks or months. For example, according to stories in the media, children raised in Belfast and Beirut taken to the USA have no problems in adapting within a few months to the US norm, and then readapting when they return home. Group traits also adjust, but in the medium-run of years to decades. The capacity for vengeful behaviour v^{max} can be thought of as mainly genetic, and thus it too can adjust in the long run, over several generations.

The dynamics are trivial in this case, because in each time scale only a single scalar variable is adjusting, the fitness functions are single peaked, and the direction of change is immediate from the definition of fitness. First, individual values of v converge to the level that maximizes individual fitness given v^n and t. Then v^n adjusts (for t fixed) to the level that maximizes the group average fitness given the error and noise rates and v^{max}; the individual v's trail along with the adjustments in v^n. (To be a bit more sophisticated, one could let t adjust at the same time, or separately, and possibly also allow the error and noise rates to evolve.) Finally, if the values of v are constrained by v^{max}, then it too evolves, with the other variables moving in its wake.

Of course, time-scales are not in fact so hierarchical, and there may be non-trivial co-evolution of individual v (social regulation of emotions), group traits, and emotional capacity. We conjecture that such co-evolution would not affect the relevant evolutionary equilibria nor alter their stability in the present case, although it certainly can in more general settings.

Results

We shall now sketch how efficient norms of vengeance might evolve in our setting. We use the extended Trust game with observational and behavioural errors, as in Figure 3.2, and assume that Self and Other belong to different groups. For reasons discussed above in connection with Figure 3.3, we assume a two-point distribution of types for Self: they can either have vengeance parameter 0 or $v > c$. We study a separating Perfect Bayesian Equilibrium (PBE). As shown in Friedman and Singh (2003b), this requires that the proportion of vengeful types of Self encountered by Other is neither too small nor too large. The intuition is that, if there are *too few* vengeful types, then Other has an insufficient incentive ever to co-operate, whereas if there are *too many* vengeful types, there is a pooling equilibrium with only trust and co-operation.

The fitness payoffs and probabilities in separating equilibrium are summarized in Table 3.1. Note that the probability α combines two error possibilities: an accurate observation followed by a behavioural error, and an

Table 3.1 Fitness and probabilities in separating PBE

Choice	Fitness payoff Self, Other	Equilibrium probability Strategies: (NT, DC)
$v > 0$ (N, .)	0, 0	e
(T, C)	1, 1	$(1-e)(1-\alpha)$
(T, D)	$-(1+v), 2-v/c$	$(1-e)\alpha$
$v = 0$ (N, .)	0, 0	$1-e$
(T, C)	1, 1	$e\alpha$
(T, D)	$-1, 2$	$e(1-\alpha)$

Notes: Other observes $s = 1$ with probability a in $(0, \frac{1}{2})$ when $v = 0$, and observes $s = 0$ with probability a when $v > 0$. Other chooses his/her less-preferred action with probability $\alpha = a(1-e) + e(1-a) = e + a - 2ae$.

observation error followed by intended behaviour. To the fitness payoffs in Table 3.1, we add the consequences of the social norm, the group trait. If the individual vengeance parameter, v, deviates from the group norm, v^n, then the individual suffers a fitness loss, through loss of status, given by $\rho(x; t) = x^2/(2t)$, where $x = v - v^n$. Incorporating this additional term, then, using the payoffs and probabilities in Table 3.1, the vengeful Self's expected fitness is given by:

$$W(v; v^n) = 0.e + 1.(1-e)(1-\alpha) - (1+v).(1-e)\alpha + \rho(v - v^n).(1-e)\alpha$$

The short-run dynamics push the individual's vengeance parameter towards the value that maximizes individual expected fitness at the given social norm. A simple calculation yields the first-order condition $\rho'(v - v^n) = 1$. In the quadratic case, the condition reduces $v = v^n - t$. Thus, in short-run (hence also in medium- and long-run) equilibrium, groups enjoin an exaggerated version of the optimal v, but the individually optimal v prevails. That optimum is as in Figure 3.3, since group reputations have only a small observational error.

Note the comparative statics: the punishment technology for out-group interactions is parameterized by the relevant c, and the prevailing v tracks the optimum given that value of c. Thus the model implies that easier detection and punishment of culprits will lower people's taste for the amount of punishment in medium- and long-run equilibrium. Also, higher tolerance t in a group correlates with higher v^n, although there is not really a causal relationship either way.

In the analysis to this point, we can assume that everyone in Self's group is identical, so that $v^n - t$ is also the group average vengeance parameter. But this group average evolves in the medium run. To see how, first note that status losses represented by the function $\rho(x; t)$ net out to 0 for the group, and so average fitness is:

$$W^g(\bar{\nu}) = 0.e + 1.(1 - e)(1 - \alpha) - (1 + \bar{\nu}).(1 - e)\alpha$$

The subtlety for medium-run dynamics is that the observational error amplitude is negatively related to the level of the group average vengeance parameter. That is, $a = A(\bar{\nu})$ for some decreasing function A. For the functional form $A(\nu) = 0.5 \exp(-\nu/b)$,[12] and many other specifications, $W^g(\bar{\nu})$ is single-peaked at some optimal level ν^0 of the group average vengeance parameter $\bar{\nu}$.

The group optimum ν^0 is characterized by first order condition $A'(\nu^0)$ $(2 + \nu^0) + A(\nu^0) = 0$. This expression can be solved explicitly for the given parametric versions of A and ρ to yield the simple expressions $\nu^0 = b - 2$ and $\nu^n = t + b - 2$. In general, evolution in the medium run pushes the group trait (ν^n, ρ) so as to increase $W^g(\bar{\nu})$. Without corner solutions or multiple peaks in the group fitness function,[13] the group trait will evolve so that it supports an optimal level of vengefulness in interactions outside the group.

We close this section with some caveats. Our result is partial equilibrium in that it takes as given the behavioural error rate e, the observational error function A, the marginal punishment cost c, and a sufficiently large upper bound ν^{max} on vengefulness to avoid corner solutions. We have already pointed out, in the previous section, that the upper bound is not a problem in the long run. Also, we have not worked out how the entire distribution of vengeance parameters over different groups might evolve. Various groups may differ in their environments and the frequency of their interactions with each other. For example, pastoral and agricultural groups might end up with different equilibrium levels of vengeance (Nisbett and Cohen, 1996, and references therein).

Related literature

Ours is not the only analysis of vengefulness. Elster (1989) was perhaps the first to highlight vengeance as a problematic economic issue, and to suggest the importance of social norms in overcoming this problem. Since then, several authors have encountered the viability problem in one form or another, and have found ways to finesse it.

Rosenthal (1996) considers a limited form of vengeance in which a player can detect culprits and shun them after the first encounter. The payoffs of such players (called TBV for 'trust but verify') are all reduced by verification costs. Rosenthal begins with a basic-stage game like ours and then modifies it by expressing payoffs as present values of the continuing relationship. The harm a TBV player inflicts on a culprit is the present value of payoffs the culprit forgoes after the initial temptation payoff. The punishment cost is the present value of verification less the present value of the avoided (sucker payoff) loss, which for relevant parameter values is negative. Thus punishment brings a net personal *benefit* and the all-C strategy (corresponding to our $\nu = 0$ player) does not weakly dominate the TBV strategy. Rosenthal finds several NE for his 3×3 symmetric game, and all-D need not be the only stable equilibrium. For certain parameter configurations, there is an interior NE that is stable under some (but not all) monotone dynamics. Unfortunately, no such stable equilibrium would exist under our maintained assumption that vengeance is costly and cannot reduce the sting of the sucker payoff.

Huck and Oechssler (1999) deal with the problem in a richer context than ours. In the 'ultimatum game' they study, players interact in small groups and have two roles, each played for half the time. In one role ('responder') they can pursue a costly vengeance strategy. Since there are only two possible offers, shading of punishments is not possible. With finite populations (or infinite populations interacting in small groups), delivering punishments may increase an individual's relative fitness, although it lowers absolute fitness. As the dynamics in their model are driven solely by relative fitness, the vengeful trait survives. However, there is no continuous evolvable trait in their model, which would be analogous to our vengeance parameter, ν.

Sethi and Somanathan (1996) offer two attempts to get around the viability problem. First, they define stability to include neutral stability, not requiring convergence back to an equilibrium point following a small perturbation (that is, they do not require local asymptotic stability). In their model there is a continuum of neutrally stable equilibria with no culprits. Following a perturbation (a small invasion of culprits) the state moves along the continuum away from a vertex. Eventually, following sufficiently many such perturbations, the state leaves the equilibrium set and ultimately converges back to the all-D equilibrium. Thus, from a long-run evolutionary viewpoint, their other equilibria really are not stable, and their vengeful strategies are not viable. Implicitly recognizing this problem, Sethi and Somanathan refer in an appendix to a second approach, from Binmore and Samuelson (1999), in which the evolutionary dynamic is perturbed by a continuing stream of mutants in fixed positive proportions. The perturbed dynamic has a single asymptotically stable equilibrium point instead of the continuum of neutrally stable equilibria, but it has a very shallow basin of attraction and is supported by an arbitrary convention on the composition of mutants.

The solution we have proposed to the viability problem is related to the two-level model for the evolution of co-operation as exposited in Sober and Wilson (1998) and Frank (1998). These authors note that, using a tautology known as the Price equation (Price, 1970),[14] one can demonstrate the possibility that a socially beneficial but dominated strategy (call it C) might survive in evolutionary equilibrium when group interactions are important. The idea in their analysis is that groups with a high proportion of C players have higher average fitness and thus grow faster than groups with a smaller proportion, and this effect may more than offset C's decline in relative prevalence within each particular group. The necessary conditions for C to survive (it can never eliminate D but may be able to coexist in equilibrium) are rather stringent. Besides the obvious condition that the group effect favouring C must be stronger than the individual effect favouring the dominant strategy D, it must also be the case that the groups dissolve and remix sufficiently often, and that the new groups have sufficiently variable proportions of C and D players. These special conditions may be met for some parasites, but seem quite implausible as a genetic explanation of human co-operation. Richerson and Boyd (1998) point out that genetic group selection in humans is implausible because of relatively rapid cross-group gene flow rates. Indeed, Sober and Wilson devote much of their book to discussing cultural norms for rewarding co-operative behaviour and punishing uncooperative behaviour. They avoid the viability problem by assuming in essence that c is 0 (1998; see p. 151 for the most explicit discussion of this point).

Bowles and Gintis (1998) consider the genetic evolution of vengeance in the context of a voluntary contribution game. They assume a direct tie between two discrete traits, a preference for punishing shirkers (analogous to our v) and a preference for helping a team of co-operators. Their argument is a version of two-level selection as in Sober and Wilson, and again is rather delicate. In an essay on the rise of the nation state in the last millennium, Bowles (1998) uses a version of the same model that allows for cultural and genetic coevolution. Gintis (2000) focuses on group extinction threats. In his model, strong reciprocity is favoured in between-group selection, since it increases group survival chances.

Still other approaches are possible; for example, Bowles and Gintis (2001) and Sethi and Somanathan (2001). The first of these papers shares some ideas with Bowles and Gintis (1998), in a model of team production with mutual monitoring. A sufficient proportion of 'strong reciprocators', who gain subjective payoffs from punishing shirkers, leads to a more co-operative outcome. Sethi and Somanathan use a variant of reciprocal preferences, which place negative weight on the payoffs of materialists (those with conventional selfish preferences) and positive weight on the payoffs of sufficiently altruistic individuals. Such preferences do better evolutionarily than purely altruistic or spiteful preferences.

Another way to avoid the viability problem is to assume that individuals with higher values of v encounter D play less frequently. Frank (1987) discusses this possibility informally, and formally models the evolution of a visible altruistic (rather than vengeful) trait. Under some specifications of how the frequency of co-operators depends on v it is not hard to show that there is a positive level of v that maximizes fitness. Indeed, if each individual's v were observable, then those with higher v might encounter D play less frequently (as in Frank's 1988 discussion) and thus maintain equal or higher fitness.[15] This 'green beard' solution,[16] of course, ignores the mimicry problem.

Henrich and Boyd (2001) argue that the negative gradient or threshold problem can be overcome within groups if more popular behaviour tends to be imitated, even when this conformity effect is very weak. Groups with this trait would achieve better internal co-operation, and displace other groups. The issue in Henrich and Boyd is the same as here – why people would bear the personal cost to punish defectors. The paper notes the game theory device of higher-order punishments – for example, second-order is punishing those who don't punish defectors. The modelling goal is to stabilize punishments at finite order, and the key insight is that under reasonable conditions the need (hence the cost) for higher-order punishment decreases exponentially as the order increases. If conformist transmission has a positive constant impact, then even if it is rather small it can reverse the negative payoff gradient at some sufficiently high order of punishment, and hence stabilize lower orders of punishment and co-operation. This does seem to be a possible solution, but its appeal to an economist is reduced by two considerations. First, if conformist transmission is modelled explicitly, it might be difficult to make it independent of the order. For example, if third-order punishments are relevant, an imitator would only rarely observe the difference between his/her own third-order behaviour and that of the majority. The transmission rate parameter alpha thus might also decline exponentially in the punishment order and may never reverse the negative payoff gradient. Second, economists tend to think that actual payoffs trump conformity when they point in opposite directions. (Psychologists and other social scientists are unlikely to share this prejudice.)

A variation on the repeated interaction scenario is one where co-operative acts are credibly communicated to others, who are then more likely to be co-operative in interactions with the first individual. This version is referred to as 'indirect reciprocity' (for example, Fehr and Henrich, 2003), and has been discussed or modelled by Alexander (1987) and Nowak and Sigmund (1997), for example. Nowak and Sigmund model (and simulate numerically) indirect reciprocity as 'image scoring', in which an individual's score increases when s/he helps someone who needs it, and decreases when such help is not offered. This process is thought of as taking place in social groups that are small enough to allow members to track everyone else's scores. Leimar and Hammerstein (2000) have suggested that image scoring by itself is not individually rational, and offered alternative simulations that call its evolutionary robustness into question.

Finally, we should note how this chapter fits with our own earlier work. Much of the material comes from our 1999 working paper. The underlying game there, and in our 2001a and 2003a papers, however, is a simultaneous-move prisoner's dilemma rather than the Trust game. Our 2003b paper is based on the Trust game, but considers only interactions in a large population with no group structure, and obtains equilibria with no optimality properties parallel to those obtained here. It considers alternative assumptions on the observational error function $A(v)$ and focuses on a Gaussian rather than exponential form, and analyses the resulting second-order conditions in detail. The 2003a paper allows for somewhat more complex group interactions and perceptions than in this chapter, and includes a more detailed discussion of reputation and status issues, and of relevant biological constraints on vengeance. As this chapter, it obtains an optimality result based on the marginal logic of trading off the cost of individual retaliation and the impact on status within the group. None of our previous papers treats the threshold and mimicry problems.

Discussion

We have argued that small-group interactions play a crucial role in the evolution of vengeful preferences. The most relevant and problematic interactions across groups (a) are not frequent enough to support the use of repeated game or related mechanisms for reciprocity, yet (b) are important enough in the aggregate to affect fitness. We showed how small groups can, at low cost to the group, enforce specific norms of vengeance on their members. Status is key: those who depart further from the group norm suffer greater reductions in status, which ultimately decreases their fitness. Individual adherence to group norms, while imperfect, can be strong enough in evolutionary equilibrium to sustain co-operative outcomes in inter-group encounters. Thus small groups can overcome the basic viability problem for vengeance.

Earlier we presented a simple argument on how small groups overcome the threshold problem. Within small groups, even a small degree of vengefulness can help support repeated game equilibria. We also discussed how status-mediated group enforcement can also discipline mimicry by outsiders as well as by group members.

Our approach has focused most directly on the problem of the evolution and persistence of vengefulness, and we believe that it provides some new insights. Nevertheless, our discussion has finessed many important questions. Here are two methodological questions that we have not addressed in this chapter:

(i)   Other-regarding preferences may involve a host of contingencies besides whether Other belongs to Self's own group and whether s/he is a culprit. What theoretical discipline, as well as empirical evidence, can keep such models sharp and tractable? Indirect evolution dictates that the requisite preferences must aid fitness in a variety of situations, and the answer to this question may require identifying canonical games that best capture human experience.

(ii) Introducing a group structure on interactions and allowing groups a very low-cost punishment strategy creates a huge set of possible evolutionary equilibria, larger even than in the 'folk theorem'. What selection criteria can be brought to bear on the model to narrow down the set of equilibria? Friedman and Singh (2003b) introduced the concept of Evolutionary Perfect Bayesian Equilibrium, which is one approach to answering this question.

Finally, we provide some broader perspective on our approach to modelling the appearance and persistence of vengefulness. We have used the existence of well-functioning norms within small groups to support the long-run use of vengeful behaviour in across-group interactions. The analogy we can offer is to a trellis or scaffolding, where either structure supports the growth or erection of something else. The difference between a trellis and scaffolding is that the latter is temporary, whereas the former is permanent. In that sense, group traits or norms in our model act as a trellis. Without them, the kind of behaviour that we posit would erode, as, over time, individuals would find it beneficial to shade their vengefulness.

Some aspects of within-group interactions, however, have the characteristics of scaffolding – in particular, in overcoming the threshold problem because a small amount of vengefulness increases the range of discount factors for which co-operation works in repeated settings. Once the threshold is crossed, other factors sustain the level $v > c$. Of course, the repeated interactions can still play a role in enforcing the norms that matter for sustaining vengefulness. We would like to suggest that this perspective, of one set of traits, whether cultural or biological, providing direct support for another trait to develop, is a useful idea in general discussions of coevolution. In particular, distinguishing between trellises and scaffoldings can be helpful in understanding the relationship between present and past.

Notes

1 For simplicity, we neglect here possible effects on third parties, such as customers of a cartel. Extensions of the present diagram could replace 'other' by 'average of everyone else affected' or could look explicitly at all affected types.

2 Another way to think about it is that, with positive reciprocity (or genetic relatedness), one takes a weighted average of the first outcome (in II + or IV +) and the reciprocal outcome (reflected through the 45-degree line, as Self and Other are interchanged, so now in IV + or II +). This gives an outcome in the mutual gains quadrant I if the weight δ (or r) on the reciprocal outcome is sufficiently large.

3 Indeed, in principle we could have $r > 1$ and explain inefficient altruistic behaviour. The golden rule ('love thy neighbour as thyself') value $r = 1$ seems to be a practical upper bound, however, since no evolutionary devices that we know of will tend to push it higher.

4   Other's utility function here is simply own payoff. If we were focusing on friendliness instead of vengeance, we might write Other's utility function with a positive component for Self's payoff when Self chooses T. This would also lead to an efficient Nash equilibrium if the relevant coefficient r exceeds 0.5. Güth has produced a series of papers with various co-authors that develop the evolutionary implications of such friendly (or positively reciprocal) preferences.

5   Oechssler and Riedel (2000) deal with evolutionary dynamics in continuous games, and point out some difficulties with evolutionary stability if the strategy space is continuous. In our case, it is the preference trait that is continuous, so again the situation we analyse is somewhat different.

6   See Kapan (2001) and the references therein.

7   Fehr and Henrich (2004) argue forcefully that this is the usual situation for contemporary hunter-gatherer groups as well as for our Paleolithic ancestors.

8   Across-group encounters are also frequent, but a given individual will encounter a specific non-group member only very sporadically. An individual in such encounters cannot reliably signal his/her true ν because outward signs can be mimicked at low cost, but neither (because of the large numbers of sporadic personal encounters) can s/he easily establish a reputation for his/her true ν. A specific assumption that would capture these considerations is that the perceived vengeance parameter of one's opponent ν^e is equal to the true value ν in encounters within the group, but in encounters outside the group $\nu^e = \lambda\bar{\nu} + (1-\lambda)E\bar{\nu} + \varepsilon$, an idiosyncratic error plus the weighted average of the partner's group average $\bar{\nu}$ and overall population average E $\bar{\nu}$, with the weight λ on the group average an increasing function of group size. The idea is that ν^e is a Bayesian posterior, with sample information on any individual overwhelming priors for internal matches, and sample information on the relevant group being important for external matches. Implicit in this formulation is a theory of group size. Very large groups would violate the assumptions that everyone knows everyone well, and that everyone monitors the all-C equilibrium, so there are diseconomies of scale. At the margin, these diseconomies should balance the economies arising from the dependence of λ on group size. We shall not attempt to develop such a theory here, but will simply assume the existence of moderate-sized groups.

9   M. N. Srinivas termed this process 'Sanskritization' – see, for example, Srinivas (2002).

10   See Akerlof (1983) for several seminal essays that model enforcement in this context, as well as Henrich and Boyd (2001) for a more recent contribution. In such cases, the costs of enforcement are a major concern. Fines are an enforcement mechanism whose cost to a group is near zero (or perhaps negative). Elinor Ostrom, in a communication with the first author, offered the example of cow jails in Nepal. Cows grazing in the wrong places are 'jailed' and the owner has to pay a fine. Until the owner does so, the community gets the cow's milk. Hence the enforcement cost is negative, that is, the community (apart from the owner) gets a small net benefit from punishing the norm violator.

11   In these experiments, the relevant dimension of organizational culture is a specialized homemade language developed by organization members to complete a task efficiently. This kind of group trait is not relevant for encounters with outsiders, but only for within-group interactions. Corporate dress codes would matter for outsiders, but are copied very easily. However, it is easy to think of being 'hard-nosed' as a corporate trait that might be valuable in dealings with outsiders, difficult to imitate, and enforced by internal norms of status.

12 Note that the factor 0.5 ensures that, as vengefulness goes to 0, the observation becomes completely noisy, which is as it should be.

13 Our papers discussed at the end of the permultimate section show that corner solutions disappear in the long-run equilibrium and that multiple peaks do not arise in the interesting cases.

14 The Price equation uses the definition of covariance to decompose the change in prevalence of a trait into two components, for example, the direct effect from individual fitness and an indirect effect incorporating the spillovers within the group.

15 We have offered a somewhat more complex resolution of the viability problem because we believe that the relationship between v and the frequency of encountering co-operators arises mainly at the group level rather than at the individual one. We have argued that, within well-functioning groups, D behaviour is rare, and dealing with it is not an important source of fitness differences. Presumably D behaviour is encountered more frequently with partners outside one's own group, and we believe that here group reputations are the key, not individual signals or individual reputations. We have also suggested how within-group mechanisms might control the Viceroy problem.

16 This term is from Dawkins (1976), and is a used as a fanciful but striking example of identifiability. A certain type of individual is identified by a green beard, and no other type of individual is able to mimic that, even when it is strongly in their evolutionary interest.

References

Akerlof, George (1983) *An Economist's Book of Tales*, New York, Cambridge University Press.

Akerlof, George, and Rachel Kranton (2000) 'Economics and Identity', *Quarterly Journal of Economics*, 115, 715–53.

Alexander, R. D. (1987) *The Biology of Moral Systems*, New York, Aldine de Gruyer.

Becker, Gary S. (1976) *The Economic Approach to Human Behaviour*, Chicago, University of Chicago Press.

Binmore, Kenneth and Samuelson, Larry (1999), 'Evolutionary Drift and Equilibrium Selection, *Review of Economic Studies*, 66, 363–93.

Bolton, Gary E. and Ockenfels, Axel 2000 'ERC: A Theory of Equity, Reciprocity and Competition', *American Economic Review*, March, 90, 1, 166–93.

Bowles, Samuel (1998) 'Cultural Group Selection and Human Social Structure: The Effects of Segmentation, Egalitarianism and Conformism, Working paper, University of Massachusetts, Amherst.

Bowles, Samuel and Gintis, Herbert (1998) 'The Evolution of Strong Reciprocity', Working paper, Department of Economics, University of Massachusetts, Amherst.

Bowles, Samuel and Gintis, Herbert (2001) 'Social Capital and Community Governance', Working paper, Department of Economics, University of Massachusetts, Amherst.

Catanzaro, R. (1992) *Men of Respect: A Social History of the Sicilian Mafia*, New York, The Free Press.

Charness, Gary and Rabin, Matthew (2001) 'Understanding Social Preferences with Simple Tests', *Quarterly Journal of Economics*, 117, 817–69.

Cipolla, Carlo (1976) *The Basic Laws of Human Stupidity*, Bologna, The Mad Millers.

Cox, James C. and Friedman, Daniel (2002) 'A Tractable Model of Reciprocity and Fairness', Manuscript, University of California at Santa Cruz.

Dawkins, R. (1976) *The Selfish Gene*, New York, Oxford University Press.
Dekel, Eddie, Ely, Jeffrey C. and Yilankaya Okan (1998) 'The Evolution of Preferences', Working paper, Northwestern University, Evanston. See http://www.kellogg.nwu.edu/research/math/Je_Ely/working/observe.pdf.
Dufwenberg, Martin and Kirchsteiger, Georg (1999) 'A Theory of Sequential Reciprocity', Discussion paper, CentER for Economic Research, Tilburg University.
Elster, Jon (1989) 'Social Norms and Economic Theory', *Journal of Economic Perspectives*, 3, 4, 99–117.
Ely, Jeffrey C. and Yilankaya, Okan (2001) 'Nash Equilibrium and the Evolution of Preferences', *Journal of Economic Theory*, 97, 255–272.
Eshel, I. (1983) 'Evolutionary and Continuous Stability', *Journal of Theoretical Biology*, 103, 99–111.
Falk, Armin and Fischbacher, Urs (2001) 'Distributional Consequences and Intentions in a Model of Reciprocity', *Annales d' Economique et de Statistique*, 63–64 (Special issue), July–December.
Fehr, Ernst and Henrich, J. (2004) 'Is Strong Reciprocity a Maladaptation?' *Genetic and Culture Evolution of Cooperation* edited by Peter Hammerstein. Cambridge, MIT Press.
Fehr, Ernst and Schmidt, Klaus M. (1999) 'A Theory of Fairness, Competition, and Cooperation', *Quarterly Journal of Economics*, 114, 3, 817–68.
Frank, Robert (1987) 'If *Homo economicus* Could Choose His Own Utility Function, Would He Want One with a Conscience?', *American Economic Review*, 77, 593–604.
Frank, Robert (1988) *Passions Within Reason: The Strategic Role of the Emotions*, New York, W.W. Norton.
Frank, Steven (1998) *Foundations of Social Evolution*, Princeton NJ, Princeton University Press.
Friedman, Daniel (1991) 'Evolutionary Games in Economics', *Econometrica*, 59, 637–66.
Friedman, Daniel and Singh, Nirvikar (1999) 'On the Viability of Vengeance', Manuscript, University of California, Santa Cruz, May. See http://econ.ucsc.edu/~dan/.
Friedman, Daniel and Singh, Nirvikar (2001) 'Evolution and Negative Reciprocity', in Y. Aruka (ed.), *Evolutionary Controversies in Economics*, Tokyo, North-Holland.
Friedman, Daniel and Singh, Nirvikar (2003a) 'Negative Reciprocity: The Coevolution of Memes and Genes', Working Paper, University of California, Santa Cruz.
Friedman, Daniel and Singh, Nirvikar (2003b) 'Equilibrium Vengeance', Working paper, University of California, Santa Cruz.
Friedman, Daniel and Yellin, Joel (1997) 'Evolving Landscapes for Population Games', Manuscript, University of California, Santa Cruz.
Fudenberg, Drew and Maskin, Eric (1986) 'The Folk Theorem in Repeated Games with Discounting or with Incomplete Information', *Econometrica*, 54, 3, 533–54.
Geanakoplos, John, Pearce, David and Stacchetti, Ennio (1989) 'Psychological Games and Sequential Rationality', *Games and Economic Behaviour*, 1, 60–79.
Gintis, H. (2000) 'Strong Reciprocity and Human Sociality', *Journal of Theoretical Biology*, 206, 169–79.
Güth, Werner and Kliemt, Hartmut (1994) 'Competition or Cooperation: On the Evolutionary Economics of Trust, Exploitation and Moral Attitudes', *Metroeconomica*, 45, 2, 155–87.
Güth, Werner and Yaari, Menachem (1992) 'An Evolutionary Approach to Explaining Reciprocal Behaviour', in U. Witt (ed.), *Explaining Process and Change – Approaches to Evolutionary Economics*, Ann Arbor, Mich., University of Michigan Press.
Güth, Werner, Kliemt, Hartmut and Peleg, Bezalel (2001) 'Co-evolution of Preferences and Information in Simple Games of Trust', Manuscript, Humboldt University, Berlin.

Hamilton, W.D. (1964) 'The Genetical Evolution of Social Behaviour', *Journal of Theoretical Biology*, 7, 1–52.

Henrich, J. and Boyd, R. (2001) 'Why People Punish Defectors: Weak Conformist Transmission Can Stabilize Costly Enforcement of Norms in Cooperative Dilemmas', *Journal of Theoretical Biology*, 208, 79–89.

Huck, S. and Oechssler, J. (1999) 'The Indirect Evolutionary Approach to Explaining Fair Allocations', *Games and Economic Behaviour*, 28, 13–24.

Kandori, M., Mailath, G. J. and Rob, R. (1993) 'Learning, Mutation, and Long Run Equilibria in Games', *Econometrica*, 61, 29–56.

Kapan, Durrell D. (2001) 'Three-Butterfly System Provides a Field Test of Müllerian Mimicry', *Nature*, 409, 338–40.

Kaufman, S. (1993) *The Origins of Order: Self-Organization and Selection in Evolution*, New York, Oxford University Press.

Kockesen, Levent, Ok, Efe A. and Sethi, Rajiv 2000 'The Strategic Advantage of Negatively Interdependent Preferences', *Journal of Economic Theory*, June 92, 2, 274–99.

Krueger, Alan B. and Mas, Alexandre (2004) 'Strikes, Scabs and Tread Separations: Labor Strife and the Production of Defective Bridgestone/Firestone Tires', forthcoming, *Journal of Political Economy*.

Leimar, O. and Hammerstein, P. (2000) 'Evolution of Cooperation through Indirect Reciprocity', *Proceedings of the Royal Society of London* B, 268, 745–53.

Levine, David K. (1998), 'Modeling Altruism and Spitefulness in Experiments', *Review of Economic Dynamics*, 1, 593–622.

Nelson, Richard R. and Winter, Stanley G. (1982) *An Evolutionary Theory of Economic Change*, Cambridge, Mass., Belknap Press of Harvard University Press.

Nisbett, R. E. and Cohen, D. (1996) *Culture of Honor: The Psychology of Violence in the South*, Boulder, Col., Westview Press.

Nowak, M. A. and Sigmund, K. (1998), 'Evolution of Indirect Reciprocity by Image Scoring', *Nature*, 393, 573–7.

Oechssler, Jorg and Riedel, Frank (2001) 'Evolutionary Dynamics on Infinite Strategy Spaces', *Economic Theory*, 17, 141–62.

Possajennikov, Alex (2002a) 'Two-Speed Evolution of Strategies and Preferences in Symmetric Games', Discussion paper 02–03, National Research Center 504 'Rationality Concepts, Decision Behaviour, and Economic Modeling', University of Mannheim, January.

Possajennikov, Alex (2002b) 'Cooperative Prisoners and Aggressive Chickens: Evolution of Strategies and Preferences in 2 × 2 Games', Discussion paper 02–04, National Research Center 504 'Rationality Concepts, Decision Behaviour, and Economic Modeling', University of Mannheim, January.

Price, G. R. (1970) 'Selection and Covariance', *Nature*, 227, 5257, 1 August, 520–21.

Rabin, Mathew (1993) 'Incorporating Fairness into Game Theory and Economics', *American Economic Review*, 88, 5, 1281–302.

Richerson, Peter J. and Boyd, Robert 1998 'The Evolution of Ultrasociality', in I. Eibl-Eibesfeldt and F. K. Salter (eds), *Indoctrinability, Ideology and Warfare*, New York, Berghahn.

Romer, Paul (1995) 'Preferences, Promises, and the Politics of Entitlement', in Victor R. Fuchs (ed.), *Individual and Social Responsibility: Child Care, Education, Medical Care, and Long-Term Care in America*, Chicago: University of Chicago Press.

Rosenthal, R. W. (1996) 'Trust and Social Efficiencies', Manuscript, Boston University.

Rubin, Paul H. and Paul, C. W. (1979) 'An Evolutionary Model of Taste for Risk', *Economic Inquiry*, 17, 585–96.

Samuelson, Larry and Swinkels, Jeroen (2001) 'Information and the Evolution of the Utility Function', Mimeo, University of Wisconsin.

Sethi, R. and Somanathan, E. (1996) 'The Evolution of Social Norms in Common Property Resource Use', *American Economic Review*, 86, 766–88.

Sethi, R. and Somanathan, E. (2001) 'Preference Evolution and Reciprocity', *Journal of Economic Theory*, 97, 273–97.

Sethi, R. and Somanathan, E. (2003), 'Understanding Reciprocity', *Journal of Economic Behaviour and Organization*, 50, 1–27.

Sobel, Joel (2000) 'Social Preferences and Reciprocity', Mimeo, University of California, San Diego.

Sober, E. and Wilson, D. S. (1998) *Onto Others: The Evolution and Psychology of Unselfish Behaviour*, Cambridge, Mass., Harvard University Press.

Srinivas, Mysore N. (2002) *Collected Essays*, New Delhi: Oxford University Press.

Sugden, R. (1986) *The Economics of Rights, Co-operation and Welfare*, New York, Basil Blackwell.

Trivers, Robert L. (1971) 'The Evolution of Reciprocal Altruism', *Quarterly Review of Biology*, 46, 35–57.

Weber, Roberto A. and Camerer, Colin F. (2003) 'Cultural Conflict and Merger Failure: An Experimental Approach', *Management Science*, 49, 4, 400–15.

Wright, Sewall (1949) 'Adaption and Selection', in L. Jepsen, G. G. Simpson and E. Mayr (eds), *Genetics, Paleontology, and Evolution*. Princeton, NJ, Princeton University Press, 365–89.

Young, H. Peyton (1993) 'The Evolution of Conventions', *Econometrica*, 61, 57–84.

4
Network Formation and Co-ordination Games

*Siegfried K. Berninghaus and Bodo Vogt**

Introduction

Co-ordination games attracted many theoretically and experimentally orientated economists during the 1990s (see, for example, van Huyck *et al.*, 1990; Cooper *et al.*, 1992; Berninghaus and Schwalbe, 1996a; Young, 1998). In our paper we consider simple symmetric normal form 2×2 games which are characterized by having *two* equilibria in pure strategies. If such a 2×2 game is played in large populations with players who are randomly matched pairwise an equilibrium selection problem may arise. We know from the theoretical (for example, Boyer and Orleans, 1992) and the experimental literature (for example, Cooper *et al.*, 1992) that in case of pure co-ordination games both symmetric equilibria may be candidates for strategy selection. In co-ordination games with two asymmetric equilibria still less is known about strategy choice in experimental games. Conventions might sometimes help to solve co-ordination problems (Lewis, 1969; Young, 1993; Berninghaus, 2003). By following conventions, players are guided to select a particular equilibria and therefore avoid co-ordination failures. In real-world societies, conventions will not arise spontaneously but rather result from a long-run evolutionary process.

The problem of the evolution of conventions in large populations has often been considered under a particular assumption concerning neighbourhood structures or local interaction structures in the populations (see, for example, Blume, 1993, Ellison, 1993, Berninghaus and Schwalbe, 1996a, 1996b; Eshel *et al.*, 1998). In such a framework, a member of the population is not supposed to be randomly matched with any other member of the population, but is only matched with members of the member's neighbourhood that is a proper subset of the whole population. The neighbourhoods

---

*   The essential ideas in this chapter were developed during a research stay at the Max Planck Institute in Jena by both authors. We are grateful to Werner Güth, the Director of this institute, for his generous intellectual and financial support.

of the players constitute a local interaction structure or sometimes called a network structure on the population. Much of the research in this field has been devoted to populations with exogenously fixed local interaction structures imposed on the population.[1] In recent research this restrictive assumption has been relaxed, and players allowed to choose their neighbours in each period for themselves (see, Jackson and Wolinsky, 1996; Bala and Goyal, 2000; Goyal and Vega-Redondo, 2002). In these models, local interaction structures are regarded as resulting from individual decisions and not as being exogeneously imposed.

In this chapter we consider two types of 2×2 bimatrix games played in a population; that is, we consider both pure co-ordination games and Hawk/Dove (H/D) games as well. In pure co-ordination games, equilibria in pure strategies are characterized by the requirement that players choose the same strategy, while Hawk/Dove games equilibria in pure strategies are characterized by the requirement that both players choose different strategies (aymmetric equilibria). Hawk/Dove games have a long tradition in evolutionary game theory. Maynard Smith and Price (1973) developed their famous equilibrium concept, the evolutionary stable state (ESS), for these types of game. In an evolutionary framework, one has the following interpretation in mind. Two members of a species are randomly matched to compete for the same territory. If both members choose the Hawk strategy this results in territory fighting, with serious wounds for both. If they choose the Dove strategy they share the territory after some kind of ritual fighting. The only Nash equilibria in pure strategies are the asymmetric strategy configurations (Hawk, Dove) resp. (Dove, Hawk). The only symmetric equilibrium is the mixed strategy equilibrium, which can be shown to be the unique ESS of the game. Successful co-ordination in large populations is much more interesting in Hawk/Dove games than in pure co-ordination games, since each player wants to be matched only with players who employ only the opposite strategy.

Co-ordination in large populations with pure co-ordination games has been studied extensively since the early 1990s (see, for example, the experimental literature of van Huyck *et al.* 1990; Cooper, 1999). A survey on experimental results has been written by Ochs (1995). A recent survey on the experimental literature, in which co-ordination problems are considered in populations with network formation, can be found in Kosfeld (2003). However, we do not know of comparable studies for Hawk/Dove games. We argue that successful co-ordination in large populations is much more interesting in Hawk/Dove games than in pure co-ordination games, since each player wants to be matched only with players employing only the opposite strategy (in a 2×2 H/D game).

It is the main aim of this chapter to analyse which types of network and distributions of strategies chosen in the bimatrix game will be compatible in an equilibrium when players are allowed simultaneously to select their neighbours in the population *and* the strategy in the base game. Decision-making is supposed to be deterministic. Opening a new link to a member of the population is supposed to generate constant connection costs per link for the agent who initiates the link. It will be shown that the relative size of linking costs compared with the payoffs of the 2×2 game has an impact on the resulting equilibrium network in the population. We know only of similar work on this topic by Goyal and Vega-Redondo (2002). They concentrate on pure co-ordination games in which equilibrium selection problems arise. The main contribution of Goyal/Vega-Redondo is to analyse stochastic stable states of the process of network formation for pure co-ordination games by allowing mutation at the individual level of strategy and partner choice. It is well known from the literature on stochastic stability (for example, Kandori *et al.*, 1993; Young, 1993) that the equilibrium selection problem can be solved by letting the mutation rate approach zero.

Our model is a purely static one. We consider simultaneous network linking choice and action choice in the co-ordination game to be elements of an appropriated formalized one-shot game. It is the main aim of our study to analyse the impact of the particular underlying 2×2 base game on the resulting equilibrium network structure. The resulting equilibrium networks are characterized by non-directed graphs. Depending on the particular value of linking costs, we obtain different graphs for pure co-ordination and for Hawk/Dove games. Our work can be considered to be an extension of pure network formation approaches (for example, Bala and Goyal, 2000) in which pure network decision-making is considered abstracting from any strategic decision-making in 2×2 base games. And it can also be regarded in some sense as an extension of some aspects of Goyal and Vega-Redondo's results (2002) since additionally we consider Hawk/Dove games as a further class of co-ordination games.[2]

Model description and results

Hawk/Dove games

We consider a set $I = \{1, \ldots, n\}$ of n agents who are engaged in playing a Hawk/Dove game with all their neighbours. If two players i and j are linked with each other they play the Hawk/Dove game as a one-shot game. The Hawk/Dove game is a symmetric 2×2 normal form game $\Pi_{HD} = \{\Sigma, H(\cdot)\}$ with $\Sigma := \{X, Y\}$ which is characterized by the payoff table with $a > b > c > d > 0$, that is, Y is called the 'Dove strategy' and X is called the 'Hawk strategy'.

	X	Y
X	d, d	a, c
Y	c, a	b, b

Figure 4.1 Payoff *matrix* of a Hawk/Dove game

We do not impose a fixed network structure on the population of players, but assume that networks can be built up by individual decision-making. More precisely, we assume that all players participate in a *network game*. An individual strategy in the network game of player i is a vector of ones and zeros $g_i \in \{0, 1\}^{n-1}$. We say that player i wants to establish a link to player j if $g_{ij} = 1$, otherwise it is equal to zero. A link between two players allows both players to play the simple Hawk/Dove game Π_{HD}. Note that a bilateral connection between two players is supposed to be already established if at least *one* player wants to open it.[3]

Each strategy configuration $g = (g_1, \ldots, g_n)$ generates a directed graph denoted by $\mathcal{G}_g$, where the vertices represent players and a directed edge between i and j, that is, $g_{ij} = 1$, signals that i plans to open a link with j. The neighbours of player i, given a network $\mathcal{G}_g$ is defined to be the set of players to whom i wants to open a link (*direct neighbors*, $g_{ij} = 1$) and the players who want to open a link with i (that is, $g_{ji} = 1$). By utilizing the notation $\bar{g}_{ij} := \max\{g_{ij}, g_{ji}\}$ we simply define the set of neighbours as follows:

$$N_i(\mathcal{G}_g) := \{ j \mid \bar{g}_{ij} = 1 \}$$

The cardinality of this set is given by $n_i(\mathcal{G}_g) := |N_i(\mathcal{G}_g)|$.

Obviously, the set of neighbours need not concide with the set of direct neighbours that depends only on i's strategy network strategy vector g_i. We define the set of direct neighbours of i as follows:

$$N_i^d(g_i) := \{ j \mid g_{ij} = 1 \}$$

The cardinality of the set of direct neighbours is defined by $n_i^d(g_i) := |N_i^d(g_i)|$.

We suppose that it is not costless to establish a link with another player. Therefore, the total payoff of player i is composed of the aggregate payoff player i can extract from playing with his/her neighbours and the costs of establishing links to her direct neighbors. Let k denote the constant linking costs. Since the payoff player i can extract from playing the Hawk/Dove game depends on his/her own strategy choice, the strategy choice of his/her (direct) neighbours and the network generated by g, we define:

$$\Pi_i^X(\sigma_{-i}, g) := d \sum_{j \in N_i(\mathcal{G}_g)} 1_{\sigma_j = X} + a \sum_{j \in N_i(\mathcal{G}_g)} 1_{\sigma_j = Y} - k n_i^d(g_i)$$

$$\Pi_i^Y(\sigma_{-i}, g) := c \sum_{j \in N_i(\mathcal{G}_g)} 1_{\sigma_j = X} + b \sum_{j \in N_i(\mathcal{G}_g)} 1_{\sigma_j = Y} - k n_i^d(g_i)$$

where $\Pi_i^X(\cdot)$ resp. $\Pi_i^Y(\cdot)$ denotes the payoff a player choosing X resp. Y can gain and $\sigma = (\sigma_{-i}, \sigma_i)$ denotes the vector of actions[4] $\sigma_i \in \{X, Y\}$ for the H/D game. We assume that each player selects the same action against all other players in his neighbourhood.[5] An important consequence of our payoff definition is player i may benefit from a connection to j although s/he does not have to pay for it (that is, $g_{ij} = 0$, but $g_{ji} = 1$).

Furthermore, we utilize the following payoff convention:

CONVENTION: When g generates the empty network – that is, a network in which all players are isolated, we suppose that the relationship holds.

$$\prod_i^{(\cdot)}(\cdot, g) \equiv 0$$

We model the strategic situation of a player in a population as a non-cooperative game in which individual strategies are composed of the simultaneous choice of neighbours $i \in I$ and actions $\sigma_i \in \{X, Y\}$ in the bilateral H/D game. This non-cooperative game in normal form is a tupel:

$$\Gamma = \{S_1, \ldots, S_n, P_1(\cdot), \ldots, P_n(\cdot)\}$$

with strategy sets $S_i := \{0, 1\}^{n-1} \times \{X, Y\}$ and payoff functions $P_i : S_1 \times \ldots \times S_n \longrightarrow IR$ where we have $P_i(s) := \prod_i^{\sigma_i}(\sigma_{-i}, g)$. Each strategy configuration $s = (s_1, \ldots, s_n)$ in Γ induces a network represented by a directed graph $\mathcal{G}_g$.

It remains to consider which network structures $\mathcal{G}_g$ and action configurations σ in Γ will prove to be stable? Our notion of stability is purely non-cooperative. We utilize a canonical extension of the Nash concept.

DEFINITION 1 The strategy configuration $s^* = (g^*, \sigma^*)$ in Γ is an equilibrium if

$$\forall i : \quad P_i(s_{-i}^*, s_i^*) \geq P_i(s_{-i}^*, s_i) \text{ for } s_i \in S_i$$

In an equilibrium, no player has an incentive either to change his/her neighbours or to change his/her action choice σ_i^* or to change both unilaterally.

It follows immediately from the equilibrium definition that we need not consider as equilibrium candidates configurations $g = (g_1, \ldots, g_n)$ in which two players simultaneously have a bilateral link with each other; that is, if $g_{ij} = g_{ji} = 1$ holds for a pair of players $i, j \in I$. Then either i or j could improve their payoffs by dropping the link and saving linking costs k. Strategy configurations s that do not exhibit this property will be called *simple*.

DEFINITION 2 A strategy configuration $s = (g, \sigma)$ is called simple if the following relation holds:

$$\forall i, j : \quad \bar{g}_{ij} = 1 \implies g_{ij} \cdot g_{ji} = 0$$

Obviously, a simple strategy configuration *s* always generates a simple graph (which has no loops, and two vertices are connected by at most one directed edge). In the following theorems we refer only to simple networks without mentioning them explicitly.

In Theorem 1 below we characterize equilibrium network structures and action configurations *s** in a population playing the Hawk/Dove game.

THEOREM 1   *Given a HD-game* Γ, *then the following statements hold:*

(a) *If* $k > a$ *then the unique equilibrium network* $\mathcal{G}_{g^*}$ *is the empty network and the action choice of each player in the Hawk/Dove game is not determined.*

(b) *If* $k < d$ *then the unique equilibrium network* $\mathcal{G}_{g^*}$ *is the complete graph. In the complete graph no uniform choice of either X or Y is possible as an equilibrium action choice. Let* n_X^* *resp.* n_Y^* *denote the number of players choosing X resp. Y as an equilibrium choice in the complete network then these numbers are determined by the relationship*:

$$\frac{n(a-b)-(a-b)}{a-b+c-d} < n_X^* < \frac{n(a-b)+(c-d)}{(a-b+c-d)} \tag{4.1}$$

and $n_Y^* = n - n_X^*$.

(c) *If the relationships* $d < k < c$ *hold, then an equilibrium network* $\mathcal{G}_{g^*}$ *is a graph whose vertices can be partitioned into two non-empty sets* I_1 *of X players and* I_2 *of Y players, such that all vertices in* I_1 *are connected with all vertices in* I_2 *but not with each other while all vertices in* I_2 *are also connected with each other. Again uniform action choice is not possible in equilibrium and the number of players choosing X* (n_X^*) *has to satisfy the condition*:

$$\frac{(n-1)(a-b)+n_j^X(k-d))}{a-b+c-d} < n_X^* < \frac{n(a-b)+(c-k)}{a-b+c-k} \tag{4.2}$$

and $n_Y^* = n - n_X^*$, *where* n_j^X *denotes the number of direct links to X players of a Y player indexed by j.*

(d) *If the relation* $c < k < b$ *holds, then* $\mathcal{G}_{g^*}$ *is a graph where each vertex in a set* I_1 *(X players) is connected with all vertices in* I_2 *(Y players) but not with vertices in* I_1, *while all vertices in* I_2 *are also connected with each other. Furthermore, Y players do not have direct links with X players. Again, no uniform action choice can be part of an equilibrium.* n_X^* *has to satisfy the condition*

$$n_X^* > \frac{(n-1)(a-b)}{a-b+c-d}. \tag{4.3}$$

(e) *If* $b < k < a$ *then an equilibrium network* $\mathcal{G}_{g^*}$ *is characterized either by a bipartite graph with* $n_X^*, n_Y^* > 0$ *which is characterized by the following property: only X players in* I_1 *have direct links to Y players in* I_1, *that* *is, $g_{ij}^* = 1$ for $i \in I_1$ and $j \in I_2$ while we have $g_{ji}^* = 0$ for $j \in I_2, i \in I_1$ and, furthermore, $g_{jm}^* = 0$ for $j, m \in I_2$ resp. $j, m \in I_1$. Or the equilibrium network is the empty graph where $n_X^* = n$.*

PROOF:

(a) Suppose there exists at least one link between two players i and j, that is, $\bar{g}_{ij}^* = 1$. Since k is supposed to be larger than the maximum payoff a player can gain from the Hawk/Dove game, the net payoff from each link is negative irrespective of the individual action choices in the Hawk/Dove game. Therefore, establishing no link to any other player results in maximum individual payoff equal to 0 according to our convention on payoffs. Payoff is independent of a player's action choice and, therefore, action choice is not determined.

(b) Since opening a new connection to either an X player or a Y player results in positive net payoffs, it pays to open as many links as possible where it has to be taken into account that a player i should only open a link to j if $g_{ji} = 0$, otherwise payoffs do not have the Nash property.

Obviously, no uniform action configuration ($n_X^* = n$ or $n_Y^* = n$) is in equilibrium, since any player i could switch to the opposite action and increase his/her payoff. Now let us consider the case $n_X^*, n_Y^* > 0$. For a player choosing X, the payoff has to be higher than for choosing Y. Then the following condition has to be satisfied:[6]

$$(n_X^* - 1)d + n_Y^* a > (n_X^* - 1)c + n_Y^* b \iff (n_X^* - 1)(c - d) < n_Y^*(a - b).$$

Analogously, for a player choosing Y, the following inequality holds:

$$n_X^* d + (n_Y^* - 1)a < n_X^* c + (n_Y^* - 1)b \iff n_X^*(c - d) > (n_Y^* - 1)(a - b).$$

By substituting $n_Y^* = n - n_X^*$ we obtain from these inequalities the relationships:

$$n_X^* < \frac{n(a - b) + (c - d)}{a - b + c - d}$$

and

$$n_X^* > \frac{n(a - b) - (a - b)}{a - b + c - d}$$

which are equivalent to condition (4.1).

(c) Suppose inequality $d < k < c$ holds. Then it will not pay for an X player to be connected with other X players, since it will give him/her negative payoff. However, a Y player may open as many connections as possible provided there is not already another player who opened a link to him/her. The resulting graph can therefore be partitioned into two sets of vertices I_1 (X players), whose elements are connected with each vertex of I_2 (Y players), and each element of I_2 is, moreover, connected with each other member of I_2. Note that we only consider connections that are initiated by exactly one player (because of the required simplicity of a Nash network).

Uniform action choice is not possible in equilibrium since any player could benefit from either switching from X to Y resp. from Y to X. To determine the equilibrium number n_Y^* of Y resp. n_X^* of X players we first consider the decision problem of an X player i who can switch from X to Y *and* open new connections to the remaining n_X^* X players. This will not be profitable for an X player if the following condition holds:

$$n_Y^* a - n_i^Y k > n_Y^* b - n_i^Y k + [(n_X^* - 1)c - (n_X^* - 1)k]$$

where n_i^Y denotes the number of direct links of X player i and the expression in square brackets denotes the net benefits of i from opening (as a Y player) new links to the remaining X players.[7] This condition is equivalent to:

$$n_X^* < \frac{n(a-b)+c-k}{a-b+c-k}$$

When a Y player j changes his/her action to X s/he will simultaneously have at the same time to drop all direct links to X players, otherwise, these links would result in negative net benefits. Let us denote by n_j^Y (resp. n_j^X) the number of direct links of j to Y players (resp. X players) then a Y player j will not change his/her action choice when the following relationship holds:

$$(n_X - n_j^X)d + (n_Y - 1)a - n_j^Y k < n_X c + (n_Y - 1)b - n_j^Y k - n_j^X k$$

$$\Longleftrightarrow n_X > \frac{n(a-b)-(a-b)+n_j^X(k-d)}{a-b+c-d}$$

Both inequalities for n_X^* together imply Condition (4.2).

(d) Now suppose that $c < k < b$ holds. Then it follows from the arguments of the proof of part (c) that each X player will only connect to all Y players. Y players, however, will only connect with Y players, since any other link will give them negative payoffs. Therefore, the graph generated by g^* is a graph where the whole set of vertices can be partitioned into two non-empty sets I_1 (X players) and I_2 (Y players) such that all elements of I_1 are connected only to all elements of I_2 and all elements of I_2 are connected with one another.[8] Obviously, no uniform action choice can be part of an equilibrium.

For an X player it is not profitable to deviate to Y, since the relationship:

$$n_Y^* a - n_Y^* k > n_Y^* b - n_Y^* k$$

holds by assumption ($a > b$). For a Y player j the following condition has to be satisfied:[9]

$$n_X^* c + (n_Y^* - 1)b - n_j^Y k > n_X^* d + (n_Y^* - 1)a - n_j^Y k,$$

where n_j^Y denotes the number of direct links of player j to other Y players. This condition can easily be transformed into condition (4.3).

(e)   Consider a Y player. His/her maximum payoff is equal to b which can be reached by playing with other Y players. However, because of $k > b$, a Y player cannot extract positive net payoffs from any connection (with either an X or a Y player). Therefore, in an equilibrium network, Y players cannot have direct neighbours. The maximum payoff of an X player is equal to a, which can only be reached by being matched with a Y player. Furthermore, X can extract only a positive net payoff from being linked to a Y player (as his/her direct neighbour). By being linked to another X player s/he obtains a negative net payoff. Therefore, a candidate for an equilibrium network is a graph where the vertice set can be partitioned into two sets I_1 (of X players) and I_2 (of Y players) such that each X player is actively connected with all Y players. But there are non other direct links in the population. The number of action choices $n_X^*, n_Y^* > 0$ is not determined in equilibrium since it does not pay for any player in the equilibrium network to switch actions.[10]

Obviously, uniform choice of Y is not possible in equilibrium, since each player can benefit from switching to X and build up links to at least one of the remaining Y players. Uniform choice of X is only possible when the resulting equilibrium network is the empty graph.

Q.E.D

REMARKS

(i)   Note that in part (b) of the theorem, a complete graph may be generated by many different individual Nash configurations g which only have to be simple.[11] As an extreme case one could consider a complete graph in which there is one player who need not open any link with the remaining $(n-1)$ players, since they all want to be linked with him/her. Indeed, one can easily check that this is also an equilibrium network. Similar conclusions hold for the equilibrium networks in parts (c) and (d) of the theorem. When a link between two players exists it may not matter which player opened the link. Consequently, in equilibrium networks, payoffs may vary significantly from one player to the next even if they choose similar actions. However, there exist some configurations (for example, the direct links of X players with Y players in part (d) of the theorem) which induce unique directions of links between players.

(ii) The result of (d) shows that all Dove players are connected with each other but do not have direct links to Hawk players. This is an interesting situation, in which a 'Dove network' is stabilized by some Hawk 'invaders'. Note that the Doves in this network are playing non-equilibrium strategies of the one-shot H/D game. Such a constellation can only occur when decision making in H/D games is embedded in a network formation problem.

(iii) Our result in part (e) has an interesting economic interpretation. We see that part of the population (Y players in I_2) is subsidized by the remaining part of the population (X players in I_1). Y players benefit from playing with X players without bearing the linking costs.

(iv) It is well known in the theory of network formation (see Bala and Goyal, 2000; Goyal and Vega-Redondo, 2002) that so called *star-shaped* network structures may be stable equilibria in a particular model of a strategic neighbour's choice. In a star-shaped structure, one player (the 'centre' of the star) is connected with the rest of the population, and the remaining players are linked exclusively to the 'centre' player. Our bipartite graph (in part (e)) can be interpreted as a generalization of the star-shaped structure such that in our structure we have finitely many (≥ 1) 'centre' players.

We illustrate the main results on network formation by the drawing in Figure 4.2. In (c) and (d), the population is partitioned into two subpopulations $I_1 = \{1, 4\}$ and $I_2 = \{2, 3, 5, 6\}$, where either links between these populations are possible in both directions (part (c)) or links are only directed from I_1 to I_2 players (part (d)). Part (a) shows an example of an empty network, while part (b) shows an example of a complete network.

Pure co-ordination games

In this section we assume that if two players, i and j, are linked with each other they play a pure co-ordination game. The symmetric 2×2 pure co-ordination game is a symmetric normal form game $\Pi_C = \{\Sigma, H(\cdot)\}$ with $\Sigma := \{X, Y\}$ which is characterized by the payoff table with $a > b > c > d > 0$ and $(b - d) > (a - c)$, that is (Y, Y) is the payoff-dominant equilibrium and (X, X) is the risk-dominant equilibrium.

As in the previous section, we do not impose a fixed network structure on the population of players, but assume that networks can be built up

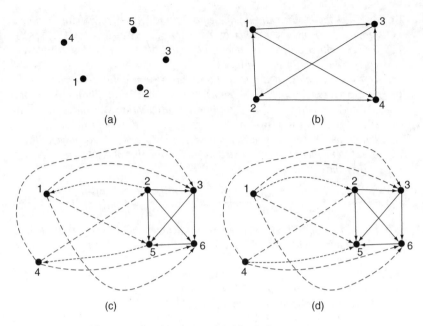

Figure 4.2 Equilibrium networks in parts (a)–(d) of Theorem 1

	X	Y
X	b, b	c, d
Y	d, c	a, a

Figure 4.3 Payoff matrix of a pure co-ordination game

by individual decisions. All members of the population are supposed to be players of the network game Γ, introduced in previous section. It has the same formal structure as before, with the restrictions concerning the numerical payoffs of the 2×2 base game being changed. In Theorem 2 we shall show that the resulting equilibrium network structures will be changed completely by altering the payoff structure of the underlying base game from a Hawk/Dove type game to a pure co-ordination game.

THEOREM 2   *Given a network game* Γ *where the underlying* 2×2 *game is a pure coordination Game.*

(a) *If* $k > a$, *then the unique equilibrium network* $\mathcal{G}_{g^*}$ *is the empty network and the action choice of each player in the pure co-ordination game is not determined.*

(b) *If $k < d$ holds, then the unique equilibrium network $\mathcal{G}_{g^*}$ is the complete graph and σ^* is given either by $\sigma^* = (X, \ldots, X)$ or by $\sigma^* = (Y, \ldots, Y)$. That is, either all players choose the equilibrium strategy X or all players choose the equilibrium strategy Y.*

(c) *If $d < k < c$ holds, then the equilibrium is the one obtained in part (b).*

(d) *If the relationship $c < k < b$ holds, then either the equilibrium is the one obtained in part (b) or an equilibrium configuration s^* induces a disconnected graph $\mathcal{G}_{g^*}$ with two components, where each component is a complete graph and players in one component (I_1) choose action X, players in the other component (I_2) choose action Y. The number of X players n_X^* has to satisfy the condition:*

$$\frac{n(a-k)+(b-d)+n_i^X(k-d)}{a+b-k-d} < n_X^* < \frac{n(a-c)-(a-c)+n_j^Y(c-k)}{a+b-k-c} \quad (4.4)$$

for all $i \in I_1$ and $j \in I_2$, where n_i^X (n_j^Y) denotes the number of direct links of X player i with other X players (direct links of Y player j with other Y players).

(e) *If the relationship $b < k < a$ holds, then an equilibrium network $\mathcal{G}_{g^*}$ is the complete graph with all players choosing Y. An alternative equilibrium is the empty graph with all players choosing X.*

Proof:

(a) The same argument as in Theorem 1.

(b) The arguments concerning equilibrium network formation for $k < d$ are analogous to Theorem 1. Since each connection to another player increases a player's payoff, s/he tries to open as many links as possible, provided the resulting network is a simple one. Therefore, the complete graph is the only candidate for an equilibrium network. If all players choose X or Y, this is obviously an equilibrium action choice. It can easily be seen that $n_X > 0$ and $n_Y > 0$ is not compatible with an equilibrium configuration in the network game. Suppose $n_X, n_Y > 0$, then for an X player resp. Y player, the following relationships must hold:

$$(n_X^* - 1)b + n_Y^* c > (n_X^* - 1)d + n_Y^* a \iff (n_X^* - 1)(b-d) > n_Y^*(a-c)$$

$$\iff \frac{n_X^* - 1}{n_Y^*} > \frac{(a-c)}{(b-d)}$$

$$n_X^* b + (n_Y^* - 1)c < n_X^* d + (n_Y^* - 1)a \iff n_X^*(b-d) < (n_Y^* - 1)(a-c)$$

$$\iff \frac{n_X^*}{n_Y^* - 1} < \frac{(a-c)}{(b-d)}$$

which implies

$$\frac{n_X^*}{n_Y^* - 1} < \frac{n_X^* - 1}{n_Y^*}$$

a contradiction.

(c) Suppose that $d < k < c$ holds. It is obvious that the equlibria of part (b) are also equilibria in this case. If some players select X and some select Y, the following holds. A Y player makes positive profits when s/he is connected with another Y player while s/he extracts negative payoffs $(d - k < 0)$ when opening a link to an X player. However, s/he benefits if an X player wants to open a link with him/her. On the other hand, an X player benefits from opening as many links as possible (with X and Y players as well).

The incentives of an X and a Y player to deviate are: For an X player i it is not profitable for him/her to switch to Y, dropping his/her links with X players if the following inequality holds:

$$(n_X^* - 1)b + n_Y^*c - n_i^X k - n_Y^* k > (n_X^* - 1 - n_i^X)d + n_Y^* a - n_Y^* k$$

where n_i^X is the number of direct links an X player i has established to other X players. This inequality is equivalent to:

$$n_X^* > \frac{n(a - c) + (b - d) + n_i^X(k - d)}{(a - c + b - d)}$$

For a Y player the following condition has to be satisfied:

$$n_X^* d + (n_Y^* - 1)a - n_j^Y k > n_X^* b + (n_Y^* - 1)c - n_j^Y k$$

which is equivalent to the inequality

$$n_X^* < \frac{n(a - c) - (a - c)}{a - c + b - d}$$

Both restrictions on n_X^* obviously cannot be satisfied simultaneously. We only have equilibria of the type obtained in part (b).

(d) As in cases (b) and (d), the complete graph with all players selecting either X or Y is an equilibrium. If some players select X and some select Y, the following holds. For X players it is profitable to build up as many links as possible with other X players. The same argument holds for Y players. All other links result in a payoff loss (either $(c - k) < 0$ or $(d - k) < 0$). In order to have no incentive for an X player to deviate it suffices to consider the effects of an action switch from X to Y together with opening links to all Y players. Such a deviation is not profitable when the inequality[12]

$$(n_X^* - 1)b - n_i^X k > n_Y^* a - n_Y^* k + (n_X^* - 1 - n_i^X)d$$

is satisfied, which is equivalent to

$$n_X^* > \frac{n(a - k) + (b - d) + n_i^X(k - d)}{a + b - k - d}$$

For a Y player, an analogous restriction holds:

$$(n_Y^* - 1)a - n_j^Y k > n_X^* b - n_X^* k + (n_Y^* - 1 - n_j^Y)c$$

which can be transformed equivalently into

$$n_X^* < \frac{n(a-c) - (a-c) + n_j^Y(c-k)}{a+b-k-c}$$

Both restrictions on n_X^* imply condition (4.4).

(e) Since $a > k > b$, it only pays for an individual player to choose Y and to look for as many Y player connections as possible. However, this argument only works when at least one player in the population selects Y. If all players choose X, the best reply of each individual player is to shut down all connections with the remaining players.

Q.E.D

REMARKS

(i) The result in parts (d) and (e) of Theorem 2 seems to have some features in common with the literature on 'equilibrium selection by migration' (for example, Ely, 1995; Bhaskar and Vega-Redondo, 1996). In these models, players can move to different locations where they play a simple 2×2 co-ordination game with each player at the same location. As a main result it can be demonstrated that all players move to the same location where the payoff-dominant equilibrium will be played provided the migration costs are low enough. In our framework, we have the opposite implications of communication costs. Moderate communication costs prevent players from co-ordination failure and let players build up isolated groups in which they choose the same action. If communication costs are large enough to make coordination on X unprofitable, then players select the payoff-dominant equilibrium in one completely connected group.

(ii) We know from theoretical and experimental work on equilibrium selection in co-ordination games that there exist many situations in which players do *not* select the payoff-dominant equilibrium (see, for example, Kandori *et al.*, 1993; Berninghaus and Schwalbe, 1996a; Cooper, 1999). In our framework it can be guaranteed that the payoff-dominant equilibrium in the co-ordination game is selected if connection costs are 'large enough'. This is made precise by the condition on k in part (e) of Theorem 2.

(iii) Our results in Theorem 2 are close to the results of the 'static' part of the paper by Goyal and Vega-Redondo (2002 – see proposition 1 and proposition 2). In addition we point explicitly to the conditions that equilibrium action distributions have to satisfy. This is important particularly for disconnected networks, where some players belong to the group of X players while the remaining ones belong to the group of Y players that is isolated from the former group. Obviously, such an action distribution cannot be stable if there is at least one player who can gain higher payoffs by joining the other group and switching strategies. Our inequality (4.4) is a sufficient condition which just prevents players from group and action switching.

We illustrate the equilibrium network of part (d) (Theorem 2) in Figure 4.4. The empty network and the complete network (see parts (a) and (b) of Theorem 2) have already been illustrated in Figure 4.3. Part (d) represents the disconnected networks where both subpopulations (of X resp. Y players) are completely isolated. This result (part (d)) follows from the size of communication costs that make co-ordination failures very costly.

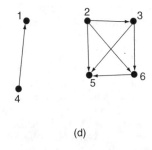

(d)

Figure 4.4   Equilibrium network in part (d) of Theorem 2

Concluding remarks

Our results in Theorems 1 and 2 show clearly the impact of communication costs and the type of the base game on network formation. As a main conclusion we state that one cannot separate the network linking decision from the action decision in a particular base game. Strategy choice in a population of players depends crucially on the communication (=linking) costs *and* the numerical payoff constellations in the base game.

Our framework is still rather simple and completely static. It can be extended in many ways. First, it is certainly not very realistic to assume that players will find an equilibrium in such a complicated one-shot network game Γ in one period. We need to find an extension of our simple static model to a dynamic strategy adaptation process in which players change their decisions (network and actions decisions) according to some well-defined adaptation rules. One could also incorporate into such a dynamic process the more or less plausible assumption that players show a lower speed of adaptation in their network decisions than in their action decisions (in the base game). For example, one could model that neighbourhood decisions are revised every 5th period, while action decisions may be changed in every period.[13]

In communication network games it is often assumed that one player can reach many players by one direct link to another player provided this player is connected with many other players via direct or indirect links. It is not easy to justify such an assumption in game networks, where all relationships are only bilateral (at least for two-person games). Nevertheless, it seems to be interesting to 'experiment' with this assumption. One could assume, for example, that opening a link to a player would guarantee access to his/her 'club' of direct and indirect neighbours. A similar assumption has been made in pure communication networks (Berninghaus *et al.*, 2003). Many other variants of access to a player's neighbourhood are possible.

Finally, one can think of substituting the simple 2×2 symmetrical co-ordination games by more complex ones, either by asymmetric co-ordination games or by $N \times N$ coordinations games (for $N > 2$). Which equilibrium networks could be expected in such models?

Notes

1 For example, players are supposed to be located on a circle or a grid.

2 Note, however, that the main aim of Goyal and Vega-Redondo is to analyse stochastic stable states, which is different from our framework.

3 This assumption and its extensions have been discussed extensively in the literature on network formation (see Jackson and Wolinsky, 1996; and Bala and Goyal, 2000). At a first glance it seems to be strange that a player has to accept the offer of any other player to play. However, this assumption simplifies the model considerably. Moreover, Berninghaus and Vogt (2003) show that by this assumption one obtains results that do not differ significantly from a model in which both players have to agree before a link is opened. Goyal and Vega-Redondo (2002) argue that positive payoffs in the base game suffice to induce a rational player to agree to play with a partner who just opened a link to him/her. By refusing to play, he/she would in fact lose payoff.

4 To avoid notational confusion, we follow the convention to call 'strategies' in the H/D game as 'actions' in the more general network game.

5 Clearly, this is a simplifying assumption which makes our results tractable.

6 Since the equilibrium network we consider is the complete graph, we simplify notation and omit total connection cost in calculating total payoffs.

7 Note that it is required to check additional payoffs of opening new links to *all* remaining X players. Because of the assumption $(c - k) > 0$ (in part (c)) each direct link of a Y player to an X player generates a positive net benefit.

8 Remember that each connection between members in I_2 in a Nash network is only generated by exactly one direct link.

9 Note that $n_j^Y k$ denotes the linking costs with the remaining Y players. A Y player j need only consider action switching from Y to X. It does not pay to change the links with the remaining players.

10 Indeed, the payoff of an X player is equal to $n_y^*(a - k) > 0$. By switching to Y s/he can gain either negative or zero net benefits. The same conclusion holds for a Y player.

11 When we talk about unique equilibrium networks (for example, in part (b) of Theorem 1) these networks are unique in being regarded as non-directed graphs. In fact, such non-directed graphs may be generated by many different simple directed graphs.

12 In calculating the payoff generated by deviation, note that (a) building up new links to all Y players generates communication costs equal to $n_Y^* k$; and (b) a deviating X player still has $(n_X^* - 1 - n_i^X)$ X players who have direct links with him/her. From each of these players s/he will extract an individual payoff of equal to d.

13 Recently, Goyal and Vega-Redondo (2002) proposed a dynamic analysis of network formation by analysing stochastic stable states in the population of players. In the tradition of Kandori et al. (1993) resp. Young (1993) they introduce a stochastic mutation term into the network formation decision problem. By letting the mutation term converge to zero the authors determine (very) long-run equilibria of action and network choice. This is an important contribution in evolutionary game theory; however, we believe that this method is more appropriate to study equilibrium selection problems than dynamic strategy adaptation processes.

References

Bala, V., and Goyal, S. (2000) A Non-Cooperative Model of Network Formation, *Econometrica*, 68, 5, 1181–230.

Berninghaus, S. K. (2003) 'The Evolution of Conventions', Hartmut Kliemt (ed.), *Jahrbuch f. Neue Politische Ökonomie*, Vol 21, Mohrsiebeck: Tübinger 101–132

Berninghaus, S. K. and Vogt, B. (2003) 'Network Formation in symmetric 2×2 Games', Mimeo, University of Karlsruhe, WIOR Institute.

Berninghaus, S. K. and Schwalbe, U. (1996a) 'Conventions, Local Interaction, and Automata Networks', *Journal of Evolutionary Economics*, 6, 297–312

Berninghaus, S. K. and Schwalbe, U. (1996b) 'Evolution, Interaction, and Nash-Equilibria', *Journal of Economic Behavior and Organization*, 29, 57–85.

Berninghaus, S. K., Ott, M. and Vogt, B. (2003) 'On Networks and "Stars": Recent Experimental Evidence on Network Games, Mimeo, University of Karlsruhe, WIOR Institute.

Bhaskar, V. and Vega-Redondo, F. (1996) 'Migration and the Evolution of Conventions', Discussion paper, University of Alicante.

Blume, L. (1993) 'The Statistical Mechanics of Strategic Interaction', *Games and Economic Behavior* 4, 387–424.

Boyer, R. and Orleans, A. (1992) 'How Do Conventions Evolve?', *Journal of Evolutionary Economics* 2, 165–77.

Cooper, R. W. (1999) 'Coordination Games', *Cambridge University Press*, Cambridge Mass

Cooper, R. W., De Jong, D.V. Forsythe, R. and Ross, T.W. (1992) 'Forward Induction in Coordination Games', *Economics Letters* 40, 167–72.

Ellison, G. (1993) 'Learning, Local Interaction, and Coordination', *Econometrica*, 61, 1047–71.

Ely, J. (1995) 'Local Conventions', Mimeo, University of California, Berkeley.

Eshel, I., Samuelson, L., and Shaked, A. (1998) 'Altruists, Egoists and Hooligans in a Local Interaction Model', *American Economic Review*, 88, 157–79.

Goyal, S. and Vega-Redondo, F. (2002) 'Learning, Network Formation and Coordination', Mimeo, University of Essent

Jackson, M. and Wolinsky, A. (1996) 'A Strategic Model of Economic and Social Networks', *Journal of Economic Theory*, 71, 44–74.

Kandori, M, Mailath, G. J., and Rob, R. (1993) 'Learning Mutation and Long Run Equilibria in Games', *Econometrica*, 61, 29–56.

Kosfeld, M. (2003) 'Economic Networks in the Laboratory: A Survey', Mimeo, University of Zürich.

Lewis, D. K. (1969) *Convention – A Philosophical Study*, Cambridge Mass., Harvard University Press.

Maynard Smith, J. and Price, G. R. (1973) 'The logic of animal conflict, *Nature*, (London) 246, 15–18.

Ochs, J. (1995) 'Coordination Problems', in J. H. Hagel and A. E. Roth Handbook of Experimental Economics, eds., Princeton, NJ, Princeton University Press.

van Huyck, J., Battalio, R. C. and Beil, R. O. (1990) 'Tacit Coordination Games, Strategic Uncertainty and Coordination Failure', *American Economic Review*, 80, 234–48.

Young, H. P. (1993) The Evolution of Conventions, *Econometrica*, 61, 57–84.

Young, H. P. (1998) *An Evolutionary Theory of Institutions*, Princeton, NJ, Princeton University Press.

5
Specific Skills, Imperfect Information and Job Rationing

*Helmut Bester** *

Introduction

This chapter is about job rationing and segmented labour markets. It analyses labour market equilibrium when job applicants cannot observe directly the amount of specific training that is provided by different firms. But employees become aware of the quality of training and their productivity as the employment relationship evolves. This is important because specifically-trained workers enjoy a bargaining advantage; an employer would incur a loss in output if he replaced his trained workforce with new employees. As a result of their bargaining position, the trained workers' wage is related positively to their productivity. Job applicants are therefore interested in identifying those firms that provide a large amount of specific training. It is shown that a firm's wage offer to untrained workers may signal information about the training it provides. This informational role of wages is the reason why they may not adjust to equilibrate demand and supply for all jobs. Under imperfect information, different jobs may be associated with different present values of lifetime income. In such an equilibrium, there is job rationing and unequal treatment among identical workers.

According to Becker's (1962) argument, employer and employee will share the costs of, and the returns to, specific training.[1] This solution gives both parties an incentive not to terminate the relationship unilaterally and impose a capital loss on the other party. This arguments reasoning is that long-term wage contracts have to be self-enforcing. Wages have to be compatible with the incentives and the bargaining position of each party, period by period. This is the case if the surplus from co-operation is shared at each stage. As a result, the wage paid for trained labour depends on its productivity within the firm. Under perfect information, employees

* I wish to thank two anonymous referees for helpful suggestions.

know their future wages, so untrained workers accept a lower wage if they are to receive higher incomes after being trained. In equilibrium they are indifferent between the lifetime income of different jobs. Rationing or unequal treatment is not compatible with competition under perfect information.

It appears reasonable, however, to assume that employees are not perfectly informed about the quality of a job when applying for it.[2] In this situation, firms may use their wage offers to untrained workers as a signal to reveal information about themselves. The resulting equilibrium involves high-productivity firms rationing workers. These firms cannot lower their wages because this would make them indistinguishable from low-productivity firms. Workers who are rejected at jobs with a large amount of specific training have to accept employment in firms with less training and lower incomes. These features resemble the distinction between 'primary' and 'secondary' sectors in Doeringer and Piore's (1971) dual labour market theory.[3]

As in the present model, asymmetric information plays a key role also in two other explanations of job rationing.[4] Imperfect information about the firm's technology has been incorporated into the implicit contract model of Azariadis (1975) and Bailey (1974). Under the informational assumptions, the firm's employment strategy becomes a signal for its profitability. The optimal contract may then entail *ex post* inefficiencies like involuntary underemployment.[5] However, the implicit contract model fails to explain *ex ante* rationing; in equilibrium, the market for employment contract clears and there is no unequal treatment.[6] The second category of models assumes that employers cannot directly observe or control their employees' productivity. In these models, lower wages reduce the worker's productivity because of either adverse selection or incentive effects.[7] This 'efficiency wage' argument explains why firms may not wish to cut wages even in the presence of excess demand. This chapter differs from these approaches mainly because workers rather than firms represent the less informed side of the labour market. Productivity is fixed within each firm and is not affected by wage payments. The firm's unobservable characteristics determine the structure of equilibrium. The model not only explains the possibility of rationing but also predicts that its occurrence is closely related to the amount of specific on-the-job training in different industries.

In the next section we develop a stylized model of a labour market with firm-specific training. The third section analyses the benchmark as equilibrium under perfect information, where job applicants can observe the amount of training in each type of firm. In the fourth section we describe competition under imperfect information: and as the fifth section shows, in this situation the competitive outcome may involve job rationing and a segmented labour market. The sixth section concludes.

The model

In a two-period economy, there are L workers and two types of firms, indexed $\theta = a, b$. The number of firms of type θ is N_θ. Labour is the only input to produce homogeneous output, and each firm can employ one worker each period. The production technology of firm θ is described by $Y_\theta = (Y_{\theta 1}, Y_{\theta 2})$, where $Y_{\theta 1}$ and $Y_{\theta 2}$ is the output from employing a worker at dates 1 and 2, respectively. If no worker is hired, output is zero. Workers acquire specific skills in the first period through on-the-job training. Therefore, $Y_{\theta 2}$ is higher than $Y_{\theta 1}$ if at date 2 the same worker is employed as in the period before. But because training is specific, a new employee can produce only $Y_{\theta 1}$ at any date.

There is a perfect credit market. Workers and producers can borrow and lend at the exogenously given interest rate $(1 - \beta)/\beta$; that is, they have the common discount factor β. Let $w_\theta = (w_{\theta 1}, w_{\theta 2})$ denote the wages paid by firm θ to its employee in Periods 1 and 2. Then its overall profit is given by:

$$\Pi_\theta(w_\theta) = Y_{\theta 1} - w_{\theta 1} + \beta[Y_{\theta 2} - w_{\theta 2}] \tag{5.1}$$

The worker's lifetime income from the wage profile w_θ is:

$$U(w_\theta) = w_{\theta 1} + \beta w_{\theta 2} \tag{5.2}$$

The worker's reservation wage for employment in any of the two industries is ω per period. Therefore, the worker will accept wage offer w_θ only if $U(w_\theta) \geqslant (1 + \beta)\omega$. The alternative income ω may either represent the worker's utility from leisure or the wage paid by firms without on-the-job training.

The firms' technology is taken to satisfy:

$$Y_{\theta 1} < \omega < Y_{\theta 2}, \, Y_{\theta 1} + \beta Y_{\theta 2} > (1 + \beta)\omega \tag{5.3}$$

Moreover, it will be assumed that:

$$N_a < L, \quad N_b < L, \quad N_a + N_b > L \tag{5.4}$$

As $Y_{\theta 1} < \omega$, training is costly in the first employment period. Therefore, replacing a trained worker in the second period is not profitable for a firm. This simplifies our analysis of the second-period equilibrium because there is no 'secondary market' for trained workers outside their current employment. But, at the social discount rate β, training increases the present value of output because of its second-period return. Under assumption (5.4), all workers receive jobs with training only if both types of firms operate. Yet producers have to compete for labour and some firms will not be able to operate.

Finally, type a and type b firms differ according to the following assumption:

$$Y_{a2} > Y_{b2} \tag{5.5}$$

Firms of type a are more productive at the second date than type b firms. In the following section, which analyses the perfect information equilibrium, we do not need a condition on the first-period productivity differential $Y_{a1} - Y_{b1}$. In the fifth section, however, we shall show that job rationing occurs under asymmetric information if Y_{b1} is not too large relative to Y_{a1}.

Equilibrium under perfect information

A worker who is employed in either industry a or b, enjoys a bargaining position in Period 2. If the firm replaces the worker, then it incurs a loss in output of at least $Y_{\theta2} - Y_{\theta1}$. Hence, employer and employee find themselves in a situation of bilateral monopoly, and the wage rate $w_{\theta2}$ depends on the outcome of a bargaining process. If co-operation is not achieved, then the worker quits and receives the alternative income ω. It is not profitable for the producer to hire an untrained worker at date 2. An untrained worker would produce $Y_{\theta1}$ and require a wage of at least ω. By Equation (5.3) this would render losses to the firm. Thus the best response of the firm is to stop operating after its employee quits. Accordingly, the surplus from co-operation amounts to $Y_{\theta2} - \omega$. This surplus is shared so that the worker receives a fraction $1 - \delta$ of it. A specific value for the parameter δ might be obtained from either axiomatic or non-cooperative approaches to bargaining theory.[8] Instead of deriving an explicit solution to the bargaining problem, it will be convenient to consider δ as exogenous and to assume simply that $0 < \delta < 1$. Given δ, second-period wages in the θ-type firms are equal to:

$$\tilde{w}_{\theta2} = (1 - \delta)[Y_{\theta2} - \omega] + \omega \tag{5.6}$$

In the second period, wages are related positively to productivity; by Equation (5.5), they are higher in sector a than in sector b.

An equilibrium of the economy specifies, for each industry θ, the wage rates w_θ and the number of firms n_θ that in fact employ a worker. While second-period wages are the outcome of the bargaining process described above, first-period wages are determined competitively. Under perfect information, workers are able to distinguish the two types of prospective employers. When hired by a producer of type θ, they know that their future wage will be $\tilde{w}_{\theta2}$, as given by Equation (5.6). Given the firms' first-period wage quotations, the workers choose among jobs to maximize income. Producers act as Bertrand–Nash competitors. They take the behaviour of workers as being fixed and seek to maximize profits through their own offers. An equilibrium is attained if no firm has an incentive to change its wage contract.

It is straightforward to compute the equilibrium $\{\hat{w}_\theta, \hat{n}_\theta\}$ under perfect information. In equilibrium, it must be the case that $\hat{n}_a + \hat{n}_b = L$. Otherwise, workers would earn their reservation income only, and $U(w_\theta) = (1+\beta)\omega$ would imply $\Pi_\theta(w_\theta) > 0$ for both types' θ. This is, however, inconsistent with equilibrium: by Equation (5.4) there are some firms that do not produce. These could make a positive profit by offering slightly higher wages to attract workers. The same argument shows that profits must be zero for at least one type of θ. In equilibrium, workers will be indifferent between $\hat{w}_a$ and $\hat{w}_b$, and so the wage bill to hire a worker for two periods is the same in each industry. Accordingly, firms with a higher present value of output $Y_{\theta 1} + \beta Y_{\theta 2}$ will earn positive profits, whereas the others make zero profits. Each of the more productive firms employs a worker in equilibrium. The less efficient ones are indifferent between operating or not; they employ all workers who do not have a job in the more productive firms.

The zero-profit condition together with Equation (5.6) determines the first-period wage in the less-productive sector. From Equation (5.6) and $U(\hat{w}_a) = U(\hat{w}_b)$ it follows that:

$$\hat{w}_{b1} - \hat{w}_{a1} = (1-\delta)\beta[Y_{a2} - Y_{b2}] \tag{5.7}$$

First-period wages are inversely related to second-period productivity in each sector, and by assumption (5.5) one has $\hat{w}_{a1} < \hat{w}_{b1}$. Workers employed by type a firms have steeper age earnings profiles than their colleagues in sector b.

Competition under imperfect information

Under imperfect information, workers cannot distinguish producers of type a and b directly at the first date. However, a worker learns about his/her employer's identity while being employed. On-the-job training not only raises the employee's productivity, but it also makes him/her aware of his/her future output as a trained worker. Since the employee knows about the producer's type at date 2, bargaining about second-period wages is in no way different from perfect information, so that w_{a2} and w_{b2} are again determined by Equation (5.6).

But, condition (5.7) may no longer be incentive compatible when information is asymmetric. Because $\hat{w}_{a1}$ is lower than $\hat{w}_{b1}$, type b producers could seek to reduce their wage costs by misrepresenting their identity. This would, of course, unravel the equilibrium $\{\hat{w}_\theta, \hat{n}_\theta\}$.

The environment of the economy is common knowledge under imperfect information. In Period 1, producers compete for workers. In addition to their wage offer $w_{\theta 1}$, they may also advertise their type in order to inform the worker about his/her future wage prospects. Of course, the worker will regard such a message as informative only if no gain can be made by an untruthful announcement. Let $S = \{\underline{s}, \bar{s}\}$ denote the set of possible signalling activities.

We shall adopt the convention that, if some firm wishes to advertise that it is of type a, it will choose the signal $\bar{s}$. No firm is interested in revealing that it is type is b. Therefore, the selection of $\underline{s}$ may simply be taken to imply that the employer is not interested in making a statement about the company's type. There are no costs associated with advertising either $\underline{s}$ or $\bar{s}$.

Workers are interested in identifying firms of type a because these yield higher second-period incomes. Their job choice will depend on their *beliefs* about the identity of employers. These beliefs are specified for all possible wage offers w_1 and messages $s \in S$. Formally, they are represented by a function $\mu(\cdot)$ such that $\mu(w_1, s) \in [0, 1]$ denotes the worker's probability assessment that the wage offer w_1, together with the signal s, comes from a type a firm. In accordance with the notion of a 'sequential equilibrium', as introduced by Kreps and Wilson (1982), the beliefs μ allow us to identify the workers' optimal behaviour for any strategy chosen by the employers. Each worker selects among all available offers to maximize expected utility. Whenever two firms adopt an identical strategy, they are equally likely to attract workers.

At date 1, each firm selects a strategy (w_1, s); then the workers apply for jobs. Given a system of beliefs μ^*, the allocation $\{w_\theta^*, n_\theta^*\}_\theta$ and the firms' choices of signals $\{s_\theta^*\}_\theta$ constitute an *equilibrium under imperfect information* if no producer of either type θ can increase expected profits through another wage offer $w_1 \neq w_{\theta 1}^*$ or message $s \neq s_\theta^*$. In addition, μ^* has to be consistent with the equilibrium outcome.

If the workers can distinguish the two types of firms by the different wage-signalling choice they make, the equilibrium is *separating* and μ^* satisfies $\mu^*(w_{a1}^*, s_a^*) = 1$ and $\mu^*(w_{b1}^*, s_b^*) = 0$.[9] Otherwise, *pooling* occurs and the workers' beliefs coincide with the actual fractions of type a and b firms in the market so that $\mu^*(w_{a1}^*, s_a^*) = \mu^*(w_{b1}^*, s_b^*) = N_a/(N_a + N_b)$. In the pooling equilibrium, workers are indifferent between all job offers, and each firm has the same probability of attracting an employee.

The definition of equilibrium imposes no restrictions upon beliefs which are conditioned on firm strategies that are not chosen in equilibrium. This indeterminacy of out-of-equilibrium beliefs typically leads to multiple equilibria in games with imperfect information. In order to avoid this problem we shall impose a further, economically reasonable restriction on μ^* by adopting the 'intuitive citerion' of Cho and Kreps (1987). This criterion is based on the following argument. Let some firm select an out-of-equilibrium strategy (w_1, s). The workers should conclude that this offer is made by a type a firm if the following two conditions are satisfied:

(i)   Should (w_1, s) attract a worker, then choosing (w_1, s) is advantageous for the firm if it is of type a; and

(ii) If the firm is of type b, then it cannot gain by proposing (w_1, s) even when a worker applies for this offer.

In equilibrium, the expected profit of a type θ producer is $n_\theta^* \Pi(w_\theta^*)/N_\theta$, because his/her offer attracts a worker with probability n_θ^*/N_θ. Accordingly, μ^* satisfies the intuitive criterion if:

$$\Pi_a(w_1, w_{a2}^*) > \frac{n_a^*}{N_a} \Pi_a(w_a^*) \tag{5.8}$$

and

$$\Pi_b(w_1, w_{b2}^*) \leqslant \frac{n_b^*}{N_b} \Pi_b(w_b^*)$$

implies

$$\mu^*(w_1, s) = 1, \text{ for all } s \in S$$

In the following, the term 'equilibrium' will be used only if the associated beliefs μ^* are consistent with the intuitive criterion (5.8).

Separating equilibria and job rationing

Job rationing occurs when some workers receive certain jobs and others do not, and the latter are worse off. This may happen if wages act as a signalling mechanism and fail to equilibrate demand and supply. When several workers apply for a job, the employer selects one of them at random and rejects the others. The latter may then proceed to seek employment in another firm. Obviously, rationing can occur at most in the separating equilibrium because in the pooling equilibrium the worker is indifferent between all job offers. The separating equilibrium entails rationing if $U(w_a^*) > U(w_b^*)$. Since by assumption (5.4) only a fraction of all workers can be employed in the a industry, some of them are denied a type a job and have to console themselves with a type b job.

The purpose of our analysis is to show that these features are a characteristic of the equilibrium when the parameters of the model satisfy the following condition:

$$\beta\delta[Y_{a2} - Y_{b2}] > Y_{b1} - Y_{a1} \tag{5.9}$$

This requires that the second-period productivity differential between type a and type b firms is sufficiently large. Inequality (5.9) holds if, and only if, $\Pi_a(w_1, \tilde{w}_{a2}) > \Pi_b(w_1, \tilde{w}_{b2})$ for all w_1.

We now show that under condition (5.9) there is a separating equilibrium with the following features: Wages and employment in the first period are:

$$w_{a1}^* = w_{b1}^* = Y_{b1} + \beta\delta[Y_{b2} - \omega], \quad n_a^* = N_a, n_b^* = L - N_a \tag{5.10}$$

The firms' second-period wages (w_{a2}^*, w_{b2}^*) are determined by the bargaining solution (5.6), and their signalling choices are given by $(s_a^*, s_b^*) = (\bar{s}, \underline{s})$.[10] This outcome is supported by the following beliefs μ^*: every wage offer above $w_{\theta 1}^*$ is considered to come from type a employers; every offer below $w_{\theta 1}^*$ is believed to originate from type b producers. If $w_{\theta 1}^*$ is offered, then $\bar{s}$ is regarded as a signal of type a and $\underline{s}$ as a signal of type b. Notice that $U(w_a^*) > U(w_b^*)$ because $w_{a1}^* = w_{b1}^*$ and $w_{a2}^* > w_{b2}^*$. Therefore, a worker prefers to be employed in a type a firm. This implies that the equilibrium outcome involves job rationing.

Indeed, it is easily verified that Equation (5.10) describes an equilibrium. The wage $w_{\theta 1}^*$ is chosen such that firms of type b earn zero profits. They are indifferent between operating or not, and have no incentive to misrepresent their identity. Each type a producer operates in equilibrium; by Equation (5.9) the producer earns positive profits. Neither s/he nor a type b producer can gain by offering a different first-period wage. This is so because no worker would accept a wage below w_{a1}^*. According to the worker's beliefs, such an offer would yield even less income than $U(w_b^*)$. Finally, μ^* is consistent with the equilibrium outcome and satisfies requirement (5.8). Indeed, since $n_a^* = N_a$, a type a producer could gain the most by offering a first-period wage $w_1 < w_{a1}^*$. Consequently, $\Pi_a(w_1, \tilde{w}_{a2}) > n_a^* \Pi(w_a^*)/N_a$ implies $\Pi_b(w_1, \tilde{w}_{b2}) > n_b^* \Pi(w_b^*)/N_b$. This precludes that a type a firm can appeal to the intuitive criterion when it offers a first-period wage $w_1 \neq w_{a1}^*$. By showing that Equation (5.10) represents in fact the only allocation that can be supported as an equilibrium if Equation (5.9) holds, we establish the following result:

PROPOSITION 1 *Let condition (5.9) be satisfied. Then there exists a unique equilibrium. This equilibrium is separating and exhibits job rationing at type a firms.*

PROOF First, it will be shown that the equilibrium must be separating. Suppose the equilibrium were pooling. Then $n_a^*/N_a = n_b^*/N_b < 1$. Define w_1^o such that $\Pi_b(w_1^o, \tilde{w}_{b2}) = n_b^* \Pi_b(w_b^*)/N_b$. Note that $w_1^o > w_{a1}^* = w_{b1}^*$. Using Equation (5.9), it is easily verified that $\Pi_a(w_1^o, \tilde{w}_{a2}) > n_a^* \Pi_a(w_a^*)/N_a$. It thus follows from Equation (5.8) that $\mu^*(w_1^o, s) = 1$. With these beliefs, the worker prefers the offer (w_1^o, s) to (w_{a1}^*, s_a^*) so that (w_1^o, s) certainly attracts a worker. But then $\Pi_a(w_1^o, \tilde{w}_{a2}) > n_a^* \Pi_a(w_a^*)/N_a$ is inconsistent with expected profit maximization, a contradiction.

Second, it will be shown that $n_b^* < N_b$. Suppose $n_b^* = N_b$. Then profit maximization implies:

$$n_a^* \Pi_a(w_a^*)/N_a \geqslant \Pi_a(w_{b1}^*, w_{a2}^*), \quad \Pi_b(w_b^*) \geqslant n_a^* \Pi_b(w_{a1}^*, w_{b2}^*)/N_a \qquad (5.11)$$

Adding these inequalities yields:

$$n_a^*[\Pi_a(w_a^*) - \Pi_b(w_{a1}^*, w_{b2}^*)]/N_a \geqslant \Pi_a(w_{b1}^*, w_{a2}^*) - \Pi_b(w_b^*) \qquad (5.12)$$

By Equation (5.9), $\Pi_a(w_a^*) - \Pi_b(w_{a1}^*, w_{b2}^*) = \Pi_a(w_{b1}^*, w_{a2}^*) - \Pi_b(w_b^*) > 0$. Therefore Equation (5.12) implies that $n_a^* = N_a$. But then $n_a^* + n_b^* = N_a + N_b > L$, a contradiction to Equation (5.4).

Since the equilibrium is separating, $\mu^*(w_{b1}^*, s_b^*) = 0$. As $n_b^* < N_b$, competition among the type b firms then implies $\Pi(w_b^*) = 0$. This proves $w_{b1}^* = Y_{b1} + \beta\delta[Y_{b2} - \omega]$.

Next, it will be shown that $w_{a1}^* = w_{b1}^*$. Since $\Pi_b(w_b^*) = 0$, any $w_{a1}^* < w_{b1}^*$ would certainly be imitated by the type b firms. This proves $w_{a1}^* \geqslant w_{b1}^*$. Suppose $w_{a1}^* > w_{b1}^*$. Then any $w_{b1}^* < w_1 < w_{a1}^*$ satisfies $\Pi_a(w_1, w_{a2}^*) > n_a^*\Pi_a(w_a^*)/N_a$ and $\Pi_b(w_1, w_{b2}^*) \leqslant n_b^*\Pi_b(w_b^*)/N_b$ so that, by Equation (5.8), $\mu^*(w_1, s) = 1$. Therefore, the worker prefers (w_1, s) to (w_{b1}^*, s_b^*), and (w_1, s) certainly attracts a worker. Since (w_1, s) generates higher expected profits to firms of type a than (w_{a1}^*, s_a^*), this contradicts expected profit maximization. This proves $w_{a1}^* = w_{b1}^*$.

Finally, one must have $n_a^* = N_a$ because $U(w_a^*) > U(w_b^*)$. This shows that Equation (5.10) is the only equilibrium allocation, and that the equilibrium must be separating if Equation (5.9) holds.

Q.E.D.

The separating equilibrium displays a segmented labour market. Access to type a jobs is restricted, and the equilibrium produces a discriminatory wage differential. The informational role of wages keeps producers of type a from eliminating excess demand by lowering wages. The profit function of these firms is discontinuous at w_{a1}^*. Profits are positive for $w_{a1} = w_{a1}^*$, but vanish to zero with a small wage reduction. The reason is the change in the workers' beliefs. They consider any wage offer below w_{a1}^* as being made by type b firms and as less attractive than available offers from competing producers. Imperfect information creates a downward rigidity of wages which makes rationing compatible with equilibrium.

To compare the equilibrium outcome under perfect and imperfect information, notice that $Y_{a1} + \beta Y_{a2} > Y_{b1} + \beta Y_{b2}$ by condition (5.9). Thus type a firms are more productive than type b firms, not only in the second period but also from the perspective of overall efficiency over the two periods. Therefore, type b firms earn zero profits and offer the same first-period wage under both perfect and imperfect information. Accordingly, the workers employed by type b firms are equally well off in both situations. Yet, firms of type a earn lower profits under imperfect information because they can no longer realize the first-period wage differential in Equation (5.7). Instead, they have to offer the same wages as type b firms. As a result, it is the employees in type a firms who gain at the cost of their employers from the informational imperfection of the labour market.

Proposition 1 identifies a range of parameter constellations which must lead to a separating equilibrium with job rationing. It is worth noting that in the case of more than two firm types, rationing may also occur when some subset of types is pooled. Indeed, it may happen that all producers whose job offers create excess demand are pooled into one group. Firms within this group have no strict incentive to distinguish themselves from other members of the group; they only wish to be separated from outsiders with rather low second-period productivities. It should be clear, however, that rationing can occur only when there is at least some degree of information revelation in equilibrium. This can be ensured by conditions that are analogous to Equation (5.9).

Conclusion

This chapter has studied the labour market equilibrium in an economy with imperfect information about specific on-the-job training. We have shown that the resulting equilibrium may involve the high productivity firms rationing workers. These firms cannot lower their wage without being perceived by the workers to be low productivity firms, in which future wages are low. Because of its informational role, the wage rate fails to equilibrate the market in the traditional sense.

In the two-period model in this chapter there is no uncertainty at date 2 and rationing occurs only at date 1. But the present analysis may easily be embedded into an overlapping generations model in which each young generation is confronted with job rationing. Of course, such a model requires persistent uncertainty about the firms' future productivities. Asymmetric information of this type could be generated through successive industry-specific shocks about which the producers are better informed than the workers. After all, it seems reasonable to assume that the employer knows more about the future prospects of the business than does the employee.

Notes

1   For a formal analysis of the sharing model, see Hashimoto (1981). The interaction between general and specific investments in the sharing model is analysed in Kessler and Lülfesmann (2000), who show that the possibility of providing specific training leads the employer to invest in general human capital.

2   The literature on labour markets distinguishes between two possible sources of asymmetric information: in, for example, Hart (1983) the firms have private information about their technological characteristics, whereas in Weiss (1980), for example, the workers have private information about their productivity. Our model assumes asymmetric information of the first type.

3   A survey of segmented labour market theories is given by Cain (1976).

4   A model of asymmetric information in which firm-specific training interacts with on-the-job screening is studied in Bac (2000).

5   A survey is given by Hart (1983) and Rosen (1985). Depending on workers' preferences, these models may result in overemployment as well as underemployment; see, for example, Azariadis (1983), Green and Kahn (1983) or Cooper (1983).

6   Such an equilibrium is described in Holmstrom (1983).

7   A survey of these so-called 'efficiency wage models' is given by Yellen (1984) and Weiss (1990). An example of the adverse selection model is Weiss (1980); as an example of the incentive model, see Miyazaki (1984).

8 See, for example, Nash (1950) or Rubinstein (1982). The non-cooperative approach would incorporate the worker's alternative income ω according to the 'outside option' principle, see Sutton (1986). It is not important to assume that δ is the same in each firm. The essential condition is that $w_{\theta 2}$ and $Y_{\theta 2}$ are positively related.

9 We confine our analysis to symmetric equilibria, in which all firms of the same type select an identical strategy.

10 The firms' signals (s_a^*, s_b^*) serve to avoid the open set problem that arises because, as long as $w_{a1} > w_{b1}^*$, a type a firm can increase its profit by slightly reducing w_{a1}. The intuitive criterion requires the worker to consider the reduced wage as an offer of a type a firm because $\Pi_b(w_{a1}, w_{b2}^*) < 0$. As a result, the type a firm would like to set w_{a1} as close to w_{b1}^* as possible without becoming indistinguishable. Allowing firms to advertise their type is thus simply a way to eliminate the discontinuity in beliefs that occurs as the type a wage converges with the type b wage from above. The problem disappears if the set of possible wage offers is finite. Type b firms would then quote the highest first-period wage at which they do not lose money; type a firms would quote the next available offer above that.

References

Azariadis, C. (1975) 'Implicit Contracts and Underemployment Equilibria', *Journal of Political Economy*, 83, 1183–202.

Azariadis, C. (1983) 'Employment with Asymmetric Information', *Quarterly Journal of Economics*, 98, 157–72.

Bac, M. (2000) 'On-the-Job Specific Training and Efficient Screening', *Journal of Labor Economics*, 18, 681–701.

Bailey, M. N. (1974) 'Wages and Employment under Uncertain Demand', *Review of Economic Studies*, 41, 37–50.

Becker, G. S. (1962) 'Investment in Human Capital: A Theoretical Analysis', *Journal of Political Economy*, 70, 9–49.

Cain, G. G. (1976) 'The Challenge of Segmented Labor Market Theories to Orthodox Theory: A Survey', *Journal of Economic Literature*, 14, 1212–57.

Cho, I.-K. and Kreps, D. M. (1987) 'Signaling Games and Stable Equilibria', *Quarterly Journal of Economics*, 102, 179–221.

Cooper, R. (1983) 'A Note on Over/Underemployment in Labor Contracts under Asymmetric Information', *Economics Letters*, 12, 81–9.

Doeringer, P. B. and Piore, M. J. (1971) *Internal Labor Markets and Manpower Analysis*, Lexington, Mass, Heath.

Green, J. and Kahn, C. M. (1983) 'Wage-Employment Contracts', *Quarterly Journal of Economics*, 98, 173–89.

Hart, O. (1983) 'Optimal Labor Contracts Under Asymmetric Information: An Introduction', *Review of Economic Studies*, 50, 3–36.

Hashimoto, M. (1981) 'Firm-Specific Human Capital as a Shared Investment', *American Economic Review*, 71, 475–82.

Holmstrom, B. (1983) 'Equilibrium Long-Term Labor Contracts', *Quarterly Journal of Economics*, 98, 23–54.

Kessler, A. and Lülfesmann, C. (2000) 'The Theory of Human Capital Revisited: On the Interaction of General and Specific Investments', CEPR Discussion Paper No. 2533.

Kreps, D. M. and Wilson, R. (1982) 'Sequential Equilibria', *Econometrica*, 50, 863–94.

Miyazaki, H. (1984) 'Work Norms and Involuntary Unemployment', *Quarterly Journal of Economics*, 99, 297–311.

Nash, J. (1950) 'The bargaining problem', *Econometrica*, 28, 155–162

Nash, J. (1953) 'Two-Person Cooperative Games', *Econometrica*, 21, 128–40.

Rosen, S. (1985) 'Implicit Contracts: A Survey', *Journal of Economic Literature*, 23, 1144–75.

Rubinstein, A. (1982) 'Perfect Equilibrium in a Bargaining Model', *Econometrica*, 50, 97–110.

Sutton, J. (1986) 'Non-Cooperative Bargaining Theory: An Introduction', *Review of Economic Studies*, 53, 709–24.

Weiss, A. (1980) 'Job Queues and Layoffs in Labor Markets with Flexible Wages', *Journal of Political Economy*, 88, 526–38.

Weiss, A. (1990) *Efficiency Wages* Princeton, NJ, Princeton University Press.

Yellen, J. L. (1984) 'Efficiency Wage Models of Unemployment', *American Economic Review* (Proceedings), 74, 200–5.

6
Testing Game Theory
*Jörgen W. Weibull**

Introduction

An important development in economics is the emergence of experimental economics, and Werner Güth has been one of its pioneers. Moving from armchair theorizing to controlled laboratory experiments may be as important a step in the development of economic theory as it once was for the natural sciences to move from Aristotelian scholastic speculation to modern empirical science.[1]

The first experiments in game theory were carried out in the early 1950s. However, a new wave of game experiments began in the mid 1970s, and Güth *et al.* (1982) pioneered experimental work on so-called ultimatum bargaining situations. For surveys of such experiments, and for introductions to experimental game theory more generally, see Güth and Tietz (1990), Bolton and Zwick (1995), Kagel and Roth (1995), Zamir (2000), the special issue of the *Journal of Economic Theory* in 2002 devoted to experimental game theory, and Camerer (2003).

The present chapter discusses some methodological and conceptual issues that arise when non-cooperative game theory is used for positive analysis of human strategic interaction. In particular, in the experimental literature, it has many times been claimed that certain game-theoretic solutions – such as Nash equilibrium and subgame perfect equilibrium – have been violated in laboratory experiments.[2] While it may well be true that human subjects do not behave according to these solutions in many situations, few experiments actually provide evidence for this. Especially in the early literature,

* This chapter is a major revision of SSE WP 382 of May 2000. I am grateful for helpful comments from Ana Ania, Geir Asheim, Kaushik Basu, Larry Blume, Vincent Crawford, Martin Dufwenberg, David Easley, Tore Ellingsen, Ernst Fehr, Jean-Michel Grandmont, Thorsten Hens, Jens Josephson, Donzhe Li, Sendhil Mullainathan, Rosemarie Nagel, Al Roth, Maria Saez-Marti, Larry Samuelson, Martin Shubik, Jon-Thor Sturlason, Sylvain Sorin, Fernando Vega-Redondo and Shmuel Zamir, and to seminar participants at presentation of various drafts of the chapter.

experimentalists did not make any effort to elicit the subjects' preferences, despite the fact that these preferences constitute an integral part of the very definition of a game. Instead, it has been customary simply to assume that subjects care only about their own material gains and losses. In later studies, subjects' preferences were allowed also to depend on the 'fairness' of the resulting vector of material gains and losses to all subjects. However, recent experiments, discussed below, suggest that even this is sometimes too restrictive – subjects' ranking of alternatives may depend on other parts of the game form, a phenomenon here called 'context dependence'.

In applications of non-cooperative game theory, the game is not only meant to represent the strategic interaction as viewed by the analyst, but also as viewed by the players – it is even assumed quite frequently that the game is common knowledge to the players. Indeed, a variety of epistemic models have been built in order to analyse the rationality and knowledge assumptions involved in game-theoretic analysis – under the classical interpretation that the game in question is played exactly once by rational players. The extent and exact form of knowledge assumed on behalf of the players varies across game forms, solutions, and on the epistemic model in question, see Tan and Werlang (1988), Reny (1993), Aumann (1995), Aumann and Brandenburger (1995), Ben-Porath (1997) and Asheim (2002). Being deductive, such epistemic models of games can, of course, not be empirically falsified as such, only their assumptions, which are known to be strong idealizations. So what, then, can be tested? One can test whether the theoretical predictions are at least approximately correct in environments that approximate the assumptions. Such testing is important, because this is the way game theory is used in economics and the other social sciences.[3]

This chapter is somewhat discursive and philosophical, and contains no theorems. I hope, though, that it sheds some light on the very definition of a non-cooperative game, on the empirically relevant possibilities of context dependent preferences and interpersonal preference dependence, on backward induction and on incomplete information modelling. The interested reader is recommended to read Levine (1998), Sprumont (2000), Ray and Zhou (2001) and Binmore et al. (2002) for other discussions of some of these, and related, issues (connections to these earlier studies are commented briefly on below).

The discussion is organized as follows. The second section pins down some terminology and notation. In particular, a notion of 'game protocol' is introduced. The third section applies this machinery to a class of very simple ultimatum bargaining situations. The fourth section discusses backward induction more generally, and in particular how to reconcile it with context-dependent preferences. The fifth section discusses briefly interpersonal preference dependence, the sixth section shows how the model in Levine (1998) can be used to address the issues discussed in the two preceding sections, and the seventh section concludes.

Games and game protocols

The present discussion is focused on a slight generalization of finite games in extensive form, as defined in Kuhn (1950, 1953).[4] Such a game is a mathematical object that contains as its basic structure a directed tree, consisting of a finite number of nodes (or vertices) and branches (or edges). A *play*, τ, of the game is a 'route' through the tree, starting at its initial node and ending at one of the *end nodes*, ω. A node k' is a successor of a node k if there is play that leads first to k and then to k'. Moreover, each end node is reached by exactly one play of the game, and each play reaches exactly one end node. Let Ω denote the set of end nodes, and T the set of plays. We then have $|\Omega| = |T| < +\infty$.

The set of non-end nodes is partitioned into player subsets, and each player subset in turn is partitioned into information sets for that player role. In each information set, the number of outgoing branches from each node is the same, and the set of outgoing branches from an information set is divided into equivalence classes, the moves available to the player at that information set, so that every equivalence class contains exactly one outgoing branch from each node in the information set. A choice in an information set is a probability distribution over the moves available at the information set. In games with exogenous random moves, one of the players is 'nature', and all information sets for this 'non-personal' player are singleton sets with fixed probabilities attached to each outgoing branch.

A pure strategy for a personal player role is a function that assigns one move to each of the role's information sets. The outcome of a strategy profile is the probability distribution induced on the set Ω of end-nodes, or, equivalently, on the set T of plays.

Since humans sometimes exhibit social preferences – that is, our choices are in part driven by concern for others, it is useful to allow for the possibility of *passive players* (or *dummy* players) – that is, player roles with empty player sets, but where the player may be affected by the choices made by other, active, players. Relevant examples are the so-called dictator games, where one player is active and one is passive. More generally, an active player may be passive in certain subgames and yet influence active players' preferences in the subgame.

The ingredients described so far together make up a *game form*.

Games

A game form becomes a game when the (personal) player roles are endowed with preferences. More exactly, in standard non-cooperative game theory, each player $i = 1, 2, \ldots, n$ is assumed to have preferences over the unit simplex:

$$\Delta(\Omega) = \Delta(T) = \left\{ p \in \mathbb{R}_+^{|\Omega|} : \sum_{i=1}^{|\Omega|} p_i = 1 \right\}$$

of lotteries over end nodes, or, equivalently, plays, satisfying the von Neumann – Morgenstern axioms.[5] Hence, for each player i there exists a real-valued function π_i with domain Ω, or T, such that player i prefers one lottery over another if and only if the expected value of the function π_i is higher in the first lottery than in the second. Such a function $\pi_i : \Omega \to \mathbb{R}$ (unique up to a positive affine transformation) will here be called the *Bernoulli function* of player i. The number $\pi_i(\omega)$ is usually called the payoff to player i at end node ω.[6] If Φ is a game form, then the pair $\Gamma = (\Phi, \pi)$, where π denotes a combined Bernoulli function $\pi : \Omega \to \mathbb{R}^n$, constitutes a finite extensive-form game.

Game protocols

In virtually all applications of game theory, including laboratory experiments, each play results in well-defined material consequences for the players. In applications to economics, and in most laboratory experiments, these material consequences are monetary gains or losses, in which case these are usually called monetary payoffs – not to be confounded with game theorists' definition of payoffs as Bernoulli function values.

In order to facilitate discussions of the effects of changed material or monetary payoffs, it is useful to introduce a name for game forms with specified material consequences. Hence, by a *game protocol* is here meant a pair (Φ, γ), where γ is a function that maps end nodes $\omega \in \Omega$ (or, equivalently, plays $\tau \in T$) to material consequences $c \in C$, for some set C rich enough to represent relevant aspects of the material consequences in question. If the material consequences are monetary gains and losses to the n players in the game form, then we may thus take C to be a subset of $\mathbb{R}^n$.[7]

The formal machinery of non-cooperative game theory does not require that a player's payoff value $\pi_i(\omega)$ at an end node ω be a function of the material consequences at that node. Indeed, two plays resulting in identical material payoffs to all players may well differ in terms of information sets reached, choices made or not made during play and so on – aspects that may be relevant for players' preferences and hence influence their Bernoulli functions. Standard game theory only requires the *existence* of a Bernoulli function π_i for each (personal) player i. Indeed, several laboratory experiments have convincingly – though perhaps not surprisingly for the non-economist – shown that human subjects' preferences are not driven only by their own monetary payoffs.[8]

Mini ultimatum protocols

A class of game protocols that have been much studied in the laboratory are those associated with ultimatum bargaining protocols. These two-player game protocols represent strategic interactions where the subject in role A, the proposer, makes a suggestion to the subject in role B, the responder, about how to split a fixed sum of money. The responder may accept or reject the proposal. If accepted, the sum is split as proposed. If rejected, both subjects receive nothing.

Figure 6.1 shows the game protocol of such a simplified strategic interaction, a mini ultimatum protocol, where 100 tokens are to be divided between two parties, a proposer and a responder. The proposer has only two choices, either to keep x tokens for him/herself, the outside option, or to offer the responder $100 - y$ tokens. In the latter case, the responder has a binary choice, whether to accept or reject the division $(y, 100 - y)$. If B rejects the proposal, both players receive zero tokens. Hence, the game form has three plays: $T = \{\tau_1, \tau_2, \tau_3\}$. In play τ_1, A chooses the division $(x, 100 - x)$ and play stops at end node ω_1. In play τ_2, A proposes the division $(y, 100 - y)$, B accepts this and play stops at end node ω_2. In play τ_3, A proposes $(y, 100 - y)$, B rejects this and play stops at ω_3. The numbers x and y are fixed and given, and known by the player subjects, where $0 < x < y < 100$. At the end of the experiment, tokens are exchanged for money, at a pre-set exchange rate.

In the early experimental literature it was presumed that the payoff values $\pi_i(\omega)$ to the subjects are monotone functions of their own monetary payoffs. Hence, in strategic interactions like this, it was claimed that non-cooperative game theory predicts play τ_2, namely, that A will propose the division $(y, 100 - y)$ and that B will accept this. This is, of course, the unique subgame-perfect equilibrium of the game that defined by such preferences. The implicit hypothesis in this early literature is fivefold:

H1: the responder prefers play τ_2 over play τ_3;
H2: the responder is rational in the sense of playing optimally, according to his/her preferences;
H3: the proposer knows that H1 and H2 hold (or at least believes that they hold with a sufficiently high probability);
H4: the proposer prefers τ_2 over τ_1; and
H5: the proposer is rational in the sense of acting optimally, in accordance with his/her knowledge and preferences.[9]

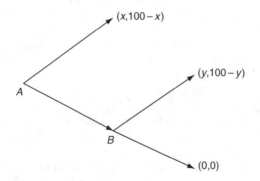

Figure 6.1 A mini ultimatum game protocol

A large number of laboratory experiments with more complex ultimatum bargaining situations of this sort have shown that many proposer subjects instead offer sizeable shares to the responder, and that many responder subjects reject small shares if offered. In the present mini ultimatum game protocol, this corresponds to plays τ_1 and τ_3. Such findings were initially interpreted as rejections of the subgame perfection solution concept. What was rejected was the combined preference-cum-knowledge hypothesis H1–H5 given above. This is not surprising, since neither hypothesis H1 nor H3 is true for all subjects.

In the present example, let $\succeq_A$ be a proposer subject's preferences over the set $\Delta(\Omega)$, and let $\succeq_B$ be a responder subject's preferences over the same set. For example, suppose $x = 50$ and $y = 90$. A subject in player role A may then have the preference $\tau_2 \succ_A \tau_1 \succ_A \tau_3$, and the subject in role B may have the preference $\tau_1 \succ_B \tau_3 \succ_B \tau_2$.[10] Indeed, such preferences are consistent with many subjects' behavior in laboratory experiments. All games in the associated game class (that is, with compatible Bernoulli functions) have the 50/50 split – that is, play τ_1, as the unique subgame perfect outcome.

In an experimental study of a variety of mini ultimatum protocols slightly more complex than the one in Figure 6.1, Falk *et al.* (2003) found that the responder rejection rate (across 90 subjects) depends not only on the current offer they faced, but also on the 'outside option' available to the proposer.[11] In the context of the present example: B's ranking of plays τ_2 and τ_3 may well depend on the material consequences of play τ_1.

As indicated above, this observation has implications for backward induction arguments: a change in one part of a game protocol may change players' preferences in another part, even if the first part cannot be reached from the second. This issue is addressed in the next section.

Backward induction

In a given game form Φ, let K_0 be the subset of nodes k such that (i) k is either a move by nature or $\{k\}$ is an information set of a personal player; and (ii) no information set in Φ contains both a successor node and a non-successor node to k. Each node $k \in K_0$ is the initial node of a subform, a game form Φ_k. For any such node k, let the associated subprotocol be defined as the game protocol (Φ_k, γ_k), where γ_k is the restriction of γ to the subset $\Omega_k \subset \Omega$ of end nodes that succeed node k. So far, so good. The subtlety arises when we try to define subgames, since for any given subprotocol there are two distinct candidates claiming that name.

First, there is the following definition of a subgame: if $\Gamma = (\Phi, \pi)$ is a game and $k \in K_0$, then the subgame at k is the game $\Gamma_k = (\Phi_k, \tilde{\pi})$, where $\tilde{\pi}$ is the restriction of π to the subset Ω_k. In other words, the Bernoulli function values in Γ_k coincide with those in Γ at all end nodes in Ω_k. This is the *context-dependent* definition of a subgame. In this definition, players' preferences in the subgame, represented by $\tilde{\pi}$, may depend on parts of the full game protocol (Φ, γ) outside the subprotocol (Φ_k, γ_k) in question. For example, $\tilde{\pi}$ may depend on choices available along the unique play leading up the node k and/or on material payoffs at end nodes not in Ω_k. This approach treats the full game protocol as the relevant context for all players' decisions at all points in the game protocol.

A second candidate for the title of 'subgame' at a node $k \in K_0$ is the game $\Gamma^o = (\Phi_k, \pi^o)$ that is obtained if the subprotocol (Φ_k, γ_k) is played in isolation – that is, beginning at node k as the initial node and without the 'context' of the rest of (Φ, γ). We shall call $\Gamma^o = (\Phi_k, \pi^o)$ the *isolated subgame* at k.

As an illustration of this distinction, consider the subform in Figure 6.1 beginning at the node k where player B has to accept or reject the proposal $(y, 100 - y)$. Viewed in isolation, this is a one-player game protocol, where the unique active player (B) has a binary choice of either (a) receiving $100 - y$ tokens while y tokens are given to a passive player A; or (b) no tokens to either of the two players. I guess an overwhelming majority of subjects in this isolated game protocol would choose the first option. However, we know that many subjects in player role B in the full game protocol in Figure 6.1 choose the second option. Taking their behaviour as indicating their revealed preference, this means that $\tilde{\pi} \neq \pi^o$. Indeed, it is an empirical question whether $\pi^o = \tilde{\pi}$. The above-mentioned observations in Falk *et al.* (2003), if taken as revealed preferences, show that $\tilde{\pi} \neq \pi^o$. Hence, the distinction between the two definitions of subgames may be critical.

Since the full game protocol is supposed to represent the relevant decision context for the players' decision-making, it is this author's opinion that backward induction should be applied to the full game protocol, with all players' preferences defined *in this protocol*. In particular, when applying subgame perfection, one should use the *context-dependent* definition of a subgame, and not that of an isolated subgame. Formally, a subgame perfect equilibrium is then a strategy profile that induces a Nash equilibrium on every subgame $\Gamma_k = (\Phi_k, \tilde{\pi})$.

However, much of the game-theoretic literature seems to ignore this distinction. For a recent example, see Ray and Zhou (2001), where it is assumed implicitly that $\tilde{\pi} = \pi^o$.[12] However, other researchers have provided experimental evidence against backward induction when carried out in terms of the isolated subgames – that is, by using π^o instead of $\tilde{\pi}$; see Binmore *et al.* (2002). Hence, what they reject is the combined hypothesis that the subjects' play is compatible with backward induction, as described here, *and* that their preferences satisfy $\tilde{\pi} = \pi^o$.

Note that the suggested approach – to base backward induction (and all other analysis) on the preferences in the full game protocol – is not a critique of 'Kuhn's algorithm', the usual way of solving finite games of perfect information by way of successively replacing each final decision node in the game tree by an end node with a payoff vector that equals the expected payoff vector achieved by some optimal move by the player at the decision node in question. All that is suggested here is that the payoffs should then be Bernoulli function values as defined from players' preferences in the full game protocol.

Context dependence in preference formation may have many causes. It may be that human subjects in player roles have opinions about actions taken and not taken on the way to the information set in question, with or without regard to the possible intentions behind those actions. It may also be that players have social preferences that depend on options available to others outside the subprotocol in question. However, for the purposes of game-theoretic analysis, it does not matter what the causes are, as long as players' preferences in the full game protocol are well defined. The analyst's task of eliciting the preferences of subjects in player roles of a given game protocol is in general not easy. It is particularly difficult if subjects have interpersonally dependent preferences – that is, if their rankings of outcomes depend in part on their expectations of other player subjects' rankings of outcomes, which depend on those other subjects' expectations of the others' rankings, and so on – which is the topic of the next two sections.

Interpersonal preference dependence

The elicitation of players' preferences raises a fundamental issue in the very definition of a game, namely whether a player's preferences may depend on (knowledge of, or beliefs about) another player's preferences, which in turn may depend on (knowledge of, or beliefs about) the first player's preferences and so on. Such potential interpersonal preference dependence is disturbing, since it makes the domain of preferences unclear – the game protocol is then not an exhaustive representation of the interactive situation – and yet such interdependence might realistically exist in some interactions.

Example

In order to illustrate this possibility, consider the game protocol in Figure 6.2.[13] Here, a mini 'dictator game' form follows the tossing of a fair coin deciding which of the two players should be the 'dictator'. The game form has four plays: τ_1, where A is the dictator and decides that they get 50 tokens each; τ_2, where A is the dictator and decides that s/he will get 70 tokens while B will get only 10; τ_3, where B is the dictator and decides that they will get 50 tokens each; and, finally, τ_4, where B is the dictator and decides that he will get 70 tokens while A will get only 10. (The token sum is thus 100 in the even splits and 80 in the uneven splits.)

Suppose there is one subject in each of the two player roles, and the experimentalist wants to elicit their preferences in order to identify the game they are playing. Suppose the experimentalist is able to find out each subject's ordinal preference ranking of the four plays in the above game

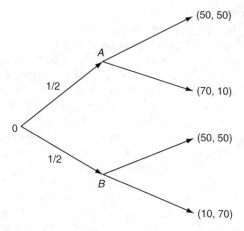

Figure 6.2 A mini random dictator game protocol

protocol. Suppose the subject in player role A ranks play τ_1 (being dictator and splitting 50/50) highest, while the subject in role B ranks τ_4 first (being the dictator and keeping 70 for him/herself). Can the experimentalist then conclude that, if the game (or more precisely, game class) were known by both subjects, the subjects would play according to their stated preferences? Not necessarily. Suppose, for example, that the subject in player role A, when learning about B's 'selfish' preference, changes his/her own ranking and now prefers to keep 70 tokens for him/herself if called upon to act as dictator.

Suppose now that the experimentalist anticipates this possibility of preference interdependence, and therefore instead proceeds as follows: the experimentalist asks subject A to state his/her ordinal ranking of the four plays for each of the 24 possible (strict) ordinal rankings that B may have, and similarly for subject B. Having done this, the experimentalists looks for matching preference orderings – that is, a pair of orderings such that each subject's ordering applies to the other subject's ordering. In this way, the experimentalist may end up with no matching pairs, one matching pair, or several matching pairs. For example, subject A may rank play τ_1 (being dictator and splitting 50/50) highest if B prefers play τ_3 (being dictator and splitting 50/50) over play τ_4, and similarly for subject B. Presumably, two such subjects would not change their own rankings even when told the other's. However, the same two subjects might also have another matching pair of rankings, such as A ranking play τ_2 (being dictator and keeping 70 tokens for him/herself) highest if subject B prefers play τ_4 (being dictator and keeping 70 tokens for him/herself) over play τ_3, and similarly for subject B. This corresponds to a behaviourally distinct class of games, so the experimentalist eventually has two distinct game classes for one and the same game protocol and pair of subjects.

Games of incomplete information

Can this kind of preference interdependence be avoided by way of modelling the interaction as a game of incomplete information, and using the
Harsanyi approach of transforming that game into a 'meta game' of complete but imperfect information, with a common prior? The feasibility of
this programme seems to be an empirical question, for two reasons. First, it
is an empirical question if such a game exists for given subjects in the player
roles of a given game protocol, since interpersonal preference dependence
may also arise in the resulting meta game of imperfect information: subjects may alter their own rankings of outcomes once they learn about the
others' rankings.[14] Second, it is doubtful if human subjects will understand
the meta game constructed in this way, since such a game is usually quite
abstract, nor is it clear that they will agree on a common prior. In view of
these difficulties, the game theorist may abandon direct preference elicitation, and work under the (falsifiable) hypothesis that the subjects act *as if*
they were players with hypothesized preferences in such a meta game. The
subsequent section illustrates that route by way of applying the simple and
yet rich incomplete-information model of interpersonal preference dependence suggested by David Levine (1998). As will be seen, a certain form of
individual preference elicitation is possible even in that setting.

Altruism-driven interpersonal preference dependence

In a two-player game protocol, such as those in Figures 6.1 and 6.2, let $x_A(\omega)$
and $x_B(\omega)$ be the monetary payoffs to the two players, A and B, at each
end-node $\omega \in \Omega$. The players are drawn from the same population (of, say,
subject in an experiment). The type space is a subset Θ of $\mathbb{R}$, and players'
types are i.i.d. draws from a c.d.f. $F : \Theta \to [0, 1]$.[15] In the game protocol of the
associated meta game, where 'nature' first chooses the two players' types, an
end node is a triplet $(\omega, a, b) \in \Omega \times \Theta^2$, where a and b are A's and B's types,
and Ω is the set of end nodes in the underlying, or 'basic', game protocol
that neglects nature's draws of player types (as in Figures 6.1 and 6.2).

The Bernoulli functions of players A and B in this meta game, when A is
of type a and B is of type b, are taken to be of the form:

$$\pi_A(\omega, a, b) = x_A(\omega) + W(a, b) x_B(\omega) \tag{6.1}$$

and

$$\pi_B(\omega, a, b) = x_B(\omega) + W(b, a) x_A(\omega) \tag{6.2}$$

where $W : \Theta^2 \to \mathbb{R}$ is a function that attaches a *relative weight* to the other
player's material payoff (compared with the unit weight attached to the

player's own material payoff). A positive weight thus represents altruism towards the other player while a negative weight represents spite. Levine (1998) uses the following functional form:

$$W(\theta, \theta') = \frac{\theta + \lambda \theta'}{1 + \lambda} \qquad \forall \theta, \theta' \in \Theta \tag{6.3}$$

where $\lambda \in [0, 1]$ is a parameter that reflects interpersonal preference dependence: the weight $W(\theta, \theta')$ placed on the other player's material payoff depends non-negatively on the other player's type θ' and is more sensitive to the other player's type the larger that λ is.

Mini ultimatum games

Let us first apply this approach to the game protocol in Figure 6.1, for $0 < x < y < 100$. Suppose thus that 'nature' first chooses the two players' types and reveals each player's type privately to the player in question. In the notation of Figure 6.1, an outcome in the meta-game protocol is a triplet (ω, a, b), where $\omega \in \{\omega_1, \omega_2, \omega_3\}$, a is A's type and b is B's type.

Let $\lambda \in [0, 1]$, and let $F : \mathbb{R} \to [0, 1]$ be a c.d.f. with finite mean value $\bar{\theta}$. Player A's Bernoulli function is then:

$$\pi_A(\omega_1, a, b) = x + \frac{a + \lambda b}{1 + \lambda}(100 - x) \tag{6.4}$$

$$\pi_A(\omega_2, a, b) = y + \frac{a + \lambda b}{1 + \lambda}(100 - y) \tag{6.5}$$

and $\pi_A(\omega_3, a, b) = 0$ for all a and b. Player B's Bernoulli function is similarly defined:

$$\pi_B(\omega_1, a, b) = 100 - x + \frac{b + \lambda a}{1 + \lambda}x \tag{6.6}$$

$$\pi_B(\omega_2, a, b) = 100 - y + \frac{b + \lambda a}{1 + \lambda}y \tag{6.7}$$

and $\pi_B(\omega_3, a, b) = 0$ for all a and b.

Note, however, the informational asymmetry in the two player's expectation formation at their respective decision nodes in Figure 6.1. While player A has not yet seen B take any action when making her choice, player B, when making his choice, has observed that A has chosen not to take her outside option $(x, 100 - x)$.

The above description specifies a meta game of complete but imperfect information, defined by the following data: x, y, λ and F.

Suppose we have laboratory data for given monetary payoffs x and y, and for human subjects who have been matched randomly and anonymously to play the two player roles. Let $p = (p_1, p_2, p_3) \in \Delta(\{\omega_1, \omega_2, \omega_3\})$ be the empirical outcome in the underlying game protocol; that is, the observed population frequencies of the three plays of the game protocol in Figure 6.1. What such outcomes are consistent with sequential equilibrium in the meta game?[16] Clearly the empirical outcome is a random variable and hence cannot be expected to agree exactly with the theoretically predicted outcome. However, for a sufficiently large subject pool, it should not differ too much, and, moreover, when changing material payoffs x and y, one may study econometrically if the empirical outcome moves in the theoretically predicted direction.

In order to answer such questions we now turn to an investigation of the sequential rationality and consistency conditions that constitute the definition of sequential equilibrium. It is sequentially rational for player B to reject the offer $(y, 100 - y)$ if and only if this choice gives at least the same conditionally expected utility as accepting the offer. Hence, rejection is sequentially rational if and only if

$$100 - y + \frac{b + \lambda \mathbb{E}_B[a]}{1 + \lambda} y \leq 0 \tag{6.8}$$

where $\mathbb{E}_B[a]$ is B's conditional expectation of A's type $a \in \Theta$, given that A has chosen to offer $(y, 100 - y)$ instead of taking the outside option $(x, 100 - x)$. Inequality (6.8) may equivalently be written as $b \leq b^*$, where

$$b^* = -(1 + \lambda)\frac{100 - y}{y} - \lambda \mathbb{E}_B[a] \tag{6.9}$$

In other words, it is sequentially rational for player B to reject the offer if B is sufficiently spiteful, given B's belief about A's type.

Let q denote the rejection rate of subjects in player role B, that is, the conditional equilibrium probability that a subject in role B will reject the offer $(y, 100 - y)$ if made. We then have $q = F(b^*)$, or, more explicitly:

$$q = F\left[-(1 + \lambda)\frac{100 - y}{y} - \lambda \mathbb{E}_B[a]\right] \tag{6.10}$$

Secondly, it is sequentially rational for player A to forego the outside option if and only if the expected utility of doing so is no less than the expected utility to A of instead offering $(y, 100 - y)$. In the latter case, the offer is accepted with probability $1 - q$. Hence, the outside option is optimal for A if and only if

$$x + \frac{a + \lambda \mathbb{E}_A^0[b]}{1 + \lambda}(100 - x) \geq (1 - q)\left[y + \frac{a + \lambda \mathbb{E}_A^1[b]}{1 + \lambda}(100 - y)\right] \tag{6.11}$$

where $\mathbb{E}^0_A[b]$ is A's unconditional expectation of B's type, and $\mathbb{E}^1_A[b]$ is A's conditional expectation of B's type, given that B accepts the offer $(y, 100 - y)$ if made. Inequality (6.11) is equivalent with $a \le a^*$, where

$$a^* = (1 + \lambda) \frac{(1 - q)y - x + \lambda[(1 - q)(100 - y)\mathbb{E}^1_A[b] - (100 - x)\mathbb{E}^0_A[b]]}{y - x + (100 - y)q} \quad (6.12)$$

Having considered sequential rationality, let us now turn to the consistency condition in the definition of sequential equilibrium. This condition determines the three expectations above. First, it requires that B's conditional expectation of A's type, if B is called upon to make a move, equals the conditional mean-value of A's type a, given that $a \le a^*$:

$$\mathbb{E}_B[a] = G(a^*). \quad (6.13)$$

where $G: \mathbb{R} \to \mathbb{R}$ is the truncated mean-value function associated with F, defined by

$$G(t) = \frac{1}{F(t)} \int_{-\infty}^{t} s \, dF(s) \quad (6.14)$$

Secondly, consistency requires player A's unconditional expectation of B's type to equal the unconditional mean- value under F:

$$\mathbb{E}^0_A[b] = \bar{\theta} \quad (6.15)$$

If A chooses the outside option, then A's expectation of B's type should equal the subject pool's average type.

Thirdly, consistency requires A's conditional expectation of B's type if B accepts the offer $(y, 100 - y)$ to equal the conditional mean-value of B's type b, given that $b > b^*$:

$$\mathbb{E}^1_A[b] = \frac{\bar{\theta} - qG(b^*)}{1 - q} \quad (6.16)$$

In order to facilitate the exposition, suppose that $\Theta = \mathbb{R}$ and that F is continuous and strictly increasing.[17] Let

$$\Psi(x, y, q) = G\left[\frac{(1 + \lambda)(y - x - yq) - \lambda[(100 - y)G[F^{-1}(q)]q + (x - y)\bar{\theta}]}{y - x + (100 - y)q}\right] \quad (6.17)$$

Equation (6.10) may then be re-written as

$$q = F\left[-(1 + \lambda)\frac{100 - y}{y} - \lambda\Psi(x, y, q)\right] \quad (6.18)$$

a fixed-point equation in the rejection rate q. Given the meta-game data x, y, λ and F, the right-hand side defines a continuous function of q that maps the closed unit interval into itself, and thus has at least one fixed point. This equation is necessary for a rejection rate q to be compatible with sequential equilibrium in this model. Conversely, if q is a fixed point of equation (6.18), then $\mathbb{E}_B[a]$ and $\mathbb{E}_A^1[b]$ are uniquely determined by equations (6.10) and (6.16), and these in turn uniquely determine the critical types a^* and b^*, hence determining a sequential equilibrium.

Figure 6.3 below shows the graphs of the left-hand side (the 45-degree line) and the right-hand side (the three curves) in equation (6.18), in the special case when F is the uniform distribution on the interval $[-1, 1]$, $\lambda = 0.5$ and $y = 80$, for $x = 20$ (the lowest curve), 35 (the middle curve) and 50 (the highest curve), respectively.[18] We see that the associated equilibrium rejection rates are approximately $q = 0.34$, 0.38 and 0.46, respectively. This

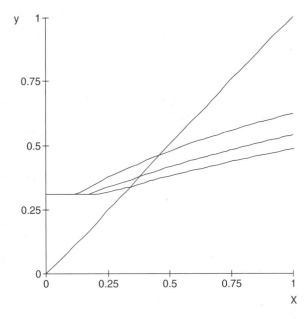

Figure 6.3 The fixed-point equation for the rejection rate

monotonicity is in qualitative agreement with the empirical findings in Falk *et al.* (2003): the more 'generous' the outside option is, in comparison with the offer $(y, 100 - y)$, the more likely it is that the responder will accept the latter. It is as if the responder more easily 'forgives' the proposer for not offering $(x, 100 - x)$ the smaller x is (recall that $x < y$ by hypothesis).[19]

By varying the material payoffs x and y and recording the associated empirical rejection rates, $\tilde{q}(x, y)$, equation (6.18) can be used to pin down λ and F

for a given subject pool—under the hypothesis, of course, that the model is valid. In this sense, indirect aggregate preference elicitation is possible. Indeed, also certain individual preference elicitation is possible. To see this, suppose estimates $\tilde{\lambda}$ and $\tilde{F}$ have been obtained as mentioned above. If records of individual subjects' actions have been kept, then further estimation and testing of the model at an individual level can be done by way of inequalities (6.9) and (6.12), as applied to the proposer and responder roles for different values of x and y. For each subject j, Let $\Theta_j \subset \Theta$ be the subset of parameter values θ that satisfy these conditions for subject j, for all values of x and y in the experimental data. The set Θ_j is either empty or a non-empty interval (determined by j's choices for different values of x and y). If $\Theta_j = \phi$ for some subject j, then the model together with its estimates ($\tilde{\lambda}$ and $\tilde{F}$) is empirically rejected. If, however, $\Theta_j \neq \phi$ for all subjects j, then interval-valued estimates of all subjects' types have been obtained.[20] In the latter case it would be interesting to see whether such individual estimates have predictive power for the same subject's behavior in other game protocols.

Mini random dictator games

Turning to the game protocol in Figure 6.2, suppose that 'nature' not only chooses who will be the dictator, but also both players' types. In the notation of Figure 6.2, an outcome in the meta-game protocol is thus a triplet (ω, a, b), where $\omega \in \{\omega_1, \omega_2, \omega_3, \omega_4\}$, a is A's type and b is B's type. Let $\lambda \in [0, 1]$, and let $F : \mathbb{R} \to [0, 1]$ be a c.d.f. with finite mean value $\hat{\theta}$. Player A's Bernoulli function is then defined by:

$$\pi_A(\omega_1, a, b) = \pi_A(\omega_3, a, b) = 50 + 50\frac{a + \lambda b}{1 + \lambda} \tag{6.19}$$

$$\pi_A(\omega_2, a, b) = 70 + 10\frac{a + \lambda b}{1 + \lambda} \tag{6.20}$$

$$\pi_A(\omega_4, a, b) = 10 + 70\frac{a + \lambda b}{1 + \lambda} \tag{6.21}$$

and analogously for B.

Suppose, first, that both players' types are common knowledge. Then player A, if selected to be the dictator, will choose 50/50 if and only if $a + \lambda b \geq (1 + \lambda)/2$. Similarly, player B, if selected to be the dictator, will choose 50/50 if and only if $b + \lambda a \geq (1 + \lambda)/2$. Hence each player's decision will depend in part on the other's type. In the context of the present example, it seems reasonable to constrain players' types to lie between zero and one; that is, $\Theta = (0, 1)$. Figure 6.4 shows how the unit square Θ^2 is divided into four regions by the two players' indifference lines (the diagram has been drawn for $\lambda = 1/2$). Note, in particular, that there are type combinations (a, b) for which one player will choose 50/50 when selected to be the dictator, despite

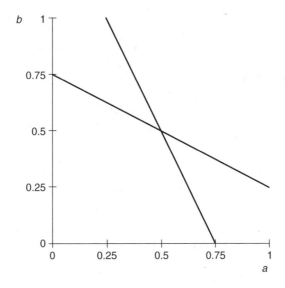

Figure 6.4 Players' indifference lines in the space of type pairs

the fact (known by that player) that the other player, if called upon, would have chosen 70 for him/herself. (These are the regions NW and SE of the intersection of the two indifference lines.)

Second, suppose that each player's type is his or her private information. In sequential equilibrium, either of the two players, if selected to be the dictator, will choose 50/50 if and only if his or her own type θ is at least $(1+\lambda)/2 - \lambda\bar{\theta}$, where $\bar{\theta}$ is the the mean value under F. If $\lambda = 0$, then this condition is independent of the other's expected type $\bar{\theta}$, while for positive λ, the condition depends on the other's type and it is met more easily the higher is the other's expected type.

Concluding remarks

The methodological issues discussed here are relevant for an array of other game protocols than the few examples discussed here. For example, one may ask if prisoner's dilemma game protocols indeed are prisoner's dilemma games for all pairs of human players. For example, in the light of the many experiments based on such protocols, it is not impossible that some individuals in fact prefer the play (C, C) to the play (D, C) even in the role of Player 1, although their own material payoff is then lower.[21]

A related empirical question is whether a repeated game protocol results in a repeated game, since the latter requires that preferences over plays are additively separable in the material payoffs in each round – a stringent requirement on preferences. Suppose, for example, that a prisoner's dilemma protocol is repeated ten times in a laboratory setting, and that the subjects are paid the sum of their material gains after the last round. Will they behave as if they were striving to maximize the sum of their material gains, even if they were exhibiting such preferences in the one-shot prisoner's dilemma protocol?

It was shown above how Levine's (1998) model allows for certain inter-personal preference dependence in game protocols. In particular, it can explain why responders' rejection rate in ultimatum game protocols may depend on the proposer's outside options. While players' preferences in this approach are driven by altruism and spite, there may well be other reasons why some subjects reject small offers. A responder subject may, for example, want to punish the proposer's action, irrespective of the proposer's possible motives, because the action violates some 'norm' supported by the responder. This would not be a case of interpersonal preference dependence, but of context-dependent preferences, as discussed in the fourth section above. Indeed, Fehr and Gächter (2003) reported empirical evidence, in the context of a public-good provision game protocol, that suggests such explanations. Further analysis of preferences of this type seems highly relevant for our understanding of much social behaviour.

An even more basic issue, not discussed here, but yet of great import-ance for the relevance of game theoretic analysis for predictive purposes, is whether human subjects reason in a way that is consistent with any form of backward induction. Johnson *et al.* (2002) provide evidence that a significant fraction of human subjects in laboratory experiments do not even care to inform themselves of the material consequences in distant parts of the game protocol, although such information would be necessary for backward induction reasoning (and despite the fact that the subject can inform him/herself at no other cost than that of touching a computer key).[22] (See Costa-Gomes *et al.* (2001) for similar evidence concerning normal-form games.) Such behaviour is clearly at odds with current game theory – another major challenge for future research.

To sum up, I would like to thank Werner Güth for his pioneering experi-mental studies of the predictive power of non-cooperative game theory. For economic theorists, the huge amount of experimental work done since the 1980s should be good news: while many theoretical presumptions have been challenged, new theoretical ideas can now be tested in the many labora-tories around the world, hopefully leading us to better models of economic behaviour.

Notes

1 The likelihood of success, however, may be smaller, in view of the complexity of human choice behaviour and strategic interaction.
2 The number of citations that could be made here is so large that any selection would be arbitrary.

3   In recent years, evolutionary alternatives to these epistemic models have been developed. However, those models are not discussed here.

4   See Ritzberger (2002) for a rigorous definition and analysis of finite extensive-form games.

5   Standard game theory can thus be criticized for its reliance on the von Neumann – Morgenstern axioms – an empirically valid critique that will not be discussed here, however.

6   By contrast, a payoff function usually means the induced mapping from strategy profiles to Bernoulli function values.

7   This is similar to consumer theory, where the consumption space is supposed to be rich enough to represent all relevant aspects of consumption alternatives.

8   For prominent examples, see Güth, *et al.* (1982); Binmore *et al.* (1985), Ochs and Roth (1989), Fehr and Schmidt (1999), Bolton and Ockenfels (2000), Brandts and Solà (2001), Binmore *et al.* (2002), Falk *et al.* (2003).

9   In more complex games, hypotheses H2 and H5, which here seem innocuous, may in fact be highly implausible. For example, in chess we know that H2 and H5 do not hold: no human player knows how to play optimally (presuming a strict preference for winning) from all game positions on the board.

10 Here each play is shorthand notation for the lottery that places unit probability mass on it.

11 In Falk *et al.* (2003), the 'outside option' was also subject to the responder's acceptance.

12 Similarly, Sprumont (2000) presumes preferences in normal-form games to be context independent in the same way: a player's ranking of pure strategies in a subset remains the player's ranking in the reduced game in which these subsets are the full strategy sets.

13 I am grateful to Al Roth for suggesting this game protocol, which is simpler than the one I originally used.

14 This existence problem is distinct from the related existence problem analysed in Mertens and Zamir (1985).

15 Levine (1998) sets $\Theta = (-1, +1)$, and assumes F to have finite support.

16 Levine (1998) sets $\Theta = (-1, +1)$, and assumes F to have finite support (see note above), while we here allow for the possibility that F has infinite support. We thus here deal with a potentially infinite extensive form. Despite this, I will speak of sequential equilibrium rather than perfect Bayesian equilibrium, since the extension is here evident.

17 If these assumptions are not made, a fixed-point equation for the critical type b^* can be likewise derived, and this equation can be used to determine the rejection rate q and other observables.

18 The algebraic operations performed above apply also to this case, although this distribution function does not have full support.

19 In a slightly more complex game protocol, where also the proposer's "outside option" was subject to the responder's approval, Falk *et al.* (2002) found that the rejection rate q to a given offer $(y, 100 - y)$ increases when the proposer's share x of the outside option $(x, 100 - x)$ was increased. More exactly, they found this for $y = 80$ and $x = 50$ and $x = 20$, respectively.

20 One further test is to see whether this collection of interval estimates, one for each subject, is consistent with $\tilde{F}$ in the sense that, for every $\theta \in \Theta$, the number of subjects j with intervals Θ_j that intersect $(-\infty, \theta]$ approximates $N\tilde{F}(\theta)$, where N is the total number of subjects in the experiment.

21 Empirical support for this hypothesis has been found in preliminary experimental work by M. Kosfeld, E. Fehr and the author.
22 Indifference to material consequences at distant nodes makes sense if a subject does not care at all about material payoffs, or holds beliefs that they differ so little across plays that it is better to save effort by not hitting the computer key than to find out these material payoffs.

References

Asheim, G. (2002) 'On the Epistemic Foundation for Backward Induction', *Mathematical Social Sciences*, 44, 121–44.
Aumann, R. (1995) 'Backward Induction and Common Knowledge of Rationality', *Games and Economic Behavior*, 8, 6–19.
Aumann, R. and Brandenburger, A. (1995) 'Epistemic Conditions for Nash Equilibrium', *Econometrica*, 63, 1161–80.
Ben-Porath, E. (1997) 'Rationality, Nash Equilibrium, and Backwards Induction in Perfect Information Games', *Review of Economic Studies*, 64, 23–46.
Binmore, K., Shaked, A. and Sutton, J. (1985) 'Testing Noncooperative Bargaining Theory: A Preliminary Study', *American Economic Review*, 75, 1178–80.
Binmore, K., McCarthy, J., Ponti, G., Samuelson, L. and Shaked, A. (2002) 'A Backward Induction Experiment', *Journal of Economic Theory*, 104, 48–88.
Bolton, G. and Ockenfels, A. (2000) 'ECR: A Theory of Equity, Reciprocity and Competition', *American Economic Review*, 90, 166–93.
Bolton, G. and Zwick, R. (1995) 'Anonymity versus Punishment in Ultimatum Bargaining', *Games and Economic Behavior*, 10, 95–121.
Brandts, J. and Solà, C. (2001) 'Reference Points and Negative Reciprocity in Simple Sequential Games', *Games and Economic Behavior*, 36, 138–157.
Camerer, C. (2003) *Behavioral Game Theory*, Princeton, NJ, Princeton University Press.
Costa-Gomes, M., Crawford, V. and Broseta, B. (2001) 'Cognition and behavior in Normal-form Games: An Experimental Study', *Econometrica*, 69, 1193–235.
Falk A., Fehr, E. and Fischbacher, U. (2003) 'On the Nature of Fair Behavior', *Economic Inquiry*, 41, 20–6.
Fehr, E. and Gächter, S. (2003) 'Altruistic Punishment in Humans', *Nature*, 415, 137–40.
Fehr, E. and Schmidt, K. (1999) 'A Theory of Fairness, Competition and Cooperation', *Quarterly Journal of Economics*, 114, 817–68.
Glaser, E., Laibson, D., Scheinkman, J. and Soutter, C. (2000) 'Measuring trust', *Quarterly Journal of Economics*, August, 811–46.
Güth, W. and Tietz, R. (1990) 'Ultimatum Bargaining Behavior: A Survey and Comparison of Experimental Results', *Journal of Economic Psychology*, 11, 417–49.
Güth, W., Schmittberger, R. and Schwarze, B. (1982) 'An Experimental Analysis of Ultimatum Bargaining', *Journal of Economic Behavior and Organization*, 3, 376–88.
Johnson, E., Camerer, C. Sen, S. and Rymon, T. (2002) 'Detecting Failures of Backward Induction: Monitoring Information Search in Sequential Bargaining', *Journal of Economic Theory*, 104, 16–47.
Kagel, J. and Roth, A. (eds) (1995) *The Handbook of Experimental Economics*, Princeton, NJ, Princeton University Press.
Kuhn, H. (1950) 'Extensive Games', *Proceedings of the National Academy of Sciences*, 36, 570–6.
Kuhn, H. (1953) 'Extensive Games and the Problem of Information', *Annals of Mathematics Studies*, 28, 193–216.

Levine, D. (1998) 'Modelling Altruism and Spitefulness in Experiments', *Review of Economic Dynamics*, 1, 593–622.

Mertens, J.-F. and Zamir, S. (1985) 'Formulation of Bayesian Analysis for Games with Incomplete Information', *International Journal of Game Theory*, 10, 619–32.

Ochs, J. and Roth, A. (1989) 'An Experimental Study of Sequential Bargaining', *American Economic Review*, 79, 355–84.

Rabin, M. (1993) 'Incorporating Fairness into Game Theory and Economics', *American Economic Review*, Vol. 83, 1281–302.

Ray, I. and Zhou, L. (2001) 'Game Theory via Revealed Preferences', *Games and Economic Behavior*, 37, 415–24.

Reny, P. J. (1993) 'Common Belief and the Theory of Games with Perfect Information', *Journal of Economic Theory*, 59, 257–74.

Ritzberger, K. (2002) *Foundations of Non-Cooperative Game Theory*, Oxford University Press.

Sprumont, Y. (2000) 'On the Testable Implications of Collective Choice Theories', *Journal of Economic Theory*, 93, 205–32.

Tan, T. and Werlang, S. (1988) 'The Bayesian Foundations of Solution Concepts of Games', *Journal of Economic Theory*, 45, 370–91.

Zamir, S. (2000) 'Rationality and Emotions in Ultimatum Bargaining', Mimeo, The Hebrew University and LEI/CREST, Paris.

7

A Dialogue Concerning the Nature of Rationality

*Hartmut Kliemt and Axel Ockenfels**

Introduction

In the spirit of David Hume's 'dialogues concerning natural religion' this dialogue presents several views on the nature of rationality and its role in economics. Definite solutions of problems are not on offer. There may, however, be a range of persuasive reasons to prefer some views over others.

Dramatis personae

MAX: An orthodox economist who believes that rationality depends on whether the consequences of choice can be viewed as (if) the outcome of maximization under constraints according to given preferences.

PSYCHE: An economic psychologist and adherent of bounded rationality who insists on 'true' explanations in terms of behavioural laws and rejects 'as if' accounts of rationality.

BORA: An 'enlightened' adherent of bounded rationality who believes in the 'rational' in 'boundedly rational' and tries to steer a middle course between the more extreme positions of MAX and PSYCHE.

SCENE: *Two economists and a psychologist sit together in the autumn of 2002 when the breaking news is that the Nobel Prize for Economics has been awarded to Daniel Kahneman and Vernon Smith.*

Economists and psychologists

MAX: Have you heard that the Nobel Prize in Economics has been awarded to Daniel Kahneman and Vernon Smith?

---

* We thank Anthony de Jasay, Björn Frank and two anonymous referees for very helpful comments.

PSYCHE: No, but that is great news! Not unexpectedly, and clearly not undeservedly, it has been awarded to Vernon Smith. And I am particularly delighted that the committee chose to include a psychologist. Nevertheless, it is a pity that none of the several runners-up have been included.

MAX: Indeed, when it comes to traditions and to putting a premium on seminal work in experimental economics, the Sauermann–Selten school should be held in the highest esteem. Reinhard Selten would at least have been a real economist who knows what rationality stands for. Including a psychologist sent the wrong signals.

PSYCHE: But Reinhard Selten and his followers were themselves demanding that economics should be based on behavioural laws. Even though they were in the forefront of working out the full rationality approach, they were often more radical critics of the prevailing rational choice paradigm of economics than many psychologists. In fact, the Sauermann–Selten school delivered the strongest arguments for including a psychologist.

BORA: It should not be forgotten that Reinhard Selten has always insisted on working 'both sides of the street', so to speak. On the one hand, there is a philosophical interest in explicating what it means to participate in a strategic interaction as a rational actor along with other rational actors, and, on the other, the interest of explaining actual behaviour in real-world interactions between human actors.

PSYCHE: I do not object to separating the two issues. However, it seems puzzling to me what the role of a theory of fully rational behaviour could be in the explanation of real behaviour. The two seem to be so much apart that I cannot see how the gulf – rather than the street – between them could conceivably be bridged.

MAX: Of course, the model of fully rational behaviour is not realistic in the strict sense of the term. Like all models of science, it is an abstraction or useful idealization. As the idealized laws of physics apply only approximately in, say, the presence of friction, so do the idealized laws of rationality apply only approximately in the presence of cognitive constraints. In that sense, rationality as utility maximization under constraints should be regarded an approximately true idealization of real behaviour.

BORA: I do not think that the analogy between 'laws of physics' and 'laws of rationality' really works. Think of game theory as an example here. The game theorist must assume that game theory itself is 'absorbed' by the players. The players follow its precepts and each knows that the other does so. Their predictions of the behaviour of other players are not based on behavioural laws but rather on game theoretic inferences, assuming that the other players are acting on the basis of the same *theory* they themselves apply. In a strict sense, the term 'game theory' is not about behaviour but rather about the reasoning of human beings who are deemed to be rational by the theory if and only if they themselves apply that theory.

PSYCHE: Pure game theory, as you rightly characterize it, adopts what philosophers such as P. F. Strawson used to call a participant's attitude, as opposed to an objective one. That is, it analyses from the internal point of view of a participant rather than from the external point of view of an onlooker – what K. R. Popper called 'the logic of the situation'. All actors use the prescriptions of the theory of rational behaviour to develop predictions of what other actors will do.

MAX: We should not forget here, too, that the economist Oskar Morgenstern was a thoroughly subjectivist Austrian economist who was particularly interested in how our theories about the world might affect what goes on in that world. He was 'absorbed' by the idea of 'theory absorption'. This idea, rather than any approximation of real behaviour, is the key to understanding classical game theory.

PSYCHE: But classical game theory has no model of the mental processes in which its logic is applied, and affects social reality. What matters is the correctness of inferences drawn by rational individuals rather than the inference process itself. In that sense, pure game theory is logical rather than psychological or behavioural. The immediate consequence is that the inferences from rationality models like those of classical game theory cannot serve as approximations of behavioural laws.

MAX: Nevertheless, pure game theory can define, or rather 'explicate', what rationality (or more precisely *rational choice-making*) in interactive situations ideally means. The concept of rationality in interactive situations is theory-dependent. An absorbable theory of rationality is prescribing how to maximize utility in interactive situations simultaneously to all actors. And it does so without providing a reason to deviate from the theory or its prescriptions if all actors act according to that theory and share common knowledge that they do. This is a minimum requirement of coherence that must be met by any theory of rationality. Unless it is fulfilled, no reflective equilibrium can be reached.

PSYCHE: But this, as the concept of a reflective equilibrium indicates, is a purely philosophical exercise. Theory absorption of this kind is concerned with 'logical' and conceptual coherence rather than with behaviour. It is completely unrelated to real human behaviour. We need to learn how such theories of ideal rationality relate to real behaviour.

BORA: Indeed, if we want to understand economic decision-making we need to bring economics closer to psychology, or to build psychology into economics, so to speak. Experimental economics is a step in that direction.

MAX: I partly agree with this. But, unlike psychology, economics is not concerned so much with the process of making choices *per se*, but rather with the outcome of such processes. It appears to me that, especially within economic institutions, the predictions of game theory often work sufficiently well without addressing the underlying complicated and messy mental processes. Taking them into account creates so much chaos that the psychologist might indeed find new clients among formerly remarkably healthy and sober-minded economists.

BORA: If we become as preoccupied with all the minute facts and details of cognitive processes as the psychologists we shall in the end break down economics into bits and pieces without greatly improving its predictive value. The strength of economics is its unity of perspective. Economics avoids the chaos of realm-dependent psychological theories, and thereby stands out in the crowd of disparate psychological and social theories.

PSYCHE: I have heard the latter argument again and again – and I disagree. In practice, economic theories are not less disparate than psychological theories, although the reasons are quite different. In psychology, the different camps have their own hypotheses about the reasoning processes that affect behaviour in a given environment. In economics, the different camps typically have their own hypotheses about the environmental factors that affect behaviour, given rational decision processes. There are probably hundreds, if not thousands, of economic theories on bargaining, growth, unemployment and so on, each coming to different, often opposite, solutions for very similar problems, but all under the same umbrella – called rationality or rational choice.

MAX: But unlike cognitive processes, environmental aspects of a theory are in principle observable objectively.

PSYCHE: You sound a bit like an old-fashioned behaviourist who does not believe in the legitimacy of using theoretical terms. All interesting theories contain terms with an extension that is unobservable directly.

BORA: Also, unfortunately, looking at environmental parameters does not tell the whole story. Economic theories based on rationality typically also depend on, say, individual knowledge and beliefs, and unobservable individual preference parameters such as risk aversion and impatience that further add to the plethora of rationality models. Thus, in fact, economics shares more drawbacks with psychology than many economists are willing to concede. It is an illusion brought about by putting everything into the single mould of maximization that economics describes a single animal while psychology takes us to the zoo.

MAX: Yes, both economics and psychology must deal with the subtleties of the behaviour of human beings who are endowed with complex brains. No theory will ever be in a position to deal in full with these complexities. My claim is not that economics does this perfectly, but rather that it fares better than its competitors.

BORA: As always, the proof of the pudding will be in the eating, and as far as this is concerned I suspect that some variety in our theoretical diet may serve us best. For example, when we are dealing with non-market behaviour, a greater dose of psychology may be necessary, while a somewhat stricter rational choice and maximization under constraints approach may lead to a better understanding of market processes.

Market behaviour

MAX: The economists' understanding of market equilibrium is cast exclusively in terms of external constraints and observable responses of market participants.

BORA: Experiments tend to show that, under proper conditions, market equilibria emerge from a process in which traders behave 'as if' led by an invisible hand. Market equilibria are typically unintended results of human action that are not to be explained as the outcomes of human design. Therefore, when it comes to markets, I'd shift the focus on the 'rationality' of results rather than the rationality of the decision-making processes.

MAX: I prefer to restrict the primary extension of the term 'rational' to choice-making and the behaviour stemming from it. To refer to results of choice-making as 'rational' is a secondary use of the term. Results are individually rational if they cannot be improved by individual maximizing behaviour.

PSYCHE: But what about the choice-making leading to the results? If we expect that results cannot be rational unless brought about by rational choice-making, a closer look reveals that market equilibrium in these environments is reached *despite* the fact that actors are not fully rational. Even so-called 'zero-intelligence traders' could do it. Rationality as a behavioural premise is unnecessary, and we do not need such smuggled-in dubious figures such as *Homo oeconomicus* when it comes to an explanation of how markets work. The explanation in the last resort relies mostly on insights into the workings of institutions rather than into human motives.

MAX: My requirement that rational results cannot be improved on by individually rational behaviour is impervious to such criticisms.

BORA: How can there be an institutional explanation without minimal behavioural rules? In fact, even the zero-intelligence traders Psyche is talking about respond to incentives, because if they don't and instead execute obviously non-rational strategies, equilibrium is not reached. Far-sighted maximization with perfect information about all constraints, consequences and options is an unnecessary assumption, but we must assume that human beings seize opportunities once they 'stumble' on them. And, as Adam Smith already had it, the desire to improve our situation is with us from the cradle to the grave even though we might not always know how to fulfil that desire optimally.

MAX: I see it from a slightly different perspective. For markets to work, you only need individuals at the margins who respond properly to incentives. Market outcomes are not affected so much by intra-marginal irrationalities. Game theory analyses incentive structures, and since most humans – both in the lab and in the field – respond to incentives, game theory is in most circumstances a useful tool to capture behaviour.

BORA: So, if an economist approaches markets s/he can use the rational choice model as a useful instrument for predicting those 'intelligent outcomes'. Because I know that markets work in general as if populated by fully rational individuals, I can rely on the model of fully rational behaviour to predict what will happen in a specific market. In that sense, *Homo oeconomicus* and economics as a predictive science are going hand in hand.

MAX: And, for that matter, they are going stronger than other social sciences that are allegedly more empirically minded but not very well-integrated bodies of knowledge.

PSYCHE: But your confidence that your predictions will work depends on the fact that you can explain why markets work in terms other than the rational choices of all participants. Markets lead to rationally desirable results because they are robust against many deficiencies in individual rationality as well as against, let's say 'anomalies' – like fully rational or fully selfish behaviour.

MAX: Still, it is true that one can use the rational choice model for predictions. The possibility of prediction is all that matters when it comes to acting in markets and to improving markets themselves by acts of institutional design.

BORA: With respect to markets, I agree. But I hesitate when we try to go beyond markets. There, *Homo oeconomicus* may not help any more. Think, for example, of the repeated prisoner's dilemma game. The prediction, by backward induction, is that there will be no co-operation at all as long as the number of rounds to be played is finite and known to be finite. To my knowledge, there is not a single experiment that supports this prediction.

PSYCHE: But are we not adherents of the methodologically individualist view that, in all market as well as non-market contexts, the same rational individuals interact? And would that not rule out the use of one model of behaviour for market contexts and a different model for non-market contexts?

BORA: I can in principle think of models that include non-rational and probably non-selfish elements. For example, there are models that capture seemingly disparate sets of behavioural patterns across institutional environments, such as markets and public goods' settings. But this involves, of course, a departure from standard economic modelling. So I am slightly reluctant here, as are many economists. But it is at least good to see that more and more economists care about the robustness of their models with respect to the rationality and motivational requirements.

PSYCHE: The analysis of repeated prisoner's dilemma games, for example, is clearly not robust as it rests on common knowledge of rationality and selfishness. Even, say, a single 'crazy' decision-maker in a large population of fully rational and selfish players would imply co-operative behaviour across the board until the last rounds of play.

MAX: What you deem to be crazy is in the last resort part of a wonderful idealized model. Without claiming to be realistic, the model teaches us a lesson about the principles that may be expected in some form or other to rule the social world. The ideal model informs us that in social reality there may be a rather delicate interplay of incomplete information and motivational types. Even completely selfish and rational players have good reasons to behave co-operatively in a world with a few fair-minded players.

PSYCHE: Well, this sounds exciting, but the reality is that, if you allow non-standard preferences to enter the model, folk theorems show that basically everything can happen, even in finite games. Game theory has nothing to offer here.

BORA: In fact, these models quickly become so complex that undergraduate economics students cannot understand them, given their cognitive limitations as boundedly rational individuals. So they are typically taught the simple message that, in a population of rational decision-makers it is best never to co-operate. This is terribly misleading.

Motivations

PSYCHE: Let's get a bit more focused here. Outside market contexts much evidence seem to speak loudly against classical *Homo oeconomicus*. For example, think of the one-off ultimatum game – the most intensely studied lab game besides the prisoner's dilemma game – how would you explain the results of this *'experimentum crucis'* on opportunistic motivation? Why in all the world would *Homo oeconomicus* in the responder role reject positive sums of money if s/he is assured s/he will never meet the proposer again, and perhaps even knows that the experiment is performed according to what experimenters call the 'double blind' procedure?

MAX: I agree that we as economists cannot ignore such evidence. The original findings of Werner Güth and his collaborators that positive offers deemed too low by the responder in ultimatum game experiments are rejected turned out to be very robust, although many colleagues tried hard to rescue the standard model's prediction. There is no way to explain ultimatum game behaviour in terms of rational choice as long as one sticks to monetary rewards as the crucial motivating factor.

BORA: It sometimes appears to me that almost everything we know about bounded rationality comes from very simple lab games where arguments in favor of truly *boundedly* rational decision-making have the least bite, and where reduction of complexity, as postulated by advocates of bounded rationality, can hardly be observed. In the ultimatum and many other experimental games, the set of alternatives is as clear as it can be. It is hard to imagine a responder rejecting a positive offer just because of a cognitive misperception of the pecuniary consequences. I guess, if this were the case, there would be no room at all for the term 'rational' in 'boundedly rational behaviour'. Motivations other than the maximization of monetary rewards must play a crucial role in such rational choice-making.

MAX: You are fighting a straw man here. I do not deny that ever since Spinoza there have been some social theorists who seemed to suggest that selfish opportunism would guide behaviour in every instance. But even Spinoza's godfather, Hobbes, violated that assumption, and so did all later founding fathers of economics – including, of course, the Adam of economics himself: Adam Smith.

BORA: *Homo oeconomicus* is characterized merely by rationality assumptions in the sense of consistency rather than by being motivated exclusively by selfish motives. *Homo oeconomicus*, as I would perceive him/her, can be selfish or unselfish, an idealist, a materialist or whatever. As long as his/her preferences are given, and comply with certain axioms of consistency, economics can run its course.

MAX: In fact, if we include motives other than maximization of pecuniary rewards we can very well explain the results of ultimatum game experiments and the like in a maximization paradigm. Human individuals are no doubt motivated by aims such as fairness. As Lionel Robbins rightly insisted, economics is about the rational choice of means to given ends, while on the rationality of the ends themselves it must remain silent. Whatever the ends, as long as they are treated as given, economics can explain observed behaviour within its means – ends framework. In fact, we can often identify motives by asking ourselves what would be maximized by observed behaviour.

PSYCHE: But that smells of '*ad hoc*ery'. The results of the ultimatum game experiments are not coherent with the standard theory, and so some additional motives are wheeled in to fix the problem. You observe less unequal results than would be expected according to standard rational choice models based on selfish motives of maximizing pecuniary payoffs, and then introduce an additional preference for equity and so on. This seems dangerously close to explaining the sleepiness of opium-smoking individuals by the soporific power of opium.

BORA: Any theory with some predictive value, be it a rational or a psychological theory, needs to deal with and to specify motivational forces. As far as this is concerned, the often heard '*ad hoc*ery' protest applies across the board, and cannot serve to disqualify economic approaches in particular. Other measures for the quality of a model are needed.

Constructing preferences

PSYCHE: The economists' concept of rationality is relying on the old idea of given preferences. There is no room for a model of the decision-making apparatus of the individual. The decision-making individual is, at the last resort, represented by his or her preferences and the assumption that the individual tries to maximize the satisfaction of those preferences. Once s/he knows the preferences, the economist allegedly can predict choices as maximizing preference satisfaction under knowledge and other constraints. This explains why economists quite mistakenly believe that they do not need a psychological theory of how preferences are formed and decisions are in fact made.

BORA: If I want to predict the effects of a bullet fired by a gun I do not need to rely on a description at the level of the molecules forming the bullet. If I want to predict the effects of human rational decision-making I do not need to go down to the level of psychological processes all the time. I should rather choose the adequate level leading to the most fruitful approach. And starting from preferences is more often than not a useful approach. As in all good empirical science, we try to formulate theories that are as general as possible and as specific as necessary.

MAX: Maximization according to given preferences is and will always be at root of the economic enterprise because it is at the root of rational behaviour. We just need to use richer utility functions than the traditional ones. The latter were dominated unduly by preferences for monetary rewards. But after appropriate enlargement of the motivational basis, maximization of utility functions explains behaviour. *At least this is what should explain behaviour. Sound economic explanations of interaction in terms of rational choice rely on maximization.*

PSYCHE: The methodological individualism of economists like you is a strange thing. Think about it: in economic theory there is no model of the individual as a decision-maker, no model of his/her deliberation process. There is only the allegedly given preference order that serves as a shorthand symbol representing the individual and his/her decision-making process as maximization.

MAX: Indeed, since we are interested chiefly in human interaction we treat the human individual more or less as a black box.

BORA: But as economists we make quite a lot of assumptions about the motives that are operative inside that box, so to speak, and are guiding its behaviour. We do factor in empirical information about the human individual, his or her motives and preferences, and the influence of these on overt behaviour.

PSYCHE: This has nothing to do with the real decision-making processes that lead to individual choice-making. Preferences are not given but rather construed in the decision-making process. People sometimes don't exactly know what they want. Sometimes there may be no rates of substitution between the level of fulfilment of multiple goals. A theory of bounded rationality deals with the cognitive *process of analysis*, including the emergence of, say, aspirations and the construction of the model and the alternatives as they are perceived. Bounded rationality is not just a combination of objective rules plus some inner constraints as it has been modelled quite often recently.

MAX: You seem to say that preferences are construed by the decision-maker *at the same time* as s/he is analysing a situation. But if preferences are not given independently of the decision-making process itself it becomes impossible for an external observer to predict individual choices in a general way.

BORA: If preferences and models of specific situations cannot be disentangled, how can there be general theories at all? The advantage of economics is that it formulates general theories of social interaction that detract from the specific individual and from what is dependent on the minute details of situations. Leaving certain things unaddressed is of the essence of generalization.

PSYCHE: We cannot avoid forming theories that in some way try to look into the black box of human motivation, and of cognitive processes that generate actions, because cognitive limits affect behaviour.

BORA: I see your point. We should not, however, underestimate the flexibility of non-cooperative game theory here. In the end, the concept of a non-cooperative game model boils down to the assumption that everything that is not subject to individual choice-making is modelled explicitly as part of the rules of the game. If it is not on paper, so to speak, it is not there. On the other hand, whatever is written on paper by the non-cooperative game theorist is assumed to be in the minds of the players and nothing more than that.

PSYCHE: But that means that, apart from the simplest cases of games, the game model must be based on a far-fetched and indeed absolutely unrealistic knowledge assumption.

BORA: Though this is true, the ignorance of an actor can itself be modelled explicitly as part of the rules of the game. Ignorance of the preferences or aims of other players can be modelled by fictitious moves of nature determining the types of other players. Which type of player is in fact playing a game need not be part of the common knowledge of the decision-makers. The assumption of common knowledge must apply only at some ultimate or basic level. At that level players must commonly know what they are ignorant of, as otherwise the game would not be well defined.

PSYCHE: But imagine how complicated such a game tree must necessarily become. It seems outrageously unrealistic if we ascribe the knowledge of such a tree to a human individual.

BORA: Sure, but rationality is a theoretical, or rather a philosophical, concept. Realism is not necessarily among its merits. If we tried to apply the model directly to the world, very subtle differences in knowledge assumptions could render any seemingly irrational behaviour rational in an appropriately specified non-cooperative game model. And vice versa, hardly any behavioural observation can be ruled out by non-cooperative game theory as such. Precisely for this reason the language of non-cooperative game theory may be used for a reconstruction of boundedly rational behaviour by modelling explicitly all phenomena of boundedness.

PSYCHE: I have even heard that in this philosophical account personal players can, or rather must, be split up into agents.

BORA: Yes, at each decision node there is another new one. The agent is simply identified with the preference order that is prevalent at that node. All assumptions we make about the decision-making of the person, including the internal mental processes of that personal decision-maker are represented as part of the rules of the game.

PSYCHE: But even at the philosophical level I would object because in the last resort preferences are still treated as given. They are exogenous to the model but should become endogenous.

MAX: In the standard account, preferences are part of the rules of the game. But even if you do not treat preferences as part of the rules of the game you could model the process in which the preferences are formed as part of the rules of the game.

BORA: Still, within an economic approach, the ultimate aims or ends of the actors must be exogenous to our modelling efforts and in that sense be given.

MAX: In any model we must treat something as being exogenous. As economists we must stick to the principle that in normative as well as descriptive analyses we treat ends as being given.

BORA: Perhaps a psycho-biological theory of the formation of aims ends or values may be used to inform us about factors that must be included in the rules of any of the games we analyse. There is a general part of the rules of any game that humans play, and human nature itself brings that, so to speak, into play.

MAX: Whatever we want to assume, we can represent it as part of the rules of the game and then analyse the game according to maximization assumptions.

PSYCHE: I do not deny that such possibilities may conceivably exist. But this only implies that we first need to form a sound empirical theory of how individual preferences are formed. Once we have such a theory we could introduce additional assumptions about constraints, such as to model behaviour *as if* it were maximizing. However, I do not see what this additional step could be good for except to serve the methodological prejudices of economists.

BORA: A theory of preference formation would be desirable. Once you have such a theory you can build it into the rules of any game. This is, no doubt, an ambitious and complex endeavour, regardless of whether a full rationality or bounded rationality approach is taken.

MAX: In any case, desirability is not the issue here – but practicability and fruitfulness in theory formation are. For all practical purposes of economic theory formation it seems best to start from preferences that may be expected to prevail among actors in some context. Experimental economics, observations in the field and so on provide some information about prevailing preferences and prevailing cognitive constraints on the maximization of preference satisfaction. Once such information is available we can represent those preferences by utility functions and proceed with a conventional economic analysis in terms of utility maximization under constraints.

A good theory

PSYCHE: Is there any methodological reason why one *should* explain all behaviour in terms of utility maximization? Are there any ends of theorizing that can be reached if we comply with such a 'technological norm' of doing economic science?

MAX: There are several reasons why we should not give up lightly the framework of utility maximization that has served us so well. First, there is a body of economic theory that has been developed within this framework. The unity and continuity of argument is a value in itself for science. It facilitates the incorporation of new knowledge into a body of old, determines what questions should be asked, and how answers should be evaluated. Second, since every theory is falsified in one way or another, it does not make sense to throw away a theory simply because there are some falsifying experiences. Third, besides coherence with observations, the simplicity of a theory weighs heavily in its favour. As a tool of organizing experiences and predicting results, standard economic theory based on the assumption of maximization under constraints has worked, and still works, well.

PSYCHE: I believe that in the end the value of a theory depends crucially on the truth of its basic laws. As far as this is concerned it is simply untrue that human beings maximize a well-defined utility function.

BORA: Like management in general, rational self-management of the human individual is 'management by exception' rather than maximizing behaviour. Since human beings do respond to incentives and more often than not can learn how to respond best to similar incentive systems to which they are exposed repeatedly, the conclusions suggested by the maximization assumption point generally in the right direction.

PSYCHE: But sometimes they do so in ways that must seem 'perverse' from a traditional economic point of view. Just think of the example of providing monetary incentives to donate blood; as a result of this you may find less willingness to donate.

MAX: As far as I know, the evidence on this as discussed in the so-called Titmuss versus Arrow debate is not really convincing.

PSYCHE: Still, there are examples of crowding-out as documented in psychology and in economic disciplines such as management and organizational studies.

BORA: But this is the great exception rather than the rule. The overwhelming evidence outside of research searching for crowding-out effects is that people respond to external incentives – perhaps not always fully rationally, but at least in the direction that standard economic theory predicts.

MAX: Economists typically insist only on the plain fact that human behaviour as a rule is sensitive to relative prices. Most of our predictions depend qualitatively only on such simple truths. It is an interesting anomaly if we observe such phenomena as the crowding-out of intrinsic motivation by extrinsic incentives. But it should not be forgotten that it is simply what it is – an anomaly. To base a theory of human behaviour on such exceptions amounts to almost the same thing as building a theory of rational entrepreneurial behaviour on a sample of individuals who became bankrupt.

PSYCHE: Is it so anomalous to become bankrupt?

MAX: It is not. Still, we would not develop the prescriptive elements of a theory of rational behaviour by idealizing and suggesting an imitation of the rules purportedly applied by those who failed.

BORA: Yes, something may be said in favour of not abandoning economic theory prematurely. The speed of light is also an anomaly in Newtonian physics, nevertheless standard mechanics works very well for all practical earthly purposes. Therefore, for practical purposes, we had better stick to it unless we have something better to put in its place.

MAX: And just ask yourself why Newtonian physics is up to the present day regarded as such a tremendous success. This is precisely because it can subsume seemingly vastly diverse phenomena as the fall of an apple from a tree as well as the movement of planets around the sun under the same law of gravitation. That economics, like physics, provides a unified theory that can exhibit common structures of the world and our experience is the essence of good science. Where is the psychological theory that might replace traditional economics in providing such a unified account of social phenomena?

BORA: It is not by chance that classical social philosophers and classical economists were in former times frequently used the expression 'an economy of nature'. Typically, this economy emerges if there is an objective criterion of success that determines, along with some kind of inheritable characteristic, how many of some competing entities will be represented in successive generations. Only the best-adapted variants, emerging from the variation of inheritable traits, will eventually survive and things will look as if they were ordered deliberately.

MAX: A competitive market in which bankruptcy is well-defined, and so is profit, will show the same characteristics. As long as some individuals hit on the maximizing strategies by chance (or deliberately) all the others will have to follow suit or fail. Selection and adaptation are forces that operate at all levels of organization, social or natural. Their outcomes can be predicted by the economic model, since they appear as if brought about by maximization under constraints.

PSYCHE: But there is a difference between spurious unification resulting from describing things arbitrarily in ways that make them look similar, and a unity of experience based on common empirical laws. The explanation that you offer for the apparent unity of our experience is basically an evolutionary one.

BORA: I have no complaints about that except that it has nothing to do with an explanation in terms of individually rational behaviour. In such evolutionary accounts of behaviour as became popular among traditional economists, the fact that behaviour appeared to be rationally chosen is explained. Rational choice-making is not itself an explanatory factor.

MAX: Be that as it may, as a tool of making predictions in a wide variety of situations, economics fares much better than its competitors.

PSYCHE: My criterion of better or worse science seems to be quite different from that applied by economists like you. As an empirically minded psychologist I subscribe to a realistic view of science. Ultimately a theory is the better if it better represents the true facts of the world and the laws that govern it. What are your criteria of better or worse science?

MAX: Like other complicated things, theories are evaluated along several dimensions. The claim that, in evaluating theories, only their truth matters seems as single-minded to me as the claim that good explanations of human behaviour must be formulated in terms only of the fully rational pursuit of selfish motives.

PSYCHE: Still, if I grant for the sake of the argument that criteria other than truth are relevant for the evaluation of theories, what exactly do you have in mind?

BORA: All models are false, so the distinction between false and true models is not helpful. But you seem to ask whether, at least, a 'truer' model is a better model. Not necessarily! Even if the sole motivation of a model-builder is to find the truth, s/he wouldn't necessarily build a model that captures it in as many facets as possible. This is because the truth is likely to be so complex that any single model reflecting it in some detail will be incomprehensible. I guess you will agree that a 'true model' of boundedly rational decision-making, for example, needs to include cognitive, motivational, adaptive and other bounds of decision-making such as neuro-biological constraints. To my knowledge, no such model (and not even an attempt to combine all these aspects) exists, even though considerable knowledge has been accumulated in each of these fields. Such a model would simply not be able to convey a *useful* picture of the truth. So, even if truth is the ultimate aim, it is not the only measure of a good theory.

MAX: Besides, when it comes to modelling, economists tend not to be interested so much in the truth but rather in features such as simplicity, generality, elegance and predictive value.

PSYCHE: But the generality of economic theory may be bought at the price of rendering it either devoid of all content as in the case of non-cooperative game theory, or by using all sorts of *ad hoc* assumptions, as in explaining non-opportunistic behaviour. What we need are robust theories of what is going to happen in classes of situations characterized by certain general features that are sufficient for certain phenomena to occur.

MAX: Where are the robust non-*ad-hoc* psychological theories that are at the same time general? With few exceptions, disproportionately coming from the German experimental economics tradition, I have hardly seen any general account of bounded rationality. What I know of are experimental studies and the like that all lead to very strongly domain-dependent and short-range theories. There is no theory of boundedly rational behaviour; at best there are fragments of it, and programmatic visions of what might and perhaps should be done.

BORA: There are promising and interesting beginnings, but the market for publications still does not always reward these efforts. Part of the hesitant reactions of many economists to these new approaches may be a large, though sunk, investment in standard theory. But the Nobel Prize for 2002 will attract more good researchers who, I am confident, will dig the tunnel from both ends: from one end using traditional economic theory, and on the other with the help of the bits and pieces of empirical evidence derived from game experiments and psychology. I estimate that by, say, 2054, there will be a point where psychology becomes more like economic theory, and economic theory more like psychology.

8
Behavioural Game Theory: Thinking, Learning and Teaching

Colin F. Camerer, Teck-Hua Ho and Juin Kuan Chong[*]

Introduction

Game theory is a mathematical system for analysing and predicting how humans behave in strategic situations. Standard equilibrium analyses assume that all players: (1) form beliefs based on an analysis of what others might do (strategic thinking); (2) choose the best response given those beliefs (optimization); and (3) adjust best responses and beliefs until they are mutually consistent (equilibrium).

It is widely accepted that not every player behaves rationally in complex situations, so assumptions (1) and (2) are sometimes violated. For explaining consumer choices and other decisions, rationality may still be an adequate approximation even if a modest percentage of players violate the theory. But game theory is different: the players' fates are intertwined. The presence of players who do not think strategically or optimize, even if there are very few such players, can change what rational players should do. As a result, what a population of players is likely to do when some are not thinking strategically and optimizing can only be predicted by an analysis that uses the tools of (1)–(3) but accounts for bounded rationality as well, preferably in a precise way.[1]

An alternative way to define the equilibrium condition (3) is that players are never surprised when the strategies of other players are revealed. Defined

---

[*] This research was supported by NSF grants SBR 9730364, SBR 9730187 and SES-0078911. Thanks to many people for helpful comments on this research, particularly Caltech colleagues (especially Richard McKelvey, Tom Palfrey and Charles Plott), Mónica Capra, Vince Crawford, John Duffy, Drew Fudenberg, John Kagel, members of the MacArthur Preferences Network, our research assistants and collaborators Dan Clendenning, Graham Free, David Hsia, Ming Hsu, Hongjai Rhee and Xin Wang, and seminar audience members and referees (especially two for this book) too numerous to mention. Dan Levin gave the shooting-ahead military example of sophistication. Dave Cooper, Ido Erev and Guillaume Frechette wrote helpful emails.

this way, it seems unlikely that equilibrium (3) is reached instantaneously in one-shot games. Indeed, in the modern view, equilibrium should be thought of as the limiting outcome of an unspecified learning or evolutionary process that unfolds over time.[2] In this view, equilibrium is the *end* of the story of how strategic thinking, optimization, and equilibration (or learning) work, not the beginning (one-shot) or the middle (equilibration).

This chapter has three goals. First we develop an index of bounded rationality which measures players' steps of thinking and uses one parameter to specify how heterogeneous a population of players is. Coupled with best response, this index makes a unique statistical prediction of behaviour in any one-shot game. Second, we develop a learning algorithm (called a self-tuning EWA – experience-weighted attraction) to compute the path of equilibration. The algorithm generalizes both fictitious play and reinforcement models, and has shown greater empirical predictive power than those models in many games (adjusting for complexity, of course). Consequently, the self-tuning EWA can serve as an empirical device for finding the behavioural resting point as a function of the initial conditions. Third, we show how the index of bounded rationality and the learning algorithm can be extended to understand repeated game behaviours such as reputation building and strategic teaching. The idea is to present new, procedurally-rational models inspired by data, in the spirit of Werner Güth's (for example, Güth, 2000) many, many contributions of a similar kind.

Our approach is guided by three stylistic principles: precision, generality and empirical discipline. The first two are standard desiderata in game theory; the third is a cornerstone in experimental economics.

Precision

Because game theory predictions are sharp, it is not hard to spot likely deviations and counter-examples. Until recently, most of the experimental literature consisted of documenting deviations (or successes) and presenting a simple model, usually specialized to the game at hand. The hard part is to distil the deviations into an alternative theory that is as precise as standard theory and can be applied widely. We favour specifications that use one or two free parameters to express crucial elements of behavioural flexibility, because people are different. We also prefer to let data, rather than our intuition, specify parameter values.[3]

Generality

Much of the power of equilibrium analyses, and their widespread use, comes from the fact that the same principles can be applied to many different games, using the universal language of mathematics. Widespread use of the language creates a dialogue that sharpens theory and accumulates worldwide know-how. Behavioural models of games are also meant to be general, in the sense that the models can be applied to many games with minimal customization. The insistence on generality is common in economics, but is not universal. Many researchers in psychology believe that behaviour is so context-specific that it is impossible to have a common theory that applies to all contexts. Our view is that we cannot know whether general theories fail until they are applied widely. Showing that customized models of different games fit well does not mean there isn't a general theory waiting to be discovered that is even better.

It is noteworthy that, in the search for generality, the models we describe below typically fit dozens of different data sets, rather than just one or two. The number of subject periods used when games are pooled is usually several thousand. This does not mean the results are conclusive or unshakeable. It just illustrates what we mean by a general model. The emphasis on explaining many data sets with a single model is also meant to create a higher standard for evaluating models of learning (and limited cognition as well). When an investigator reports only a single experiment and a learning model customized to explain those data, it is hard to take such models seriously until their generality has been explored by applying them to other games.

Empirical discipline

Our approach is heavily disciplined by data. Because game theory is about people (and groups of people) thinking about what other people and groups will do, it is unlikely that pure logic alone will tell us what will happen. As the physicist Murray Gell-Mann supposedly said, 'Think how hard physics would be if particles could think.' It is even harder if we do not watch what 'particles' do when interacting. Or, as Thomas Schelling (1960, p. 164) wrote, 'One cannot, without empirical evidence, deduce what understandings can be perceived in a nonzero-sum game of maneuver any more than one can prove, by purely formal deduction, that a particular joke is bound to be funny.'

Our insistence on empirical discipline is shared by others, past and present. Von Neumann and Morgenstern (1944) thought that 'the empirical background of economic science is definitely inadequate . . . it would have been absurd in physics to expect Kepler and Newton without Tycho Brahe, – and there is no reason to hope for an easier development in economics'. Fifty years later, Eric Van Damme (1999) had a similar thought:

> Without having a broad set of facts on which to theorize, there is a certain danger of spending too much time on models that are mathematically elegant, yet have little connection to actual behavior. At present our empirical knowledge is inadequate and it is an interesting question why game theorists have not turned more frequently to psychologists for information about the learning and information processes used by humans.

The data we use to inform theory are experimental. Laboratory environments provide crucial control of what players know, when they move, and what their payoffs are, which is crucial because game-theoretic predictions are sensitive to these variables (see Crawford, 1997). As in other lab sciences, the idea is to use lab control to sort out which theories work well and which do not, then use them later to help understand patterns in naturally-occurring data. In this respect, behavioural game theory resembles data-driven fields such as labor economics or finance more than does analytical game theory. The large body of experimental data accumulated since the 1980s (and particularly over the late 1990s early 2000s; see Camerer, 2003) is a treasure trove that can be used to sort out which simple parametric models fit well.

While the primary goal of behavioural game theory models is to make accurate predictions when equilibrium concepts do not, it can also circumvent two central problems in game theory: refinement and selection. Because we replace the strict best-response (optimization) assumption with stochastic better-response, all possible paths are part of a (statistical) equilibrium. As a result, there is no need to apply subgame perfection or propose belief refinements (to update beliefs after zero-probability events where Bayes' rule is useless). Furthermore, with plausible parameter values, the thinking and learning models often solve the long-standing problem of selecting one of several Nash equilibria, in a statistical sense, because the models make a unimodal statistical prediction rather than predicting multiple modes. Therefore, while the thinking-steps model generalizes the concept of equilibrium, it can also be *more* precise (in a statistical sense) when equilibrium is imprecise (see Lucas, 1986).[4]

We shall make three remarks before proceeding. First, while we do believe the thinking, learning and teaching models in this chapter do a good job of explaining some experimental regularity parsimoniously, many other models are actively being explored.[5] The models in this chapter illustrate what most other models also strive to explain, and how they are evaluated.

The second remark is that these behavioural models are shaped by data from game experiments, but are intended for eventual use in areas of economics where game theory has been applied successfully. We shall return to a list of potential applications in the conclusion, but to whet the reader's appetite, here is a preview. Limited thinking models might be useful in explaining price bubbles, speculation and betting, competition neglect in business strategy, simplicity of incentive contracts, and persistence of nominal shocks in macroeconomics. Learning might be helpful for explaining evolution of pricing, institutions and industry structure. Teaching can be applied to repeated contracting, industrial organization, trust-building, and policy-makers setting inflation rates.

The third remark is about how to read this long chapter. The second and third sections, on learning and teaching, are based on published research and an unpublished paper introducing the one-parameter self-tuning EWA approach. We chose some examples to highlight what the models do and how they compare to other models, and also use this opportunity to comment briefly on methodology. The first section, on the cognitive hierarchy thinking model, is newer. Our 2002 working paper has many additional results. A short paper (Camerer *et al.*, 2003) reports some brief results.

We put all three into one paper to show the ambitions and sweep of behavioural game theory. It is no longer the case that game theory is a broad enterprise that has solved all the hard problems, and behavioural game theory has just nibbled at some minor anomalies, with no underlying theoretical principles. Formal behavioural models, sharply honed on data, have been proposed in all the central issues in non-cooperative game theory – How will people play a new game? How do they learn? Do they behave differently when they are playing against others repeatedly? A natural question is how the models of thinking, learning and teaching fit together. The short answer is that the learning model is a mathematical special case of teaching in which all players are adaptive rather than sophisticated. The thinking and teaching models are not designed to fit together, since the cognitive hierarchy model is designed to apply to one-shot games (where planning for future rounds does not matter) and teaching is clearly for repeated games. One could presumably unify the models by mapping increasing steps of strategic thinking into increasing degrees of sophistication, but we have not done so. Or calling the thinking steps 'rules' and allowing players to learn in the domain of rules is a way of unifying the two (for example, Stahl, 1996). Since the models are so parsimonious there is no great saving in degrees of freedom by unifying them, but it would be important, both scientifically and practically, to know if there is a close link.

A cognitive hierarchy thinking model and bounded rationality measure

The cognitive hierarchy (CH) model is designed to predict behaviour in one-shot games and to provide initial conditions for models of learning. The model uses an iterative process which formalizes Selten's (1998, p. 421) intuition that 'the natural way of looking at game situations ... is not based on circular concepts, but rather on a step-by-step reasoning procedure'.

We begin with notation. Strategies have numerical attractions that determine the probabilities of choosing different strategies through a logistic response function. For player i, there are m_i strategies (indexed by j) which have initial attractions denoted $A_i^j(0)$. Denote i's jth strategy by s_i^j, chosen strategies by i and other players (denoted $-i$) in period t as $s_i(t)$ and $s_{-i}(t)$, and player i's payoffs of choosing s_i^j by $\pi_i(s_i^j, s_{-i}(t))$.

A logit response rule is used to map attractions into probabilities:

$$P_i^j(t+1) = \frac{e^{\lambda \cdot A_i^j(t)}}{\sum_{k=1}^{m_i} e^{\lambda \cdot A_i^k(t)}} \tag{8.1}$$

where λ is the response sensitivity.[6]

We model thinking by characterizing the number of steps of iterated thinking that subjects do, and their decision rules.[7] In the CH model some players, using zero steps of thinking, do not reason strategically at all. (Think of these players as being fatigued, clueless, overwhelmed, uncooperative, or simply more willing to make a random guess in the first period of a game and learn from subsequent experience than to think hard before learning.) We assume that zero-step players randomize equally over all strategies.

Players who do one step of thinking *do* reason strategically. What exactly do they do? We assume they are 'over-confident' – though they use one step, they believe others are all using zero steps. Proceeding inductively, players who use K steps think all others use zero to $K-1$ steps.

It is useful to ask why the number of steps of thinking might be limited. One answer comes from psychology. Steps of thinking strain 'working memory', where items are stored while being processed. Loosely speaking, working memory is a hard constraint. For example, most people can remember only about five to nine digits when shown a long list of digits (though there are reliable individual differences, correlated with reasoning ability). The strategic question: 'If she thinks he anticipates what she will do what should she do?' is an example of a recursive 'embedded sentence' of the sort that is known to strain working memory and produce inference and recall mistakes.[8]

Reasoning about others might also be limited, because players are not certain about other players' payoffs or degree of rationality. Why should players think others are rational? After all, adherence to optimization and instant equilibration is a matter of personal taste or skill. But whether other players do the same is a guess about the world (and iterating further, a guess about the contents of another player's brain or a firm's boardroom activity).

The key challenge in thinking steps models is pinning down the frequencies of K-step thinkers, $f(K)$. The constraint on working memory suggests that the relative fraction of $K-1$ step thinkers doing one more step of thinking, $f(K)/f(K-1)$, should be declining in K. For example, suppose the relative proportions of 2-step thinkers and 3-step thinkers is proportional to 1/3, the relative proportions of 5- and 6-step thinkers is proportional to 1/6, and in general $f(K)/f(K-1) \propto 1/K$. This axiom turns out to imply that $f(K)$ has a Poisson distribution with mean and standard deviation τ (the frequency of level K types is $f(K) = \frac{e^{-\tau}\tau^K}{K!}$). Then τ is an index of the degree of bounded rationality in the population.

The Poisson distribution has three appealing properties: it has only one free parameter (τ); since Poisson is discrete it generates 'spikes' in predicted distributions reflecting individual heterogeneity (other approaches do not[9]); and for sensible τ values the frequency of step types is similar to the frequencies estimated in earlier studies (see Stahl and Wilson, 1995; Ho *et al.*, 1998; and Nagel *et al.*, 1999). When we compare Poisson-constrained distributions $f(K)$ to a 7-parameter free distribution, with $f(0), f(1), \ldots, f(7)$ each a separate parameter, the Poisson distribution fits only about 1 per cent worse in log-likelihood terms in most data sets (see Camerer *et al.*, 2002b), so it is a very close approximation to a freer distribution.

Figure 8.1 shows four Poisson distributions with different τ values. Note that there are substantial frequencies of steps 0–3 for τ around one or two. There are also very few higher-step types, which is plausible if the limit on working memory has an upper bound.

Modelling heterogeneity is important, because it allows the possibility that not every player is rational. The few studies that have looked carefully found fairly reliable individual differences, because a subject's step level or decision rule is fairly stable across games (Stahl and Wilson, 1995; Costa-Gomes *et al.*, 2001). Including heterogeneity can also improve learning models by starting them off with enough persistent variation across people to match the variation seen among actual people.

To make the CH model precise, assume players know the absolute frequencies of players at lower levels from the Poisson distribution. But since they do not imagine higher-step types, there is missing probability. They must adjust their beliefs by allocating the missing probability in order to compute sensible expected payoffs to guide choices. We assume players divide the correct relative proportions of lower-step types by $\sum_{c=1}^{K-1} f(c)$ so the adjusted frequencies maintain the same relative proportions but add up to one.

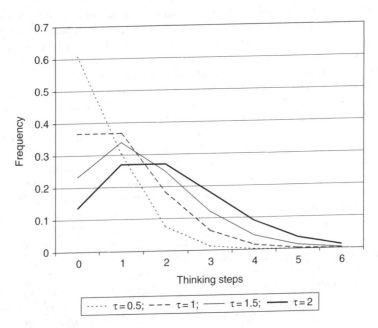

Figure 8.1 Poisson distributions for various τ

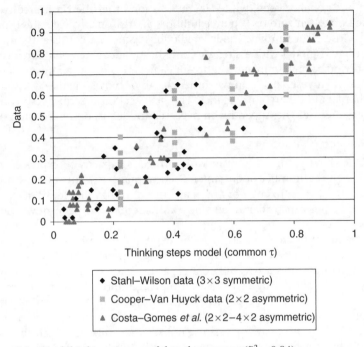

Figure 8.2 Fit of thinking-steps model to three games ($R^2 = 0.84$)

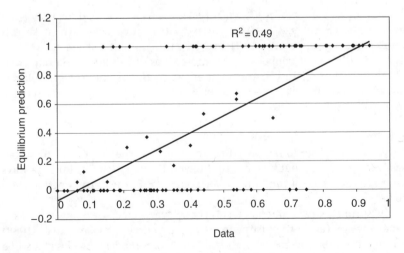

Figure 8.3 Nash equilibrium predictions versus data in three games

fits are reasonably good, while Figure 8.3 shows that the Nash predictions (which are often zero or 1, pure equilibria, are reasonably accurate, though not as close as the thinking-model predictions). Since τ is consistently around 1–2, the CH model with a single τ could be an adequate approximation to first-period behaviour in many different games. To see how far the model can take us, we investigated it in two other classes of games – games with mixed equilibria, and binary entry games. The next section describes results from entry games (see the Appendix on page 163 for details on mixed games).

Market entry games

Consider binary entry games in which there is capacity c (expressed as a fraction of the number of entrants). Each of many entrants decides simultaneously whether to enter or not. If an entrant thinks that fewer than c per cent will enter, s/he will enter; if s/he thinks more than c per cent will enter, s/he stays out.

There are three regularities in many experiments based on entry games such as this one (see Ochs, 1999; Seale and Rapoport, 2000; Camerer, 2003, ch. 7): (i) entry rates across different capacities c are closely correlated with entry rates predicted by (asymmetric) pure equilibria or symmetric mixed equilibria (that is, about c per cent of the people enter when capacity is c); (ii) players slightly over-enter at low capacities and under-enter at high capacities; and (iii) many players use noisy cut-off rules in which they stay out for most capacities below some cut-off c^* and enter for most higher capacities.

Let us apply the CH thinking model with best response. Step zero players enter half the time. This means that when $c < 0.5$, 1-step thinkers stay out, and when $c > 0.5$ they enter. Players doing 2 steps of thinking believe the fraction of zero steppers is $f(0)/(f(0)+f(1)) = 1/(1+\tau)$. Therefore, they enter only if $c > 0.5$ and $c > \frac{0.5+\tau}{1+\tau}$, or when $c < 0.5$ and $c > \frac{0.5}{1+\tau}$. To make this more concrete, suppose $\tau = 2$. Then 2-step thinkers enter when $c > 5/6$ and $1/6 < c < 0.5$. What happens is that more steps of thinking 'iron out' steps in the function relating c to overall entry. In the example, 1-step players are afraid to enter when $c < 1/2$. But when c is not too low (between 1/6 and 0.5) the 2-step thinkers perceive room for entry because they believe the relative proportion of zero-steppers is 1/3 and those players enter half the time. Two-step thinkers stay out for capacities between 0.5 and 5/6, but they enter for $c > 5/6$ because they know half of the (1/3) zero-step types will randomly stay out, leaving room even though 1-step thinkers always enter. Higher steps of thinking smooth out steps in the entry function even further.

The surprising experimental fact is that players can co-ordinate entry reasonably well, even in the first period. ('To a psychologist', Kahneman (1988) wrote, 'this looks like magic.') The thinking-steps model provides a possible explanation for this magic and can account for the other two regularities for reasonable τ values. Figure 8.4 plots entry rates from the first block of two studies for a game similar to the one above (Sundali *et al.*, 1995; Seale and Rapoport, 1999). Note that the number of actual entries rises almost monotonically with c, and entry is above capacity at low c and below capacity at high c.

Figure 8.4 also shows the CH entry function $N(all|\tau)(c)$ for $\tau = 1.5$ and 2. Both functions reproduce monotonicity and the over- and under-capacity effects. The thinking-steps models also produces approximate cut-off rule behaviour for all higher thinking steps except two. When $\tau = 1.5$, step 0 types randomize, step 1 types enter for all c above 0.5, step 3–4 types use cut-off rules with one 'exception', and levels 5 and above use strict cut-off rules. This mixture of random, cut-off and near-cut-off rules is roughly what is observed in the data when individual patterns of entry across c are measured (see, for example, Seale and Rapoport, 1999).

The model can also be used to do some simple theory. For example, in Camerer *et al.*, 2002b (and Camerer, Ho and Chong, 2003) we show that the entry function is monotonic (that is, $N(all|\tau)(c)$ is increasing in c) if $1 + 2\tau < e^{\tau}$, or $\tau < 1.25$. (The same condition guarantees that the conditional

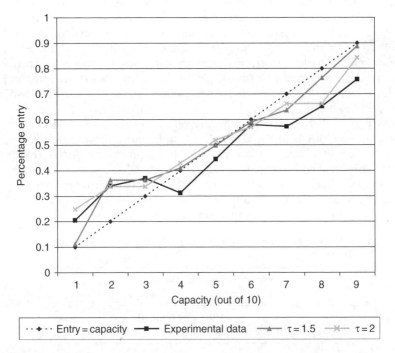

Figure 8.4 How entry varies with capacity (c), data and thinking-steps model

entry functions including only up to *K*-step players get increasingly close to the equilibrium entry as *K* rises.)

Thinking steps and cognitive measures

Since the CH model is a cognitive model, it gives an account of some treatment effects and shows how cognitive measures, such as response times and information acquisition, can be correlated with choices.

Belief-prompting

Some studies show that asking players for explicit beliefs about what others will do moves their choices, moving them closer to equilibrium (compared to a control in which beliefs are not prompted – see, for example, Costa-Gomes and Weizsacker, 2003). A simple example reported in Warglien *et al.* (1998) is shown in Table 8.2. Best-responding 1-step players think others are randomizing, so they choose *X*, which pays 60, rather than *Y*, which has an expected payoff of 45. Higher-step players choose *Y*.

Without belief-prompting, 70 per cent of the row players choose *X*. When subjects are prompted to articulate a belief about what the column players will do, 70 per cent choose the dominance-solvable equilibrium choice *Y*. Croson (2000) reports similar effects. In experiments on beauty contest games, we found that prompting beliefs also reduced dominance-violating choices modestly. Schotter *et al.* (1994) found a related display effect – showing a game in an extensive-form tree led to more subgame perfect choices.

Belief-prompting can be interpreted as increasing all players' thinking by one step. To illustrate, assume that since step zeros are forced to articulate *some* belief, they move to step 1. Now they believe others are random so they choose *X*. Players previously using one or more steps now use two or more. They believe column players choose *L* so they choose *Y*. The fraction of *X* play is therefore because former zero-step thinkers now do one step of thinking. This is just one simple example, but the numbers match up reasonably well[14] and it illustrates how belief-prompting effects could be accommodated within the thinking-steps model.

Similarly, Cooper and Kagel (2003b) report that two-person teams play signalling games more strategically (and transfer learning better) than individuals (though see Kocher and Sutter, forthcoming). This might be understood

Table 8.2 How belief-prompting promotes dominance-solvable choices by row players

Row move	Column player		Without belief prompting	With belief prompting
	L	*R*		
X	60,20	60,10	0.70	0.30
Y	80,20	10,10	0.30	0.70

Source: Warglien *et al.* (1998).

formally in terms of a model in which the highest-step player 'teaches' the lower-step player.

Information look-ups

Camerer *et al.* (1993), Costa-Gomes *et al.* (2001), Johnson *et al.* (2002), and Salmon (2003) measure directly the information subjects acquire in a game by putting payoff information in boxes which must be clicked open using a computer mouse. The order in which boxes are opened, and for how long they are open, gives a 'subject's-eye view' of what players are looking at, and should be correlated with thinking steps. Indeed, Johnson *et al.* show that how much time players spend looking ahead to future 'pie sizes' in alternating-offer bargaining is correlated with the offers they make. Costa-Gomes *et al.* show that look-up patterns are correlated with choices that result from various (unobserved) decision rules in normal-form games. These correlations means that a researcher who simply knew what a player had looked at could, to some extent, forecast that player's offer or choice. Both studies also showed that information look-up statistics helped to answer questions that choices alone could not.[15]

Summary

A simple cognitive hierarchy model of thinking steps attempts to predict choices in one-shot games and provide initial conditions for learning models. We propose a model which incorporate discrete steps of thinking, and the frequencies of players using different numbers of steps is Poisson-distributed with mean τ. We assume that players at level $K > 0$ cannot imagine players at their level or higher, but they understand the relative proportions of lower-step players and normalize them to compute expected payoffs. Estimates from three experiments on matrix games show reasonable fits for τ around 1–2, and τ is fairly regular across games in two of three data sets. Values of $\tau = 1.5$ also fits data from fifteen games with mixed equilibria and reproduces key regularities from binary entry games. The thinking-steps model also creates natural heterogeneity across subjects. When best response is assumed, the model generally creates 'purification' in which most players at any step level use a pure strategy, but a mixture results because of the mixture of players using different numbers of steps.

Learning

By the mid-1990s, it was well-established that simple models of learning could explain some movements in choice over time in specific game and choice contexts.[16] Therefore, the issue is not whether simple models of learning can capture some aspects of movement in experimental data – that issue was well-settled (the answer is Yes) by the late 1990s. The bigger challenge taken up since then is to see how well a specific parametric model can account for finer details of the equilibration process in a very wide range of games.

This section describes a one-parameter theory of learning in decisions and games called functional EWA (or self-tuning EWA for short; also called 'functional EWA' or 'EWA lite' to emphasize its simple functions). Self-tuning EWA predicts the time path of individual behaviour in any normal-form game. Initial conditions can be imposed or estimated in various ways. We use initial conditions from the thinking-steps model described in the previous section. The goal is to predict both initial conditions and equilibration in new games in which behaviour has never been observed, with minimal free parameters (the model uses one parameter, λ).

Parametric EWA learning: interpretation, uses and limits

Self-tuning EWA is a relative of the parametric model of learning called experience-weighted attraction (EWA) (Camerer and Ho, 1998, 1999). As in most theories, learning in EWA is characterized by changes in (unobserved) attractions based on experience. Attractions determine the probabilities of choosing different strategies through a logistic response function. For player i, there are m_i strategies (indexed by j) which have initial attractions denoted $A_i^j(0)$. The thinking steps model is used to generate initial attractions given parameter values τ and λ.

To avoid complications with negative payoffs, we rescale payoffs by subtracting by the minimum payoff so that rescale payoffs are always weakly positive. Define an indicator function $I(x, y)$ to be zero if $x \neq y$ and one if $x = y$. The EWA attraction updating equation is:

$$A_i^j(t) = \frac{\phi N(t-1) A_i^j(t-1) + [\delta + (1-\delta) I(s_i^j, s_i(t))] \pi_i(s_i^j, s_{-i}(t))}{N(t-1)\phi(1-\rho) + 1} \tag{8.5}$$

and the experience weight (the 'EW' part) is updated according to $N(t) = N(t-1)\phi(1-\rho) + 1$.[17] Notice that the term $[\delta + (1-\delta)I(s_i^j, s_i(t))]$ implies that a weight of one is put on the payoff term when the strategy being reinforced is the one the player chose ($s_i^j = s_i(t)$), but the weight on forgone payoffs from unchosen strategies ($s_i^j \neq s_i(t)$) is δ. (When forgone payoffs are not known exactly, averaging possible values or using historical rules can be used as proxies.[18]) Attractions are mapped into choice probabilities using a logit response function $P_i^j(t+1) = \frac{e^{\lambda \cdot A_i^j(t)}}{\sum_{k=1}^{m_i} e^{\lambda \cdot A_i^k(t)}}$ (where λ is the response sensitivity). The subscript i, superscript j, and argument $t+1$ in $P_i^j(t+1)$ are reminders that the model aims to explain every choice by every subject in every period.[19]

In implementing the model, we shall typically take strategies to be stage-game strategies. However, it is often likely that a strategy could be history-dependent or have some other form, which should be considered in future work. Furthermore, transfer of learning across games is an interesting topic we have not explored (but see Cooper and Kagel, 2003a).

Each EWA parameter has a natural interpretation.

The parameter δ is the weight placed on forgone payoffs. Presumably it is affected by imagination (in psychological terms, the strength of counterfactual reasoning or regret, or in economic terms, the weight placed on opportunity costs and benefits), or reliability of information about forgone payoffs (Heller and Sarin, 2000).

The parameter ϕ decays previous attractions because of forgetting or, more interestingly, because agents are aware that the learning environment is changing and deliberately 'retire' old information (much as firms junk old equipment more quickly when technology changes rapidly).

The parameter ρ controls the rate at which attractions grow. When $\rho = 0$, attractions are weighted averages and grow slowly; but when when $\rho = 1$ attractions cumulate. We originally included this variable because some learning rules used cumulation and others used averaging. It is also a rough way to capture the distinction in machine learning between 'exploring' an environment (low ρ), and 'exploiting' what is known by locking in to a good strategy (high ρ) (see, for example, Sutton and Barto, 1998).

The initial experience weight $N(0)$ is like a strength of prior beliefs in models of Bayesian belief learning. (Imposing $N(0) < \frac{1}{1+\phi(1-\rho)}$ guarantees that $N(t)$ is increasing, which is sensible.) It plays a minimal empirical role, so it is set to $N(0) = 1$ in our current work.

EWA is a hybrid of two widely-studied models, reinforcement and belief learning. In reinforcement learning, only payoffs from chosen strategies are used to update attractions and guide learning. In belief learning, players do not learn about which strategies work best; they learn about what others are likely to do, then use those updated beliefs to change their attractions and hence which strategies they choose (see Brown, 1951; Fudenberg and Levine, 1998). EWA shows that reinforcement and belief learning, which were often treated as being fundamentally different, are in fact related in a non-obvious way, because both are special kinds of reinforcement rules.[20] When $\delta = 0$, the EWA rule is a simple reinforcement rule.[21] When $\delta = 1$ and $\rho = 0$, the EWA rule is equivalent to belief learning using weighted fictitious play.[22]

It is important to be very clear about what the EWA formulation means and does. First, one thing EWA suggests is that general learning can be thought of as a splice of two different cognitive processes: in one process, strategies that are chosen are automatically (and fully) reinforced; and in the other, players think about the forgone payoffs they would have gained from choosing other strategies and reinforce them less strongly (with weight δ). Many kinds of behaviour are now attributable to the behaviour of two kinds of system (see, for example, Kahneman, 2003): one system is very fast, automatic, pattern-orientated and sometimes subconscious (similar to our perceptual system), and can work in parallel (for example, people can perceive sounds and images at the same time); the other is slow, deliberate, conscious, invokes abstract rules and cognition (logic, arithmetic), and works serially. Put very roughly, δ represents the strength of the second, more deliberative, process which generates counterfactual answers to the question 'Could I have done better with another strategy?' Furthermore, seen in the dual-process light, reinforcement learning can be interpreted as suggesting that the second process does not start (which may be appropriate for animals with minimal cortical apparatus for deliberation, or for people under time pressure), while belief learning suggests that the second process completely overrides the rapid instinctive response of the first process.

Second, EWA provides a way of checking whether a simple model is *too simple*. Obviously, every model is a simplification that permits counterexamples. So making imperfect predictions is no reason to abandon a model. But imperfections *are* a reason to ask whether adding features to a model can improve its predictions. EWA is one way of doing precisely this.

Forgone payoffs are the fuel that runs EWA learning. They also provide an indirect link to 'direction learning' and imitation. In direction learning, players move in the direction of observed best response (Selten and Stöcker, 1986). Suppose players follow EWA but do not know forgone payoffs, and believe those payoffs are monotonically increasing between their choice $s_i(t)$ and the best response. If they also reinforce strategies near their choice $s_i(t)$ more strongly than strategies that are further away, their behaviour will look like direction learning. Imitating a player who is similar and successful can also be seen as a way of heuristically inferring high forgone payoffs from an observed choice and moving in the direction of those higher payoffs. However, this is probably not the whole story about 'observational learning', which is a fruitful area of research (see, for example, Duffy and Feltovich, 1999; Armantier, 2004).[23]

The relationships of various learning rules can be shown visually in a cube showing configurations of parameter values (see Figure 8.5). Each point in the cube is a triple of EWA parameter values that specify a precise updating equation. The corner of the cube with $\phi = \rho = 0, \delta = 1$ is Cournot best-response dynamics. The corner $\rho = 0, \phi = \delta = 1$, is standard fictitious play. The vertex connecting these corners, $\delta = 1, \rho = 0$, is the class of weighted fictitious play rules (see, for example, Fudenberg and Levine, 1998). The vertices with $\delta = 0$ and $\rho = 0$ or 1 are averaging and cumulative choice reinforcement rules.

The biologist Francis Crick (1988) said, 'in nature a hybrid is often sterile, but in science the opposite is usually true'. As Crick suggests, the point of EWA is not simply to show a surprising relationship among other models, but also to improve their fertility for explaining patterns in data by combining the best modelling 'genes'. In reinforcement theories, received payoffs get the most weight (in fact, *all* the weight[24]). Belief theories assume implicitly that

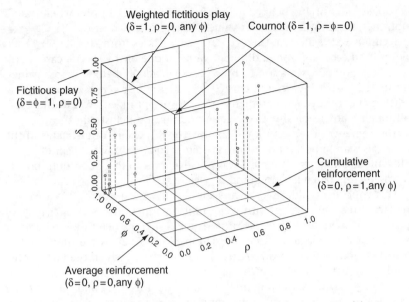

Note: The arrows indicate where the various learning models are, as special cases, in the EWA framework. The points represent empirical estimates for various games.

Figure 8.5 The EWA learning cube: learning models and empirical estimates

forgone and received payoffs are weighted equally. Rather than assuming that one of these intuitions about payoff weights is right and the other is wrong, EWA allows *both* intuitions to be true. When $0 < \delta < 1$, received payoffs can gain more weight, but forgone payoffs also get some weight.

The EWA model has been estimated by ourselves and many others on about 40 data sets (see Camerer *et al.*, 2002). The hybrid EWA model predicts more accurately in many cases than do the special cases of reinforcement and weighted fictitious play, except in games with mixed-strategy equilibrium, where reinforcement does equally well.[25] It is extremely important to emphasize that, in our model estimation and validation, we *always* penalize the EWA model in ways that are known to generally make the *adjusted* fit *worse* if a model is too complex (that is, if the data are in fact generated by a simpler model).[26] Furthermore, econometric studies show that, if the data were generated by simpler belief or reinforcement models, then EWA estimates would generally correctly identify that fact for many games and reasonable sample sizes (see Cabrales and Garcia-Fontes, 2000; Salmon, 2001), although Wilcox (2003) finds that heterogeneity in all model parameters lowers the estimate of δ in mixed-equilibrium games (which may explain why low values of δ often fit well in these games). Since EWA is capable of identifying behaviour consistent with special cases, when it does not, then the hybrid parameter values are improving in fit. Figure 8.5 also shows estimated parameter triples from 20 data sets. Each point is an estimate from a different game. If one of the special case theories is a good approximation to how people generally behave across games, estimated parameters should cluster in the corner or vertex corresponding to that theory. In fact, parameters tend to be sprinkled around the cube, although many (typically mixed-equilibrium games) cluster in the averaged reinforcement corner with low δ and ρ. The dispersion of estimates in the cube raises an important question: is there regularity in which games generate which parameter estimates? A positive answer to this question is crucial for predicting behaviour in brand new games.

This concern is addressed by a version of EWA, self-tuning EWA, which replaces free parameters with deterministic functions $\phi_i(t), \delta_i(t), \rho_i(t)$ of player i's experience up to period t. These functions determine parameter values for each player and period. The parameter values are then used in the EWA updating equation to determine attractions, which then determine choices probabilistically. Since the functions also vary across subjects and over time, they have the potential to inject heterogeneity and time-varying 'rule learning', and to explain learning better than models with fixed parameter values across people and time. And since self-tuning EWA has only one parameter that must be estimated (λ),[27] it is especially helpful when learning models are used as building blocks for more complex models that incorporate sophistication (some players think others learn) and teaching, as we discuss in the section below.

The decay rate ϕ is sometimes interpreted as forgetting, an interpretation carried over from reinforcement models of animal learning. Certainly forgetting does occur, but the more important variation in $\phi_i(t)$ across games is probably a player's perception of how quickly the learning environment is changing. The function $\phi_i(t)$ should therefore 'detect change'. As in physical change detectors (for example, security systems or smoke alarms), the challenge is to detect change when it is really occurring, but not falsely mistake noise for change too often.

The core of the $\phi_i(t)$ change-detector function is a 'surprise index', which is the difference between other players' recent strategies and their strategies in previous periods. To make exposition easier, we describe the function for games with pure-strategy equilibria (suitably modified for games with mixed equilibria, as noted below). First define a history vector, across the other players' strategies k, which records the historical frequencies (including the last period) of the choices by other players of s_{-i}^k. The vector element $h_i^k(t)$ is $\frac{\sum_{\tau=1}^{t} I(s_{-i}^k, s_{-i}(\tau))}{t}$.[28] The recent 'history' $r_i^k(t)$ is a vector of 0s and 1s which has a 1 for strategy $s_{-i}^k = s_{-i}(t)$ and 0s for all other strategies s_{-i}^k (that is, $r_i^k(t) = I(s_{-i}^k, s_{-i}(t))$). The surprise index $S_i(t)$ simply sums up the squared deviations between the cumulative history vector $h_i^k(t)$ and the immediate recent history vector $r_i^k(t)$; that is, $S_i(t) \equiv \sum_{k=1}^{m_{-i}} (h_i^k(t) - r_i^k(t))^2$. Note that this surprise index varies from zero (when the last strategy the other player chose is the one they have always chosen before) to two (when the other player chose a particular strategy 'for ever' then suddenly switches to something brand new). The change-detecting decay rate is $\phi_i(t) = 1 - 0.5 \cdot S_i(t)$. Because $S_i(t)$ is between zero and two, ϕ is always (weakly) between one and zero.

The numerical boundary cases illuminate intuition: if the other player chooses the strategy s/he has always chosen before, then $S_i(t) = 0$ (player i is not surprised) and $\phi_i(t) = 1$ (player i does not decay the lagged attraction at all, since what other players did throughout is informative). If the other player chooses a new strategy that was never chosen before in a very long run of history, $S_i(t) = 2$ and $\phi_i(t) = 0$ (player i decays the lagged attraction completely and 'starts over'). Note that, since the observed behaviour in period t is included in the history $h_i^k(t)$, $\phi_i(t)$ will typically not dip to zero. For example, if a player chose the same strategy for each of nine periods and a new strategy in period 10, then $S_i(t) = (0.9 - 0)^2 + (1 - 0.1)^2 = 2 \cdot 0.81$ and $\phi_i(t) = 1 - 0.5(2 \cdot 0.81) = 0.19$.

In games with mixed equilibria (and no pure equilibria), a player should expect other players' strategies to vary. Therefore, if the game has a mixed equilibrium with W strategies that are played with positive probability (that is, W is the cardinality of the smallest support of any Nash strategy), the surprise index defines recent history over a window of the last W periods (for example, in a game with four strategies that are played in equilibrium, $W = 4$). Then $r_i^k(t) = \sum_{k=1}^{m_{-i}} \left[\frac{\sum_{\tau=t-W+1}^{t} I(s_{-i}^k, s_{-i}(\tau))}{W} \right]$.[29]

A sensible property of $S_i(t)$ is that the surprisingness of a new choice should depend not only on how often the new choice has been chosen before, but also on how variable previous choices have been. Incorporating this property requires ϕ to be larger when there is more dispersion in previous choices, which is guaranteed by squaring the deviations between current and previous history. (Summing absolute deviations between $r_i(t)$ and $h_i(t)$, for example, would not have this property.) If previously observed relative frequencies of strategy k are denoted f_k, and the recent strategy is h, then the surprise index is $(1 - f_h)^2 + \sum_{k \neq h} (f_k - 0)^2$. Holding f_h constant, this index is minimized when all frequencies f_k with $k \neq h$ are equal. In the equal-f_k case, the surprise index is $S_i(t) = (m_{-i} - 1)/m_{-i}$ and $\phi_i(t) = (m_{-i} + 1)/2m_{-i}$, which has a lower bound of 0.5 in games with large m_{-i} (many strategies).

The opposite case is when an opponent has previously chosen a single strategy in every period, and suddenly switches to a new strategy. In this case, $\phi_i(t)$ is $\frac{2t-1}{t^2}$. This expression declines gracefully towards zero as the string of identical choices up to period t grows longer. (For $t = 2, 3, 5$ and 10, the $\phi_i(t)$ values are 0.75, 0.56, 0.36, and 0.19.) The fact that the ϕ values decline with t expresses the principle that a new choice is a bigger surprise (and should have an associated lower ϕ) if it follows a *longer* string of *identical* choices that are different from the surprising new choice. It also embodies the idea that dipping $\phi_i(t)$ too low is a mistake which should be avoided because it erases the history embodied in the lagged attraction. So $\phi_i(t)$ only dips low when opponents have been choosing one strategy reliably for a very long time, and then switch to a new one.

Another interesting special case is when unique strategies have been played in every period up to $t - 1$, and another unique strategy is played in period t. (This is often true in games with large strategy spaces.) Then $\phi_i(t) = 0.5 + \frac{1}{2t}$, which starts at 0.75 and asymptotes at 0.5 as t increases.

The calculations above show that in the first few periods of a game, $\phi_i(t)$ will not dip much below 1 (because the tth period experience is included in the recent history $r_i(t)$ vector as well as being part of the cumulative history $h_i(t)$). But in these periods players often learn rapidly. Since it makes sense to start with a low value of $\phi_i(0)$ to express players' responsiveness in the first few periods, in the empirical implementation, we smooth the $\phi_i(t)$ function by starting at $\phi_i(0) = 0.5$, and gently blending in the updated values according to $\hat{\phi}_i(t) \equiv 0.5/t + (t - 1)\phi_i(t)/t$.

The other self-tuning EWA functions are less empirically important and interesting so we mention them only briefly. The function $\delta_i(t) = \phi_i(t)/W$. Dividing by W pushes $\delta_i(t)$ towards zero in games with mixed equilibria, which matches estimates in many games (see Camerer *et al.*, 2003).[30] Tying $\delta_i(t)$ to the change detector $\phi_i(t)$ means that chosen strategies are reinforced relatively strongly (compared to unchosen ones) when change is rapid. This reflects a 'status quo bias' or 'freezing' response to danger (which is virtually universal across species, including humans). Since $\rho_i(t)$ controls how sharply subjects lock in to choosing a small number of strategies, we use a 'Gini coefficient' – a standard measure of dispersion often used to measure income inequality – over choice frequencies.[31,32]

Self-tuning EWA has three advantages. First, it is easy to use because it has only one free parameter (λ). Second, parameters in self-tuning EWA naturally vary across time and people (as well as across games), which can capture heterogeneity and mimic 'rule learning' in which parameters vary over time (see, for example, Stahl, 1996, 2000; Salmon, 2001). For example, if ϕ rises across periods from 0 to 1 as other players stabilize, players are effectively switching from Cournot-type dynamics to fictitious play. If δ rises from 0 to 1, players are effectively switching from reinforcement to belief learning. Third, it should be easier to theorize about the limiting behaviour of self-tuning EWA than about some parametric models. A key feature of self-tuning EWA is that, as a player's opponent's behaviour stabilizes, $\phi_i(t)$ goes toward 1 and (in games with pure equilibria) $\delta_i(t)$ does too. If $\rho = 0$, self-tuning EWA then automatically turns into fictitious play; and a lot is known about theoretical properties of fictitious play.

Self-tuning EWA predictions

In this section we compare in-sample fit and out-of-sample predictive accuracy of different learning models when parameters are estimated freely, and check whether self-tuning EWA functions can produce game-specific parameters similar to estimated values. We use seven games: games with unique mixed strategy equilibrium (Mookerjhee and Sopher, 1997); R&D (research and development) patent race games (Rapoport and Amaldoss, 2000); a median-action order statistic co-ordination game with several players (Van Huyck *et al.*, 1990); a continental-divide co-ordination game, in which convergence behaviour is extremely sensitive to initial conditions (Van Huyck *et al.*, 1997); a 'pots game' with entry into two markets of different sizes (Amaldoss and Ho, in preparation); dominance-solvable p-beauty contests (Ho *et al.*, 1998); and a price-matching game (called 'traveller's dilemma' by Capra *et al.*, 1999).

Estimation method

The estimation procedure for self-tuning EWA is sketched briefly here (see Ho *et al.*, 2001 for details). Consider a game where N subjects play T rounds. For a given player i of CH step-level c, the likelihood function of observing a choice history of $\{s_i(1), s_i(2), \ldots, s_i(T-1), s_i(T)\}$ is given by:

$$\Pi_{t=1}^{T} P_i^{s_i(t)}(t|c) \tag{8.6}$$

The joint likelihood function L of observing all players' choices is given by:

$$L(\lambda) = \Pi_i^{N} \left\{ \sum_{c=1}^{K} f(c) \times \Pi_{t=1}^{T} P_i^{s_i(t)}(t) \right\} \tag{8.7}$$

where K is set to a multiple of τ rounded to an integer. Most models are 'burnt in' by using first-period data to determine initial attractions. We also compare all models with burnt-in attractions with a model in which the thinking steps model from the previous section is used to create initial conditions and combined with self-tuning EWA. Note that the latter hybrid uses only two parameters (τ and λ) and does not use first-period data at all.

Given the initial attractions and initial parameter values,[33] attractions are updated using the EWA formula. Self-tuning EWA parameters are then updated according to the functions above and used in the EWA updating equation. Maximum likelihood estimation is used to find the best-fitting value of λ (and other parameters, for the other models) using data from the first 70 per cent of the subjects. Then the value of λ is frozen and used to forecast behaviour of the entire path of the remaining 30 per cent of the subjects. Payoffs were all converted to dollars (which is important for cross-game forecasting).

In addition to self-tuning EWA (one parameter), we estimated the parametric EWA model (five parameters), a belief-based model (weighted fictitious play, two parameters) and the two -parameter reinforcement models with payoff variability (Erev *et al.*, 1999; Roth *et al.*, 2000), and QRE.

Model fit and predictive accuracy in all games

The first question we ask is how well models fit and predict on a game-by-game basis (that is, parameters are estimated separately for each game). Some authors fix a set of parameters for several games and see how well predictions do based on those parameters. But it is impossible to know from such an exercise whether there is a *better* set of parameters, or how much the best-fitting parameters truly vary across games. By allowing parameter to vary across games we can both find out how well a single parameter would do (by looking at variation across games – if variation is low a single parameter would be fine, or we can restrict the parameter to have a common value and see how badly the fit degrades), and have some guidance as to which parameters fit best in which games.

An alternative method is to simulate the entire path of play and compare some statistics of the simulated path with statistics of the data (e.g., Roth and Erev 1995). We have done this in several publications and do not draw different conclusions from those drawn from the estimation method below (see, for example, Camerer *et al.*, 1998; Camerer, Hsia and Ho, 2002; and Camerer, 2003, ch. 6). Furthermore, note that if the statistic that is used is the *conditional* relative frequency (that is, frequencies conditioned on all actual histories), it makes no difference whether the conditioned histories are determined by the actual data, or paths are first simulated, then the paths which match the data histories are selected for conditional-frequency comparison. The simulation method will give exactly the same results as the method we use. Furthermore, if the frequencies that are compared are *not* conditioned on histories, then the model can be doing poorly on capturing some kind of history-dependence but appear to fit well on the unconditioned frequencies.[34]

For out-of-sample validation we report both hit rates (the fraction of most-likely choices that are picked) and log likelihood (*LL*). (Keep in mind that these results forecast a holdout sample of subjects after model parameters have been estimated on an earlier sample and then 'frozen'. If a complex model is fitting better within a sample purely because of spurious overfitting, it will in fact predict less well out of sample.) Results are summarized in Table 8.3.

Across games, parametric EWA is as good as all other theories or better, judged by hit rate, and has the best *LL* in four games. Self-tuning EWA also does well on hit rate in six out of seven games. Reinforcement is competitive on hit rate in five games, and best in *LL* in two. Belief models are often inferior on hit rate and never best in *LL*. QRE clearly fits the worst.

Combining self-tuning EWA with a thinking steps model to predict initial conditions (rather than using the first-period data), a two-parameter combination is only a little worse in hit rate than self-tuning EWA, and slightly worse in *LL*.

Table 8.3 Out-of-sample accuracy of learning models

Game	Thinking +stEWA		stEWA		EWA		Weighted fict. play		Reinf. with PV		QRE	
	%Hit	LL	%Hit	LL	%Hit	LL	%Hit	LL	%Hit	LL	%Hit	LL
Cont'l divide (7)	**45**	−483	**47**	−470	**47**	**−460**	25	−565	45	−557	5	−806
Med. action (14)	71	−112	74	−104	79	**−83**	**82**	−95	74	−105	49	−285
p-BC (1)	**8**	−2119	**8**	−2119	6	**−2042**	7	−2051	**6**	−2504	4	−2497
Pricing (0.8)	**43**	−507	**46**	−445	**43**	**−443**	36	−465	**41**	−561	27	−720
Mixed games (21)	**36**	−1391	**36**	**−1382**	**36**	−1387	**34**	−1405	33	−1392	**35**	−1400
Patents (18)	**64**	−1936	**65**	−1897	**65**	−1878	53	−2279	**65**	**−1864**	40	−2914
Pot games (50)	**70**	−438	**70**	−436	**70**	−437	66	−471	**70**	**−429**	51	−509
Pooled	**50**	−6986	**51**	**−6852**	49	**−7100**	40	−7935	46	−9128	36	−9037
KS p-BC	6		6	−309	3	−279	3	−279	4	−344	1	−346

Notes: Sample sizes are 315, 160, 580, 160, 960, 1760, 739, 4674 (pooled), 80; The best fits for each game and criterion printed in bold; hit rates statistically indistinguishable from the best (by the McNemar test) also in bold. Numbers in parentheses after each game name in col. 1 are hit rates from a random model; comparing hit rates in the '%Hit' columns indicates how much better the models are doing than random prediction.

Source: Ho *et al.* (2001).

The penultimate row of Table 8.3, 'pooled', shows results when a single set of common parameters is estimated for all games (except for game-specific λ^{35}). If self-tuning EWA is capturing parameter differences across games effectively, it should predict especially accurately, compared to other models, when games are pooled. It does so: when all games are pooled, self-tuning EWA predicts out-of-sample better than other theories, by both statistical criteria.

Some readers of our functional EWA paper were concerned that, by searching across different specifications we may have overfitted the sample of seven games we reported. To check whether we did, we announced at conferences in 2001 that we would analyse all the data people sent us by the end of that year and report the results in a revised paper. Three samples were sent and we have analysed one so far – experiments by Kocher and Sutter (forthcoming) on p-beauty contest games played by individuals and groups. The KS results are reported in the last row of Table 8.3 ('KS p-BC'). The game is the same as the beauty contests we studied (except for the interesting complication of group decision-making, which speeds equilibration), so it is not surprising that the results replicate the earlier findings: belief and parametric EWA fit best by *LL*, followed by self-tuning EWA, and reinforcement and QRE models fit worst. This is a small piece of evidence that the solid performance of self-tuning EWA (while being worse than belief learning on these games) is not entirely caused by overfitting on our original seven-game sample.

Now we shall show predicted and relative frequencies for three games that highlight differences among models. In other games the differences are minor or hard to see with the naked eye.[36]

Dominance-solvable games: beauty contests

In beauty contest games, each of n players chooses $x_i \in [0, 100]$. The average of their choices is computed and whichever player is closest to $p < 1$ times the average wins a fixed prize (see Nagel, 1999, for a review). The unique Nash equilibrium is zero. (The games get their name from a passage in Keynes (1936) about how the stock market is like a special beauty contest in which people judge who others will think is beautiful.) These games are a useful way to measure the steps of iterated thinking players seem to use (since higher steps will lead to lower number choices). Experiments have been run with exotic subject pools such as Ph.D.s and CEOs (Camerer, 1997), and in newspaper contests with very large samples (Nagel *et al.*, 1999). The results are generally robust, although specially-educated subjects (for example, professional game theorists) choose, not surprisingly, closer to equilibrium.

We analyse experiments run by Ho *et al.* (1998).[37] The data and relative frequencies predicted by each learning model are shown in Figure 8.6. Figure 8.6(a) shows that while subjects start around the middle of the distribution, they converge steadily downwards towards zero. By Period 5, half the subjects chose numbers 1–10.

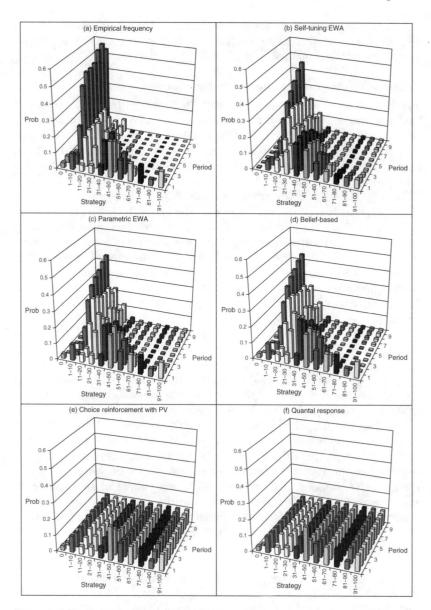

Figure 8.6 Predicted frequencies for *p*-beauty contest

The EWA, belief and thinking – self-tuning EWA model all capture the basic regularities although they underestimate the speed of convergence. (In the next section we add sophistication – some subjects know that others are learning and 'shoot ahead' of the learners by choosing lower numbers – which improves the fit substantially.) The QRE model is a dud in this game, and reinforcement also learns far too slowly because most players receive no reinforcement.[38]

Games with multiple equilibria: continental divide game

Van Huyck *et al.* (1997) studied a co-ordination game with multiple equilibria and extreme sensitivity to initial conditions, which we call the continental divide game (CDG). The payoffs in the game are shown in Table 8.4. Subjects play in cohorts of seven people. Subjects choose an integer from 1 to 14, and their payoff depends on their own choice and on the median choice of all seven players.

The payoff matrix is constructed so that there are two pure equilibria (at 3 and 12) which are Pareto-ranked (12 is the better one). Best responses to different medians are in bold. The best-response correspondence bifurcates in the middle: if the median starts at 7 virtually any sort of learning dynamics will lead players toward the equilibrium at 3. If the median starts at 8 or above, however, learning will eventually converge to an equilibrium of 12. Both equilibrium payoffs are shown in bold italics. The payoff at 3 is about half as much as at 12, so which equilibrium is selected has a large economic impact.

Figure 8.7 shows empirical frequencies (pooling all subjects) and model predictions.[39] The key features of the data are: bifurcation over time from choices in the middle of the range (5–10) to the extremes, near the equilibria at 3 and 12; and late-period choices are more clustered around 12 than around 3. There is also an extreme sensitivity to initial conditions (which is disguised by the aggregation across sessions in Figure 8.7(a)): namely, five groups had initial medians below 7 and all five converged toward the inefficient low equilibrium. The other five groups had initial medians above 7 and all five converged towards the efficient high equilibrium. This path-dependence shows the importance of a good theory of initial conditions (such as the thinking steps model). Because a couple of steps of thinking generates a distribution concentrated in the middle strategies 5–9, the thinking-steps models predicts that initial medians will sometimes be above the separatrix 7 and sometimes below. The model does not predict precisely which equilibrium will emerge, but it predicts that both high and low equilibria will sometimes emerge.

Notice also that strategies 1–4 are never chosen in early periods, but are chosen frequently in later periods. Strategies 7–9 are chosen frequently in early periods but rarely chosen in later periods. Like a sportscar, a good model should be able to capture these effects by 'accelerating' low choices quickly (going from zero to frequent choices in a few periods) and 'braking' midrange choices quickly (going from frequent choices to zero).

QRE fits poorly because it predicts no movement (it is not a theory of learning, of course, but simply a static benchmark that is tougher to beat

Table 8.4 Payoffs in 'continental divide' experiment

Choice	Median choice													
	1	2	3	4	5	6	7	8	9	10	11	12	13	14
1	45	49	52	55	56	55	46	−59	−88	−105	−117	−127	−135	−142
2	**48**	53	58	62	65	66	61	−27	−52	−67	−77	−86	−92	−98
3	**48**	**54**	**60**	**66**	70	74	72	1	−20	−32	−41	−48	−53	−58
4	43	51	58	65	**71**	77	80	26	8	−2	−9	−14	−19	−22
5	35	44	52	60	69	**77**	**83**	46	32	25	19	15	12	10
6	23	33	42	52	62	72	82	62	53	47	43	41	39	38
7	7	18	28	40	51	64	78	75	69	66	64	63	62	62
8	−13	−1	11	23	37	51	69	83	81	80	80	80	81	82
9	−37	−24	−11	3	18	35	57	**89**	**94**	91	92	94	96	98
10	−65	−51	−37	−21	−4	15	40	85	**94**	**100**	101	104	107	110
11	−97	−82	−66	−49	−31	−9	20	85	91	99	**106**	110	114	119
12	−133	−117	−100	−82	−61	−37	−5	78	91	99	**106**	*112*	*118*	**123**
13	−173	−156	−137	−118	−96	−69	−33	67	83	94	103	110	117	**123**
14	−217	−198	−179	−158	−134	−105	−65	52	72	85	95	104	112	120

Note: Best replies in bold.
Source: Van Huyck et al. (1997).

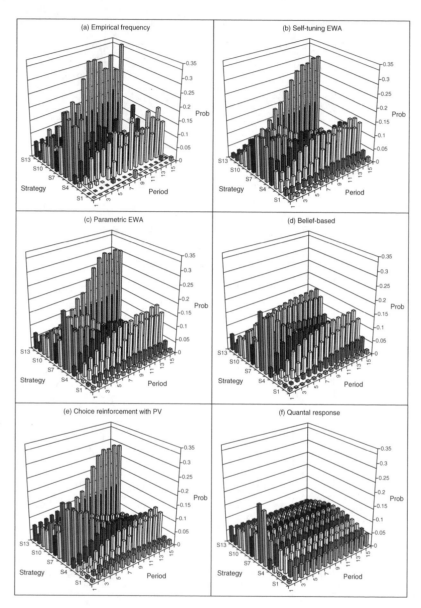

Figure 8.7   Predicted frequencies for continental divide

than Nash). Reinforcement with PV fits well. Belief learning does not reproduce the asymmetry between sharp convergence to the high equilibrium
and flatter frequencies around the low equilibrium. The reason why is dia-
gnostic of a subtle weakness in belief learning. Note from Table 8.4 that the

payoff gradients around the equilibria at 3 and 12 are exactly the same – choosing one number too high or low 'costs' $0.02; choosing two numbers too high or low costs $0.08, and so on. Since belief learning computes expected payoffs, and the logit rule means only differences in expected payoffs influence choice probability, the fact that the payoff gradients are the same means the spread of probability around the two equilibria must be the same. Self-tuning EWA, parametric EWA and the reinforcement models generate the asymmetry with low δ.[40]

Games with dominance-solvable equilibrium: price-matching with loyalty

Capra *et al.* (1999) studied a dominance-solvable price-matching game. In their game, two players simultaneously choose a price between 80 and 200. Both players earn the low price. In addition, the player who names the lower price receives a bonus of R and the players who names the higher price pays a penalty R. (If the prices they choose are the same, the bonus and penalty cancel and players just earn the price they named.) You can think of R as a reduced-form expression of the benefits of customer loyalty and word-of-mouth which accrue to the lower-priced player, and the penalty is the cost of customer disloyalty and switching away from the high-price firm. We like this game because price-matching is a central feature of economic life. These experiments can also, in principle, be tied to field observations in future work.

Their experiment used six groups of 9–12 subjects. The reward/penalty R had six values (5, 10, 20, 25, 50, 80). Subjects were rematched randomly.[41]

Figure 8.8 shows empirical frequencies and model fits for $R = 50$ (where the models differ most). A wide range of prices are named in the first round. Prices gradually fall, being 91–100 in Rounds 3–5, 81–90 in Rounds 5–6, and towards the equilibrium of 80 in later rounds.

QRE predicts a spike at the Nash equilibrium of 80.[42] The belief-based model predicts the direction of convergence, but overpredicts numbers in the interval 81–90 and underpredicts choices of precisely 80. The problem is that the incentive in the traveller's dilemma is to undercut the other player's price by as little as possible. Players only choose 80 frequently in the last couple of periods; before those periods it pays to choose higher numbers.

EWA models explain the sharp convergence in late periods by cumulating payoffs and estimating $\delta = 0.63$ (for self-tuning EWA). Players who chose 80 while others named a higher price could have earned more by undercutting the other price, but weighting that higher forgone payoff by δ means their choice of 80 is reinforced more strongly, which matches the data.

Reinforcement with payoff variability has a good hit rate because the highest spikes in the graph often correspond with spikes in the data. But the graph shows that predicted learning is much more sluggish than in the data (that is, the spikes are not high enough). Because $\phi = 1$ and players are not

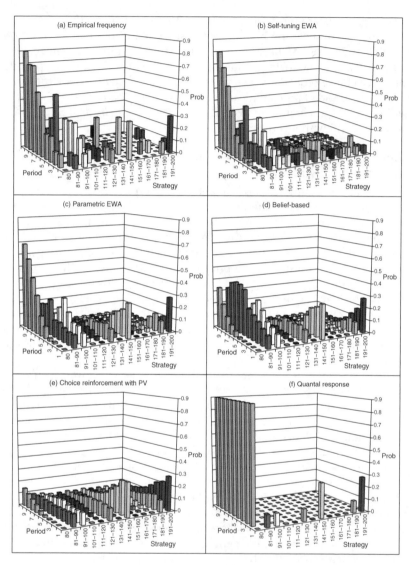

Figure 8.8   Predicted frequency for traveller's dilemma (Reward = 50)

predicted to move toward *ex post* best responses, the model cannot explain why players learn to choose 80 so rapidly.

Economic value of learning models

Since the 1980s, the concept of economic engineering has emerged as being increasingly important from its start in the late 1970s (see Plott, 1986).

Experimentation has played an important role in this emergence (see Plott, 1997; Rassenti *et al.*, 2001). For the practice of economic engineering, it is useful to have a measure of how much value a theory or design creates. For policy purposes, increases in allocative efficiency are a sensible measure, but for judging the private value of advice to a firm or consumer other measures are more appropriate.

Camerer and Ho (2001) introduced a measure called 'economic value'. Schelling (1960, p. 98) wrote, 'a normative theory must produce strategies that are at least as good as what people can do without them'. Inspired by his definition, the economic value of a theory is how much model forecasts of behaviour of other players improve the profitability of a particular player's choices. This measure treats a theory as being like the advice service that professionalssell(forexample,consultants).Thevalueofatheoryisthedifference intheeconomicvalueoftheclient'sdecisionswithandwithouttheadvice.

Besides being a businessperson's measure, economic value is a way of measuring the degree of disequilibrium in economic terms. Note that, in equilibrium, the economic value of a learning theory is zero or negative by definition (since players are already guessing perfectly accurately what others will do). A bad theory, which implicitly 'knows' less than the subjects themselves do about what other subjects are likely to do, will have negative economic value.

Furthermore, do not conclude, mistakenly, that if a learning theory has economic value it does not describe how people in fact learn. The economic value assumes that an objective observer uses the theory to make a forecast and best-responds to it – in our terms below, such a person is 'sophistic-ated'. So if the model describes accurately how adaptive players (who are *not* sophisticated) learn, it will have economic value. It *is* true, however, that a model of sophisticated players should not have economic value (since the advice it gives should already be known to the players, by definition of sophistication).

To measure economic value, we use model parameters and a player's observed experience through period t to generate model predictions about what others will do in $t + 1$. These predictions are used to compute expected payoffs from strategies, and recommend a choice with the highest expected value. We then compare the profit from making that choice in $t + 1$ (given what other players did in $t + 1$) with profit from the target player's actual choice. Economic value is a good measure because it uses the full distribution of predictions about what other players are likely to do, *and* the economic impact of those possible choices. These measures do not control for the boomerang effect of how a recommended choice would have changed future behaviour by others, but this effect is small in most games.[43]

Data from six games are used to estimate model parameters and make recommendations in the seventh game, for each of the games separately. Table 8.5 shows the overall economic value – the percentage improvement

Table 8.5   Economic value of learning theories (percentage improvement in payoffs)

Game	self-tuning EWA(%)	parametric EWA(%)	Belief-based(%)	Reinf.-PV(%)	QRE(%)
cont'l divide	5.0	**5.2**	4.6	−9.4	−30.4
median action	1.5	**1.5**	1.2	1.3	−1.0
p-Beauty contest	**49.9**	40.8	26.7	−7.2	−63.5
price matching	**10.3**	9.8	9.4	3.4	2.7
mixed strategies	**7.5**	3.0	1.1	5.8	−1.8
patent race	1.7	1.2	1.3	**2.9**	1.2
pot games	−2.7	−1.1	−1.3	−1.9	**9.9**

Note: Highest economic value for each game is displayed in **bold** type.

(or decline) in payoffs of subjects from following a model recommendation rather than their actual choices. The highest economic value for each game is printed in bold. Most models have positive economic value. The percentage improvement is small in some games because even clairvoyant advice would not raise profits much.[44]

Self-tuning EWA and EWA usually add the most value (except in pot games, where only QRE adds value). Belief learning has positive economic value in all but one game. Reinforcement learning adds the most value in patent races, but has negative economic value in three other games. (Reinforcement underestimates the rate of strategy change in continental divide and beauty contest games, and hence gives bad advice.) QRE has negative economic value in four games.

Summary

This section reports a comparison among several learning models on seven data sets. The new model is self-tuning EWA, a variant of the hybrid EWA model in which estimated parameters are replaced by functions entirely determined by data. Self-tuning EWA captures a predictable cross-game variation in parameters and hence fits better than other models when common parameters are estimated across games. A closer look at the continental divide and price-matching games shows that belief models are close to the data on average but miss other features (the asymmetry in convergence toward each of the two pure equilibria in the continental divide game, and the sharp convergence on the minimum price in price-matching). Reinforcement predicts well in co-ordination games and predicts the correct price often in price-matching (but with too little probability). However, reinforcement predicts badly in beauty contest games. It is certainly true that for explaining some features of some games, the reinforcement and belief models are adequate. But self-tuning EWA is easier to estimate (it has one parameter instead of two) and explains subtler features other models sometimes miss. It is also never fits poorly (relative to other games), which is the definition of robustness.

Sophistication and teaching

The learning models discussed in the previous section are adaptive and backward-looking: Players only respond to their own previous payoffs and knowledge about what others did. While a reasonable approximation, these models leave out two key features: adaptive players do not use information about *other* players' payoffs explicitly (though subjects in fact *do*[45]); and adaptive models ignore the fact that when the same players are matched together repeatedly, their behaviour is often different than it is when they are not rematched together, generally in the direction of greater efficiency (see, for example, Van Huyck *et al.*, 1990; Andreoni and Miller, 1993; Clark and Sefton, 1999).

In this section, adaptive models are extended to include sophistication and strategic teaching in repeated games (see Stahl, 1999; and Camerer *et al.*, 2002a, for details). Sophisticated players believe that others are learning and anticipate how those others will change in deciding what to do. In learning to shoot at a moving target, for example, soldiers and fighter pilots learn to shoot *ahead*, towards where the target *will be*, rather than shoot at the target where it is when they aim. They become sophisticated.

Sophisticated players who also have strategic foresight will 'teach' – that is, they choose current actions which teach the learning players what to do, in a way that benefits the teacher in the long run. Teaching can either be mutually beneficial (trust-building in repeated games) or privately beneficial but socially costly (entry-deterrence in chain-store games). Note that sophisticated players will use information about the payoffs of others (to forecast what others will do), and will behave differently depending on how players are matched, so adding sophistication can conceivably account for the effects of information and matching that adaptive models miss.[46]

Sophistication

Let us begin with myopic sophistication (no teaching). The model assumes a population mixture in which a fraction α of players are sophisticated and $1 - \alpha$ are adaptive. (It is possible to imagine a model with degrees of sophistication, as well, or learning to become sophisticated, as in Stahl, 1999.) To allow for possible over-confidence, sophisticated players think that a fraction $(1 - \alpha')$ of players are adaptive and the remaining fraction α' of players are sophisticated, like themselves.[47] Sophisticated players use the self-tuning EWA model to forecast what adaptive players will do, and choose strategies with high expected payoffs, given their forecast and their guess about what sophisticated players will do. Denoting choice probabilities by adaptive and sophisticated players by $P_i^j(a, t)$ and $P_i^j(s, t)$, attractions for sophisticates are:

$$A_i^j(s, t) = \sum_{k=1}^{m_{-i}} [\alpha' P_{-i}^k(s, t+1) + (1 - \alpha') P_{-i}^k(a, t+1)] \times \pi_i(s_i^j, s_{-i}^k) \qquad (8.8)$$

Note that, since the probability $P^k_{-i}(s, t + 1)$ is derived from an analogous condition for $A^j_i(s, t)$, the system of equations is recursive. Self-awareness creates a whirlpool of recursive thinking which means that QRE (and Nash equilibrium) are special cases in which all players are sophisticated and believe others are too ($\alpha = \alpha' = 1$).

An alternative structure one could study links steps of sophistication to the steps of thinking used in the first period. For example, define zero learning steps as using self-tuning EWA; one step is best-responding to zero-step learners; two steps is best-responding to choices of one-step sophisticates, and so forth. We think this model can produce results similar to the recursive one we report below, and it replaces α and α' with τ from the theory of initial conditions so reducing the entire thinking–learning–teaching model to just two parameters.

We estimate the sophisticated EWA model using data from the p-beauty contests introduced above. Table 8.6 reports results and estimates of important parameters (with bootstrapped standard errors in parentheses). For inexperienced subjects, adaptive EWA generates Cournot-like estimates ($\hat{\phi} = 0$ and $\hat{\delta} = 0.90$). Adding sophistication increases $\hat{\phi}$ and improves *LL* substantially both in and out of sample. The estimated fraction of sophisticated players is 24 per cent and their estimated perception $\hat{\alpha}'$ is zero (and is insignificant), showing over-confidence (as in the thinking-steps estimates from the previous section).[48]

Experienced subjects are those who play a second 10-period game with a different p parameter (the multiple of the average that creates the target

Table 8.6 Sophisticated and adaptive learning model estimates for the p-beauty contest game

	Inexperienced subjects		Experienced subjects	
	Sophisticated EWA	Adaptive EWA	Sophisticated EWA	Adaptive EWA
ϕ	**0.44**	0.00	**0.29**	0.22
	(0.05)*	(0.00)	(0.03)	(0.02)
δ	**0.78**	0.90	**0.67**	0.99
	(0.08)	(0.05)	(0.05)	(0.02)
α	**0.24**	0.00	**0.77**	0.00
	(0.04)	(0.00)	(0.02)	(0.00)
α'	**0.00**	0.00	**0.41**	0.00
	(0.00)	(0.00)	(0.03)	(0.00)
LL				
In sample	**−2095.32**	-2155.09	**−1908.48**	-2128.88
Out of sample	**−968.24**	-992.47	**−710.28**	-925.09

Note: *Standard errors in parentheses.
Source: Camerer *et al.* (2002b).

number). Among experienced subjects, the estimated proportion of sophist-icates increases to $\hat{\alpha} = 77$ per cent. Their estimated perceptions also increase, but are still over-confident ($\hat{\alpha}' = 41$ per cent). The estimates reflect 'learning about learning': subjects who played one 10-period game came to realize that an adaptive process is occurring; and most of them anticipate that others are also learning when they play again.

Strategic teaching

Sophisticated players matched with the same players repeatedly often have an incentive to 'teach' adaptive players by choosing strategies with poor short-run payoffs that will change what adaptive players do, in a way that benefits the sophisticated player in the long run. Game theorists have shown that strategic teaching could select one of many repeated-game equilibria (teachers will teach the pattern that benefits them) and could give rise to reputation formation without the complicated apparatus of Bayesian updat-ing of Harsanyi-style payoff types (see Fudenberg and Levine, 1989; Watson, 1993; Watson and Battigali, 1997). This section of our chapter describes a parametric model which embodies these intuitions, and tests it with experi-mental data. The goal is to show how the kinds of learning models described in the previous section can be extended parsimoniously to explain beha-viour in more complex games which are, perhaps, of even greater economic interest than games with random matching.

Consider a finitely-repeated trust game, first studied by Camerer and Weigelt (1988). A borrower B wants to borrow money from each of a series of lenders denoted L_i ($i = 1, \ldots, N$). In each period a lender makes a single lending decision (*Loan* or *No Loan*). If the lender makes a loan, the borrower either *repays* or *defaults*. The next lender in the sequence, who observed all the previous history, then makes a lending decision. The payoffs used in the experiments are shown in Table 8.7.

There are in fact two types of borrowers. As in post-Harsanyi game theory with incomplete information, types are expressed as differences in borrower payoffs which the borrowers know but the lenders do not (though the prob-ability of a given borrower is each type is commonly known). The honest (Y) types in fact receive *more* money from repaying the loan, an experimenter's way

Table 8.7 Payoffs in the borrower–lender trust game

Lender strategy	Borrower strategy	Payoffs to lender	Payoffs to borrower	
			Normal (X)	Honest (Y)
Loan	Default	−100	150	0
	Repay	40	60	60
No loan	(No choice)	10	10	10

Source: Camerer and Weigelt (1988).

of inducing preferences like those of a person who has a social utility for being trustworthy (see Camerer, 2003, ch. 3 and references therein). The normal (X) types, however, earn 150 from defaulting and only 60 from repaying. If they were playing just once and wanted to earn the most money, they would default.

In the standard game-theoretic account, paying back loans in finite games arises because there is a small percentage of honest types who *always* repay. This gives normal-type borrowers an incentive to repay until close to the end, when they begin to use mixed strategies and default with increasing probability.

Whether people in fact play these sequential equilibria is important to investigate, for two reasons. First, the equilibria impose consistency between optimal behaviour by borrowers and lenders, and Bayesian updating of types by lenders (based on their knowledge and anticipation of the borrowers' strategy mixtures); whether reasoning or learning can generate this consistency is an open behavioural question (see Selten, 1978). Second, the equilibria are very sensitive to the probability of honesty (if it is too low, the reputational equilibria disappear and borrowers should always default), and also make counterintuitive comparative statics predictions that are not confirmed in experiments (for example, Neral and Ochs, 1992; Jung *et al.*, 1994).

In the experiments, subjects play many sequences of eight periods. The eight-period game is repeated to see whether equilibration occurs across many sequences of the entire game.[49] Surprisingly, the earliest experiments showed that the pattern of lending, default and reactions to default across experimental periods within a sequence is roughly in line with the equilibrium predictions. Typical patterns in the data are shown in Figure 8.9. Sequences are combined into ten-sequence blocks (denoted 'sequence') and average frequencies are reported from those blocks. Periods 1, . . . , 8 denote periods in each sequence. The figures show relative frequencies of *no loan* and *default* (conditional on a loan). Figure 8.9(a) shows that, in early sequences, lenders start by making loans in early periods (that is, there is a low frequency of no loan), but they rarely lend in periods 7–8. In later sequences they have learned always to lend in early periods and rarely in later periods. Figure 8.9(b) shows that borrowers rarely default in early periods, but usually default (conditional on getting a loan) in periods 7–8. The within-sequence pattern becomes sharper in later sequences.

The general patterns predicted by equilibrium are therefore present in the data. But given the complexity of the equilibrium, how do players approximate it? Camerer and Weigelt (1988) concluded their paper as follows:

> the long period of disequilibrium behavior early in these experiments raises the important question of how people learn to play complicated games. The data could be fit to statistical learning models, though new experiments or new models might be needed to explain learning adequately. (pp. 27–8)

The teaching model is a 'new model' of the sort Camerer and Weigelt had in mind. It is a boundedly rational model of reputation formation in which

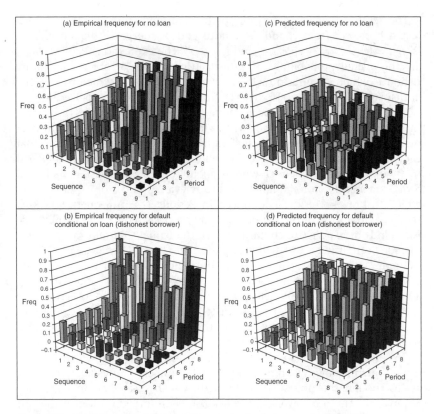

Figure 8.9 Empirical and predicted frequencies for borrower–lender trust game

the lenders learn whether to lend or not. They do not update borrowers' types and do not anticipate borrowers' future behaviour (as in equilibrium models); they just learn.

In the teaching model, a proportion of borrowers are sophisticated and teach; the rest are adaptive and learn from experience but have no strategic foresight. The teachers choose strategies that are expected (given their beliefs about how borrowers will react to their teaching) to give the highest long-run payoffs in the remaining periods.

A sophisticated teaching borrower's attractions for sequence k after period t are specified as follows ($j \in \{repay, default\}$ is the borrower's set of strategies):

$$A_B^j(s, k, t) = \sum_{j'=Loan}^{No\,Loan} P_L^{j'}(a, k, t+1) \cdot \pi_B(j, j')$$

$$+ \max_{J_{t+1}} \left\{ \sum_{v=t+2}^{T} \sum_{j'=Loan}^{No\,Loan} \hat{P}_L^{j'}(a, k, v | j_{v-1} \in J_{t+1}) \cdot \pi_B(j_v \in J_{t+1}, j') \right\}$$

The set J_{t+1} specifies a possible path of future actions by the sophist-icated borrower from round $t+1$ until end of the game sequence. That is, $J_{t+1} = \{j_{t+1}, j_{t+2}, \ldots, j_{T-1}, j_T\}$ and $j_{t+1} = j$.[50] The expressions $\hat{P}_L^{j'}(a, k, v|j_{v-1})$ are the overall probabilities of either getting a loan or not in the future period v, which depends on what happened in the past (which the teacher anticipates).[51] $P_B^j(s, k, t+1)$ is derived from $A_B^j(s, k, t)$ using a logit rule.

The updating equations for adaptive players are the same as those used in self-tuning EWAs, with two twists. First, since lenders who play in later periods know what has happened earlier in a sequence, we assume that they learnt from the experience they witnessed as though it had happened to them.[52] Second, a lender who is about to make a decision in Period 5 of Sequence 17, for example, has *two* relevant sources of experience on which to draw – the behaviour seen in Periods 1–4 in Sequence 17, *and* the behaviour seen in the Period 5s of the previous sequences (1–16). Since *both* kinds of experience could influence the lender's current decision, we include both, using a two-step procedure. After Period 4 of Sequence 17, for example, attractions for lending and not lending are first updated, based on the Period 4 experience. Then attractions are partially updated (using a degree of updating parameter σ) based on the experience in Period 5 of the previous sequences.[53] The parameter σ is a measure of the strength of 'peripheral vision' – glancing back at the 'future' Period 5s from previous sequences to help guess what lies ahead.

Of course, it is well known that repeated-game behaviour can arise in finite-horizon games when there are a small number of 'unusual' types (who act as though the horizon is unlimited), which creates an incentive for rational players to behave as if the horizon is unlimited until near the end (for example, Kreps and Wilson, 1982). But specifying why some types are irrational, and how many there are, makes this interpretation difficult to test. In the teaching approach, which 'unusual' type the teacher pretends to be arises endogenously from the payoff structure: they are Stackelberg types, who play the strategy they would choose if they could commit to it. For example, in trust games, they would like to commit to repaying; but in entry-deterrence, they would like to commit to fighting entry.

The model is estimated using repeated game trust data from Camerer and Weigelt (1988). In Camerer *et al.* (2002a), we used parametric EWA to model behaviour in trust games. This model allows two different sets of EWA parameters for lenders and borrowers. In this chapter we use self-tuning EWA to model lenders and adaptive borrowers so the model has fewer parameters.[54] Maximum likelihood estimation is used to estimate parameters on 70 per cent of the sequences in each experimental session, then behaviour in the holdout sample of 30 per cent of the sequences is forecast using the estimated parameters.

As a benchmark alternative to the teaching model, we estimated an agent-based version of QRE suitable for extensive-form games (see McKelvey and

Palfrey, 1998). Agent-based QRE is a good benchmark because it incorporates the key features of repeated-game equilibrium – strategic foresight, accurate expectations about actions of other players, and Bayesian updating – but assumes stochastic best-response. We use an agent-based form in which players choose a distribution of strategies at each node, rather than using a distribution over all history-dependent strategies. We implement agent QRE (AQRE) with four parameters – different λs for lenders, honest borrowers and normal borrowers, and a fraction θ, the percentage of players with normal-type payoffs who are thought to act as if they are honest (reflecting a 'homemade prior' which can differ from the prior induced by the experimental design[55]). (Standard equilibrium concepts are a special case of this model when λs are large and $\theta = 0$, and fit much worse than does AQRE). The implementation in our 2002 paper is itself a small contribution since it is quite complex to estimate AQRE in these games.

The models are estimated separately on each of the eight sessions to gauge cross-session stability. Since pooling sessions yields similar fits and parameter values, we report only those pooled results in Table 8.8 (excluding the λ values). The interesting parameters for sophisticated borrowers are estimated to be $\hat{\alpha} = 0.89$ and $\hat{\sigma} = 0.93$, which means that most subjects are classified as teachers and they put a lot of weight on previous sequences. The teaching model fits in-sample and predicts better out-of-sample than AQRE by a modest margin (and does better in six out of eight individual experimental sessions), predicting about 75 per cent of the choices correctly. The AQRE fits reasonably well too (72 per cent correct) but the estimated 'homemade prior' θ is 0.91, which is absurdly high. (Earlier studies estimated numbers around 0.1–0.2.) The model basically fits best by assuming that *all* borrowers simply prefer to repay loans. This assumption fits most of the data but it mistakes

Table 8.8 Model parameters and fit in repeated trust games

	Statistic	Model	
		Self-tuning EWA+ teaching	Agent QRE
In-sample Calibration (n = 5757)	Hit rate (%) log-likelihood	76.5% −2975	73.9% −3131
Out-of-sample Validation (n = 2894)	Hit rate (%) log-likelihood	75.8% −1468	72.3% −1544
Parameters		*Estimates*	
Cross-sequence learning	σ	0.93	–
Percentage of teachers	α	0.89	–
Homemade prior p(honest)	θ	–	0.91

teaching for a pure repayment preference. As a result, it does not predict the sharp upturn in defaults in Periods 7–8, which the teaching model does.[56]

Figure 8.9(c)–(d) show average predicted probabilities from the teaching model for the no-loan and conditional default rates. No-loan frequencies are predicted to start low and rise across periods, as they in fact do, though the model underpredicts the no-loan rate in general. The model predicts the increase in default rate across periods reasonably well, except for underpredicting default in the last period.

The teaching approach is a boundedly-rational alternative to type-based equilibrium models of reputation formation.[57] It has always seemed improbable that players are capable of the delicate balance of reasoning required to implement the type-based models, unless they learn the equilibrium through some adaptive process. The teaching model is one parametric model of that adaptive process. It retains the core idea in the theory of repeated games – namely, strategic foresight – and consequently respects the fact that matching protocols matter. And since the key behavioural parameters (α and σ) appear to be near 1, restricting attention to these values should make the model workable for doing theory.

Summary

In this section we introduced the possibility that players can be sophisticated – that is, they believe others are learning. (In future work, it would be interesting to link steps of iterated thinking, as in the first section, to steps of sophisticated thinking.) Sophistication links learning theories to equilibrium ones if sophisticated players are self-aware. Adding sophistication also improves the fit of data from repeated beauty-contest games. Interestingly, the proportion of estimated sophisticates is around a quarter when subjects are inexperienced, but rises to around three-quarters when they play an entire 10-period game for a second time, as if the subjects learn about learning. Sophisticated players who know they will be rematched repeatedly may have an incentive to 'teach', which provides a boundedly rational theory of reputation formation. We apply this model to data on repeated trust games. The model adds only two behavioural parameters, representing the fraction of teachers and how much 'peripheral vision' learners have (and some nuisance λ parameters), and predicts substantially better than a quantal response version of equilibrium.

Conclusion

In the introduction we stated that the research programme in behavioural game theory has three goals: (i) to create a theory of one-shot or first-period play using an index of bounded rationality measuring steps of thinking; (ii) to predict features of equilibration paths when games are repeated; and (iii) to explain why players behave differently when matched together repeatedly.[58]

The models described in this chapter illustrate ways to understand these three phenomena. There are many alternative models (especially of learning). The models described here are just some examples of the style in which ideas can be expressed, and how data are used to test and modify them.

Keep in mind that the goal is *not* to list deviations from the Nash equilibrium and stop there. Deviations are just hints. The goal is to develop alternative models which are precise, general and disciplined by data – that is, some day game theory might be taught beginning with behavioural models and *ending* with analytical concepts such as equilibrium and its many refinements. All such models deserve to be called 'game theory', except that these models are *behavioural* – rooted in psychological regularity and sharpened by having to explain data – while analytical models are simply useful fictional answers to questions about how players of varying degrees of hypothetical rationality behave.

The thinking-steps model posits a Poisson distribution (with mean τ) of numbers of thinking steps, along with decision rules for what players using each number of steps will do. Studies with simple matrix games, beauty contests (unreported), mixed games, and entry games all show that values of τ around 1.5 can fit data reasonably well (and never worse than the Nash equilibrium). The model is easy to use because players can be assumed to best-respond and the model usually makes realistic probabilistic predictions because the mixture of thinking steps types creates a population mixture of responses. The surprise is that the same model, which is tailor-made to produce spikes in dominance-solvable games, can also fit data from games with pure and mixed equilibria using roughly the same τ.

The second section compared several adaptive learning models. For explaining simple trends in equilibration, many of these models are close substitutes. However, it is useful to focus on where models fail if the goal is to improve them. The EWA hybrid was created to include the psychological intuitions behind both reinforcement learning (received payoffs receive more weight than forgone payoffs) and belief learning (both types of payoff receive equal weight). If both intuitions were compelling enough for people to want to compare them statistically, then a model that had both intuitions in it should be better still (and generally, it is). Self-tuning EWA uses one parameter (λ) and substitutes functions for parameters. The major surprise here is that functions such as the change-detector $\phi_i(t)$ can reproduce differences across games in which parameter values fit best. This means that the model can be applied to brand-new games (when coupled with a thinking-steps theory of initial conditions) without having to make a prior judgement about which parameter values are reasonable, and without positing game-specific strategies. The interaction of learning and game structure creates reasonable parameter values automatically.

In the third section we extended the adaptive learning models to include sophisticated players who believe that others are learning. Sophistication improves fit in the beauty contest game data. (Experienced subjects seem to have 'learned about learning' because the percentage of apparently sophisticated players is higher and convergence is faster.) Sophisticated players who realize they are matched with others repeatedly often have an incentive to 'teach', as in the theory of repeated games. Adding two parameters to adaptive learning was used to model learning and teaching in finitely-repeated trust games. While trustworthy behaviour early in these games is known to be rationalizable by Bayesian – Nash models with 'unusual' types, the teaching model creates unusual types from scratch. Teaching also fits and predicts better than more forgiving quantal response forms of the Bayesian – Nash type-based model. The surprise here is that the logic of mutual consistency and type updating is not needed to produce accurate predictions in finitely-repeated games with incomplete information.

Potential applications

A crucial question is whether behavioural game theory can help to explain naturally-occurring phenomena. We conclude the chapter with some speculations about the sorts of phenomena precise models of limited thinking, learning and teaching could illuminate.

Bubbles

Limited iterated thinking is potentially important because, as Keynes and many others have pointed out, it is not always optimal to behave rationally if you believe others are not. For example, prices of assets should equal their fundamental or intrinsic value if rationality is common knowledge (Tirole, 1985). But when the belief that others might be irrational arises, bubbles can too. Besides historical examples such as Dutch tulip bulbs and the $5 trillion tech-stock bubble in the 1990s, experiments have shown such bubbles even in environments in which the asset's fundamental value is controlled and commonly known.[59]

Speculation and competition neglect

The 'Groucho Marx theorem' says that traders who are risk-averse should not speculate by trading with each other even if they have private information (since the only person who will trade with you may be better-informed). But this theorem rests on unrealistic assumptions of common knowledge of rationality and is violated constantly by massive speculative trading volume and other kinds of betting, as well as in experiments.[60]

Players who do limited iterated thinking, or believe others are not as smart as themselves, will neglect competition in business entry (see Camerer and Lovallo, 1999; Huberman and Rubinstein, 2000). Competition neglect may partly explain why the failure rate of new businesses is so high. Managerial hubris, over-confidence and self-serving biases correlated with costly delay and labour strikes in the lab (Babcock *et al.*, 1995) and in the field (Babcock and Loewenstein, 1997) can also be interpreted as players not believing that others always behave rationally.

Incentives

In a thorough review of empirical evidence on incentive contracts in organizations, Prendergast (1999) notes that workers typically react to simple incentives, as standard models predict. However, firms do not usually implement complex contracts, which *should* elicit greater effort and improve efficiency. Perhaps the firms' reluctance to bet on rational responses by workers is evidence of limited iterated thinking.

Macroeconomics

Woodford (2001) notes that in Phelps–Lucas 'islands' models, nominal shocks can have real effects, but their predicted persistence is too short compared to effects in data. He shows that imperfect information about *higher-order* nominal gross domestic product (GDP) estimates – beliefs about beliefs, and higher-order iterations – can cause longer persistence which matches the data. However, Svensson (2001) notes that iterated beliefs are probably constrained by computational capacity. If people have a projection bias, their beliefs about what others believe will be too much like their own, which undermines Woodford's case. On the other hand, in the thinking-steps model, players' beliefs are not mutually consistent so there is higher-order belief inconsistency which can explain longer persistence. In either case, knowing precisely how iterated beliefs work could help to inform a central issue in macroeconomics – persistence of the real effects of nominal shocks.

Learning

Other phenomena are evidence of a process of equilibration or learning. For example, institutions for matching medical residents and medical schools, and analogous matching in college sororities and college bowl games, developed over decades and often 'unravel' so that high-quality matches occur before some agreed-upon date (Roth and Xing, 1994). Bidders in eBay auctions learn to bid late to hide their information about an object's common value (Bajari and Hortacsu, 2003). Consumers learn over time what products they like (Ho and Chong, 2003). Learning in financial markets can generate excess volatility and returns predictability, which are otherwise anomalous in rational expectations models (Timmerman, 1993). We are currently studying evolution of products in a high-uncertainty environment (electronics equipment) for which thinking-steps and learning models are proving useful.

Teaching

Teaching in repeated games may prove to be the most potentially useful tool for economics, because it is essentially an account of how bounded rationality can give rise to some features of repeated-game behaviour, where standard theory has been applied widely. The teaching model could be applied to repeated contracting, employment relationships, alliances among firms, industrial organization problems (such as pricing games among perennial rivals, and entry deterrence) and macroeconomic models of policy-maker inflation-setting. For example, the teaching model could be applied to the Kydland – Prescott model of commitment in which the public learns about inflation from past history (using the self-tuning EWA rule described in the Appendix) and unemployment is determined by an expectational Phillips curve. Since policy-makers face a temptation to choose surprisingly high inflation to lower unemployment, they can either act myopically or 'teach' the public to expect low inflation which is Pareto-optimal in the long-run (see Sargent, 1999). A model similar to self-tuning EWA applied to hyperinflations is Marcet and Nicolini (2003).

Appendix: thinking models applied to mixed games and entry games

Games with mixed equilibria

A good model of thinking steps should be able both to account for deviations from the Nash equilibrium (as in the games above), *and* reproduce the successes of the Nash equilibrium. A domain in which the Nash equilibrium does a surprisingly good job is in games with unique mixed equilibria. It is hard to beat the Nash equilibrium in these games because (as we shall see) the correlation with data is in fact very good (around 0.9) so there is little room for improvement. Instead, the challenge is to see how well a thinking-steps model which bears little resemblance to the algebraic logic of equilibrium mixing can approximate behaviour in these games.

Early tests in the 1960s and 1970s (mostly by psychologists) appeared to reject the Nash equilibrium as a description of play in mixed games. As others have noted (for example, Binmore *et al.*, 2001), these experiments were incomplete in important dimensions and hence inconclusive. Financial incentives were very low or absent; subjects typically did not play other human subjects (and often were deceived about playing other people, or were instructed only vaguely about how their computer opponents played); and pairs were often matched repeatedly so that (perceived) detection of temporal patterns permitted subjects to choose non-equilibrium strategies. Under conditions ideal for equilibration, however, convergence was rapid and sharp. Kaufman and Becker (1961), for example, asked subjects to specify mixtures and told them that a computer program would then choose a mixture to minimize the subjects' earnings. Subjects could maximize their possible gains by choosing the Nash mixture. After playing five games, more than half learned to do so. More recent experiments are also surprisingly supportive of the Nash equilibrium (see Binmore *et al.*, 2001; Camerer, 2002, ch. 2). The data are supportive in two senses: (i) equilibrium predictions and actual frequencies are closely correlated, when taken as a whole (for example, strategies predicted to be more likely are almost always played more often);

and (ii) it is hard to imagine any parsimonous theory that can explain the modest deviations.

We applied a version of the thinking model in which K-step thinkers think all others are using $K-1$ steps along with best response to see whether it could produce predictions as accurate as Nash in games with mixed equilibria. This model is extremely easy to use (just start with step-zero mixtures and compute best responses iteratively). Furthermore, it creates natural 'purification': players using different thinking steps usually choose pure strategies, but the Poisson distribution of steps generates a mixture of responses, and hence a probabilistic prediction.

Model predictions are compared with data from fifteen games with unique mixed equilibria, reported in Camerer (2002, ch. 2).[61] These games are not a random or exhaustive sample of recent research but there are enough observations that we are confident the basic conclusion will be overturned by adding more studies. Note that we use data from *all* the periods of these games rather than the first period only. (In most cases, the first-period data are rarely reported, and there is usually little trend over time in the data.)

Figure 8.A1 plots actual frequencies on the ordinate (y) axis against either mixed-strategy equilibrium predictions or thinking-steps predictions on the abscissa (x) axis. Each data point in Figure 8.A1 represents a single strategy from a different game (pooling across all periods to reduce sampling error).[62]

In Figure 8.A1, the value of τ is common across games (1.46) and minimizes mean squared deviations between predicted and actual frequencies. When values are estimated separately for each game to minimize mean squared deviations, the values across the games (in the order they are listed above) are 0.3, 0.3, 0.3, 2.2, 2.5, 0.1, 1.8, 2.3, 2.9, 2.7, 0.5, 0.8, 1.6, 1.5, 1.9. The lower values occur in games where the actual

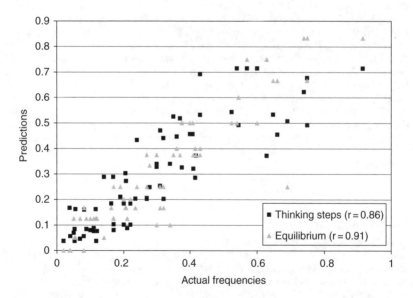

Figure 8.A1 Fit of data to equilibrium and thinking-steps model (common $\tau = 1.5$) in mixed-equilibrium games

mixtures are close to equal across strategies, so that a distribution with $\tau = 0$ fits well. When there are dominated strategies, which are usually rarely played, much higher τ values are needed, since low τ generates a lot of random play and frequent dominance violation. The simple arithmetical average across the fifteen games is 1.45, which is very close to the best-fitting common $\tau = 1.46$. Figure 8.A1 shows two regularities: both thinking-steps (circles in the plot) and equilibrium predictions (triangles) have very high correlations (above 0.8) with the data, though there is substantial scatter, especially at high probabilities. (Keep in mind that sampling error means there is an upper bound on how well any model could fit – even the true model that generated the data.) The square root of the mean squared deviation is around 0.10 for both models.

While the thinking-steps model with common τ is a little less accurate than the Nash equilibrium (the game-specific model is more accurate), the key point is that the same model that can explain Nash deviations in dominance-solvable games and matrix games fits almost as well with a value of τ close to those estimated in other games.

Table 8.A1 shows a concrete example of how the thinking model is able to approximate mixture probabilities. The game is Mookerjhee and Sopher's (1997) 4×4 game. Payoffs are shown as wins (+) or losses (−) ((2/3) + means a 2/3 chance of winning) in the upper left cells. The rightmost columns show the probabilities with which row players using different thinking steps choose each of the four row strategies. To narrate a little, zero-step players randomize (each is played with probability of 0.25); one-step players best-respond to a random column choice and choose row strategy 3 with probability of 1, and so on. First notice that the weakly dominated strategy (4) is only chosen by a quarter of zero-step players (since it is never the best response against players who randomize), which generates a small amount of choice that matches the data. Notice also that the best responses tend to lurch around as thinking-steps changes. When these are averaged over thinking-steps frequencies, a population mixture results. Furthermore, one of the quirkiest features of mixed equilibrium is that one player's mixture depends only on the other player's payoffs. This effect also occurs in the thinking steps models, because a K-step row player's payoffs affect row's best responses, which then affect a $K + 1$-step column player's best response. So one player's payoffs affect the other's strategies indirectly. Table 8.A1 also shows the MSE (mixed equilbrium) prediction, the data frequencies and overall frequencies from the thinking-steps model when $\tau = 2.2$. The model fits the data more closely than MSE for row players (it accounts for underplay of row 2 strategy and overplay of strategy 3) and is as accurate as MSE for column players. As noted in the text, the point is not that the thinking-steps model outpredicts MSE – it cannot, because MSE has such a high correlation – but simply that the model that explains behaviour in dominance-solvable, matrix and entry games also generates mixtures of players playing near-pure strategies that are close to outcomes in mixed games.

Market entry games

Analysis of the simple entry game described in the text proceeds as follows. Step 0s randomize so $f(0)/2$ level 0s enter. Define the relative proportion of entry after accounting up through level k as $N(k)$. Define a Boolean function $B(X) = 1$ if X true, $B(X) = 0$ if X false. Level 1s enter iff $1/2 < c$. Therefore, total entry 'after' Level 1 types are accounted for is $N(1) = f(0)/2 + B[N(0)/(f(0)/2) > c]f(1)$. Total entry after Level k type is therefore $N(k) = f(0)/2 + \sum_{n=1}^{k} f(n)B[N(n-1)/(\sum_{m=1}^{k} f(m)) > c]$. A given c and τ then generates a sequence of cumulated entry rates which asymptotes as k grows

Table 8.A1 How thinking steps fits mixed-game data

Row strategies	Strategies 1–4				Column thinking steps 0–5						MSE	Data	Thinking Model
	1	2	3	4	Step 0	1	2	3	4	5			
1	+	–	–	+	0.25	0	0.5	1	0	0	0.375	0.32	0.37
2	–	–	+	+	0.25	0	0	0	1	0	0.25	0.17	0.14
3	–	+	(2/3)+	(2/3)+	0.25	1	0.5	0	0	1	0.375	0.43	0.46
4	–	–	(1/3)+	+	0.25	0	0	0	0	0	0	0.08	0.03
Steps													
0	0.25	0.25	0.25	0.25									
1	0.5	0.5	0.25	0									
2	1	0	0	0									
3	0	0	1	0									
4	0	0.5	0.5	0									
5	0.5	0.5	0	0									
6	1	0	0	0									
7	0	0	1	0									
MSE	0.375	0.25	0.375	0									
Data	0.38	0.31	0.27	0.04									
Thinking	0.46	0.23	0.28	0.03									

Source: Mookerjhee and Sopher (1997) 4 × 4 game.

large. Define a function $N(all|\tau)(c)$ as the overall rate of entry, given τ, for various capacities of c.

The data reported in the text come from experiments by Sundali *et al.* (1995) and Seale and Rapoport (2000). Their game is not precisely the same as the one analyzed because, in their game, entrants earn $1 + 2(c - e)$ (where e is the number of entrants) and non-entrants earn 1. They used twenty subjects with odd values of $c(1, 3, \ldots 19)$. To compute entry rates reported in Figure 8.A1 we averaged entry for adjacent c values (that is, averaging 1 and 3 yields a figure for $c = 2$ matching $c = 0.1$, averaging 3 and 5 yields a figure for $c = 4$ corresponding to $c = 0.2$ and so on). Obviously, the analysis and data are not perfectly matched, but we conjecture that the thinking-steps model can still match data closely and reproduce the three experimental regularities described in the text; whether this is true is the subject of ongoing research.

Notes

1 Our models are related to important concepts such as rationalizability, which weakens the mutual consistency requirement, and behaviour of finite automata. The difference is that we work with simple parametric forms and concentrate on fitting them to data.

2 In his thesis proposing a concept of equilibrium, Nash himself suggested that equilibrium might arise from some 'mass action' that adapted over time. Taking up Nash's implicit suggestion, later analyses filled in details of where evolutionary dynamics lead (see Weibull, 1995; Mailath, 1998).

3 While great triumphs of economic theory come from parameter-free models (for example, Nash equilibrium), relying on a small number of free parameters is more typical in economic modelling. For example, nothing in the theory of intertemporal choice pins a discount factor δ to a specific value. But if a wide range of phenomena are consistent with a value such as 0.95, for example, then as economists we are comfortable working with such a value despite the fact that it does not emerge from axioms or deeper principles.

4 Lucas (1986) makes a similar point in macroeconomic models. Rational expectations often yield indeterminacy, whereas adaptive expectations pin down a dynamic path. Lucas writes: 'The issue involves a question concerning how collections of people behave in a specific situation. Economic theory does not resolve the question . . . It is hard to see what can advance the discussion short of assembling a collection of people, putting them in the situation of interest, and observing what they do' (p. S421).

5 Quantal response equilibrium (QRE), a statistical generalization of Nash, almost always explains the direction of deviations from Nash and should replace Nash as the static benchmark to which other models are routinely compared. Stahl and Wilson (1995), Capra (1999) and Goeree and Holt (2003) have models of limited thinking in one-shot games which are similar to ours. Jehiel (2002) proposes a concept of limited foresight in games in which analogy is used to truncate complex games. There are many learning models. Self-tuning EWA generalizes some of them (though reinforcement with payoff variability adjustment is different; see Erev *et al.*, 1999). Other approaches include rule learning (Stahl, 1996, 2000), and earlier artificial intelligence (AI) tools such as genetic algorithms or genetic programming to 'breed' rules (see Jehiel, forthcoming). Finally, there are no alternative models of strategic teaching that we know of but this is an important area others should examine.

6 Note the timing convention – attractions are defined *before* a period of play; so the initial attractions $A_i^j(0)$ determine choices in Period 1, and so on.

7 This concept was first studied by Stahl and Wilson (1995) and Nagel (1995), and later by Ho, *et al.* (1998). A 1993 working paper by Debra Holt that also pioneered this approach was published in 1999. See also Sonsino *et al.* (2000).

8 Embedded sentences are those in which subject–object clauses are separated by other subject–object clauses. A classic example is: 'The mouse that the cat that the dog chased bit ran away'. To answer the question 'Who got bitten?' the reader must keep in mind 'the mouse' while processing the fact that the cat was chased by the dog. Limited working memory leads to frequent mistakes in recalling the contents of such sentences or answering questions about them (Christiansen and Chater, 1999). This notation makes it easier: 'The mouse that (the cat that [the dog chased] bit) ran away'.

9 A natural competitor to the thinking-steps model for explaining one-shot games is quantal response equilibrium (QRE) (see McKelvey and Palfrey, 1995, 1998; Goeree and Holt, 1999a). Weiszacker (2003) suggests an asymmetric version equivalent to a thinking-steps model in which one type thinks others are more random than s/he is. More cognitive alternatives are the theory of thinking trees of Capra (1999) and the theory of 'noisy introspection' of Goeree and Holt (2004). In Capra's model, players introspect until their choices match those of players whose choices they anticipate. In Goeree and Holt's theory, players use an iterated quantal response function with a response sensitivity parameter equal to λ/t^n where n is the discrete iteration step. When t is very large, their model corresponds to one in which all players do one-step and think others do zero. When $t = 1$ the model is QRE. All these models generate unimodal distributions so they need to be expanded to accommodate heterogeneity. Further work should try to distinguish different models, or investigate whether they are similar enough to be close modelling substitutes.

10 The data are 48 subjects playing 12 symmetric 3×3 games (Stahl and Wilson, 1995); 187 subjects playing 8 2×2 asymmetric matrix games (Cooper and Van Huyck, 2003); and 36 subjects playing 13 asymmetric games ranging from 2×2 to 4×2 (Costa-Gomes *et al.*, 2001).

11 While the common-τ models have one more free parameter than QRE, any reasonable information criterion penalizing the LL would select the thinking model.

12 When λ is set to 100 the τ estimates become very regular, around 2, which suggests that the variation in estimates is caused by poor identification in these games.

13 The differences in LL across game-specific and common τ are 0.5, 49.1 and 9.4. These are marginally significant (apart from Cooper–Van Huyck).

14 Take the over-confidence $K - 1$ model. The 70 per cent frequency of X choices without belief-prompting is consistent with this model if $f(0|\tau)/2 + f(1|\tau) = 0.70$, which is satisfied most closely when $\tau = 0.55$. If belief-prompting moves all thinking up one step, then the former zero-steppers will choose X and all others choose Y. When $\tau = 0.55$ the fraction of level zeros is 29 per cent, so this simple model predicts 29 per cent choice of X after belief-prompting, close to the 30 per cent that is observed.

15 Information measures are crucial to resolving the question of whether offers that are close to equal splits are equilibrium offers which reflect fairness concerns, or reflect at limited look-ahead and heuristic reasoning. The answer is both (see Camerer *et al.*, 1993; Johnson *et al.*, 2002). In the Costa-Gomes (2001) study, two different decision rules always led to the same choices in their games, but required

different look-up patterns. The look-up data were able therefore to classify players according to decision rules more conclusively than could choices alone.

16 To name only a few examples, see Camerer (1987) in markets for risky assets (partial adjustment models); Smith *et al.* (1988) in finitely-lived asset markets (Walrasian excess demand); McAllister (1991) in co-ordination games (reinforcement); Camerer and Weigelt (1993) in stochastically-lived assets (entrepreneurial stockpiling); Roth and Erev (1995) in bargaining and best-shot public goods games (reinforcement learning); Ho and Weigelt (1996) in extensive-form co-ordination games (reinforcement and belief learning); and Camerer and Cachon (1996) in co-ordination games (Cournot dynamics).

17 The parameter ρ is different from the one used in our original paper and is equivalently called 'kappa' in our recent working papers.

18 We have also applied slight variants of the model to extensive-form centipede games (Ho *et al*, 1999), extensive-form signalling games (Anderson and Camerer, 2001); and bilateral call markets with private values (Camerer *et al.*, 2002). Adapting the model to extensive forms and incomplete information requires making very slight further assumptions, but the three studies we have done show that the same basic framework can be used successfully. In extensive-form games the model needs to estimate forgone payoffs for branches of the game tree that are not reached. In the papers listed above, we do this using historical information, some average of the minimum and maximum payoffs (in signalling and centipede games), or 'clairvoyance' (the other player's stated conditional response to the unchosen path in the tree, in centipede games). Any of these assumptions fit equally well. In the bilateral call markets we 'spill over' reinforcement from one private value to neighbouring private values (this kind of generalization is assumed in belief learning), but the strength of spillover depends on how close the private values are; see also Roth and Erev (1995).

19 Other models aim to explain choices aggregated at some level. Of course, models of this sort can sometimes be useful. But our view is that a parsimonious model that can explain very fine-grained data can probably explain aggregated data well too, while the opposite may not be true, so the harder challenge is to explain fine-grained data.

20 See also Cheung and Friedman, 1997, pp. 54–5; Fudenberg and Levine, 1998, pp. 184–5; and Ed Hopkins, 2002.

21 See Bush and Mosteller, 1955; Harley, 1981; Cross, 1983; Arthur, 1991; McAllister, 1991; Roth and Erev, 1995; Erev and Roth, 1998.

22 When updated fictitious play beliefs are used to update the expected payoffs of strategies, precisely the same updating is achieved by reinforcing all strategies by their payoffs (whether received or forgone). The beliefs themselves are an epiphenomenon that disappear when the updating equation is written expected payoffs rather than beliefs.

23 Weber (forthcoming) also finds that players' choices change over time, as if they are learning, even with *no* feedback about what others have done (that is, even when they choose strategies repeatedly with no feedback). He suggests that players are using their own previous choices as a kind of pseudo-experience.

24 Taken seriously, reinforcement models also predict that learning paths will look the same whether players know their full payoff matrix or not. This prediction is rejected in all the studies that have tested it – for example, Mookerjhee and Sopher, 1994; Rapoport and Erev, 1998; Van Huyck *et al.*, 2001.

25   In mixed games, no model improves much on the Nash equilibrium (and often does not improve on quantal response equilibrium at all, and parameter identification is poor; see Salmon, 2001).

26   Typically, we penalize in-sample likelihood functions using the Akaike and Bayesian information criteria, which subtract a penalty of one, or log(n), times the number of degrees of freedom from the maximized likelihood. More persuasively, we rely mostly on out-of-sample forecasts which will be *less* accurate if a more complex model simply appears to fit better because it overfits in sample.

27   Note that, if your statistical objective is to maximize hit rate, λ does not matter and so self-tuning EWA is a zero-parameter theory given initial conditions.

28   Note that, if there is more than one other player, and the distinct choices by different other players matter to player i, then the vector is an $n-1$-dimensional matrix if there are n players.

29   In a naturally-occurring application where the equilibrium structure is not known to the modeller, it is easy to create some statistic that proxies for W, such as the number of strategies that have been chosen more than 25 per cent of the time in the previous eight periods. Such a rule will often overestimate W in empirical application, biasing $\phi_i(t)$ upwards compared to the value created by knowing W exactly. Since letting $\phi_i(t)$ dip too low, and discarding history irretrievably is a bigger mistake than keeping too much history, such an adjustment typically should not be a big mistake.

30   If one is uncomfortable assuming subjects act as if they know W, you can easily replace W by some function of the variability of others' choices to proxy for W.

31   Formally, $\rho_i(t) = 2 \times \{\sum_{k=1}^{m_i} f_i^{(k)}(t) \times \frac{m_i - k}{m_i - 1}\}$ where $f_i^k(t)$ are ranked from the lowest to the highest.

32   In our recent work we have switched to a model in which δ is 1 for strategies that are better (or equally good) responses to the current strategy, and zero for worse responses. This in fact fits a little worse than $\delta_i(t) = \phi_i(t)/W$, but has a sensible interpretation in terms of direction learning and dual-system dynamics. We also fix $\rho = 0$ in recent work, since allowing the flexible $\rho_i(t)$ function adds little to fit.

33   The initial parameter values are $\phi_i(0) = \rho_i(0) = 0.5$ and $\delta_i(0) = \phi_i(0)/W$. These initial values are averaged with period-specific values determined by the functions, weighting the initial value by $\frac{1}{t}$ and the functional value by $\frac{t-1}{t}$.

34   For example, suppose half the players choose $ABABAB \ldots$ in a game, and half choose $BABAB \ldots$. Then a model which says that A is chosen half the time will fit perfectly in every period, even though it does not predict the transition from A to B and vice versa accurately at all.

35   The parameter λ seems to vary reliably across games in ways that are not easily adjusted for by rescaling payoffs to common inflation-adjusted currency levels.

36   More details are in Ho *et al.*, 2001, and corresponding graphs for *all* games can be seen at http://www.bschool.nus.edu.sg/ depart/mk/fbacjk/ewalite/ewalite.htm.

37   Subjects were 196 undergraduate students in computer science and engineering in Singapore. Each group played ten times together twice, with different values of p in the two 10-period sequences. (One sequence used $p > 1$ and is not included.) We analyse a subsample of their data with $p = 0.7$ and 0.9, from groups of size 7. This subsample combines groups in a 'low experience' condition (the game is the first of two they play) and a 'high experience' condition (the game is the second of two, following a game with $p > 1$).

38   Reinforcement can be speeded up in such games by reinforcing unchosen strategies in some way, as Roth and Erev (1995) did in market games, which is why EWA and belief learning do better.

39   Their experiment used ten cohorts of seven subjects each, playing for fifteen periods. At the end of each period subjects learned the median, and played again with the same group in a partner protocol. Payoffs were the amounts on the table, in pennies.

40   At the high equilibrium, the payoffs are larger and so the difference between the received payoff and δ times the foregone payoff will be larger than at the low equilibrium. (Numerically, a player who chooses 3 when the median is 3 earns \$0.60 and has a forgone payoff from 2 or 4 of \$0.58 $\cdot\delta$. The corresponding figures for a player choosing 12 are \$1.12 and \$1.10 $\cdot\delta$. The differences in received and forgone payoffs around 12 and around 3 are the same when $\delta = 1$, but the difference around 12 grows larger as δ falls (for example, for the self-tuning EWA estimate $\hat{\delta} = 0.69$, the differences are \$0.20 and \$0.36 for 3 and 12). Cumulating payoffs rather than averaging them 'blows up' the difference and produces sharper convergence at the high equilibrium.

41   They also had a session with $R = 10$ but in this session one subject sat out each round so we dropped it to avoid making an *ad hoc* assumption about learning in this unusual design. Each subject played ten times (and played with a different R for five more rounds; we use only the first ten rounds).

42   As λ rises, the QRE equilibria move sharply from smearing probability throughout the price range (for low λ) to a sharp spike at the equilibrium (higher λ). No intermediate λ can explain the combination of initial dispersion and sharp convergence at the end so the best-fitting QRE model essentially makes the Nash prediction.

43   In beauty contests and co-ordination games, payoffs depend on the mean or median of fairly large groups (7–9 except in 3-person entry games), so switching one subject's choice to the recommendation would probably not change the mean or median and hence would not change future behaviour much. In other games, players are usually paired randomly so the boomerang effect again is muted. We are currently redoing the analysis to simply compare profits of players whose choices frequently matched the recommendation with those who rarely did. This controls for the boomerang effect and also for a Lucas critique effect in which adopting recommendations would change the behaviour of others and hence the model parameters used to derive the recommendations. A more interesting correction is to run experiments in which one or more computerized subjects in fact use a learning model to make choices, and compare their performance with that of actual subjects.

44   For example, in the continental divide game, *ex-post* optimal payoffs would have been 892 (pennies per player) if players knew exactly what the median would be, and subjects in fact earned 837. EWA and self-tuning EWA generate simulated profits of 879–882, which is only an improvement of 5 per cent over 837, but is 80 per cent of the maximum possible improvement from actual payoffs to clairvoyant payoffs.

45   Partow and Schotter (1993), Mookerjee and Sopher (1994), Cachon and Camerer (1996).

46   Sophistication may also have the potential to explain why players sometimes move in the *opposite* direction to that predicted by adaptive models (Rapoport, 1999), and why measured beliefs do not match up well with those predicted by adaptive belief learning models (Nyarko and Schotter, 2002).

47   To truncate the belief hierarchy, the sophisticated players believe that the other sophisticated players, like themselves, believe there are α' sophisticates.

48 The gap between apparent sophistication and perceived sophistication shows the empirical advantage of separating the two. Using likelihood ratio tests, we can clearly reject both the rational expectations restriction $\alpha = \alpha'$ and the pure overconfidence restriction $\alpha' = 0$, although the differences in log-likelihood are not large.

49 Borrower subjects do not play consecutive sequences, which removes their incentive to repay in the eighth period of one sequence so they can get more loans in the first period of the next sequence.

50 To economize on computing, we search only paths of future actions that always have default following repay because the reverse behaviour (repay following default) generates a lower return.

51 Formally, $\hat{P}_L^{j'}(a, k, v|j_{v-1}) = \hat{P}_L^{Loan}(a, k, v - 1|j_{v-1}) \cdot P_L^{j'}(a, k, v|(Loan, j_{v-1})) + \hat{P}_L^{NoLoan}(a, k, v - 1|j_{v-1}) \cdot P_L^{j'}(a, k, v|(NoLoan, j_{v-1}))$.

52 This is called 'observational learning' (see Duffy and Feltovich, 1999; Armentier, 2004). Without this assumption, the model learns far move slowly than do the lenders, so it is clear that they are learning from observing others.

53 The idea is to create an 'interim' attraction for Round t, $B_L^j(a, k, t)$, based on the attraction $A_L^j(a, k, t-1)$ and payoff from Round t, then incorporate experience in Round $t+1$ from previous sequences, transforming $B_L^j(a, k, t)$ into a final attraction $A_L^j(a, k, t)$. See Camerer *et al.* (2002) for details.

54 We use four separate λs for honest borrowers, lenders, normal adaptive borrowers and teaching borrowers, an initial attraction for lending $A(0)$, and the spillover parameter σ and teaching proportion α.

55 See Camerer and Weigelt (1988); Palfrey and Rosenthal (1988); McKelvey and Palfrey (1992).

56 In our current extension of the Camerer *et al.* (2002) paper on strategic teaching, we impose a value of θ that was measured independently in a separate experiment on one-shot trust games. It is around 0.20, much lower than the estimate of 0.91, and when this measured number is used instead of letting θ be a free parameter, AQRE degrades rather badly.

57 One direction we are pursuing is to find designs or tests that distinguish the teaching and equilibrium updating approaches. The sharpest test is to compare behaviour in games with types that are fixed across sequences with types that are 'refreshed' independently in each period within a sequence. The teaching approach predicts similar behaviour in these two designs, but type-updating approaches predict that reputation formation dissolves when types are refreshed.

58 A fourth enterprise fits utility functions that reflect social preferences for fairness or equality. This important area is not discussed in this chapter.

59 See Smith *et al.*, 1988; Camerer and Weigelt, 1993; and Lei *et al.*, 2001.

60 See Sonsino *et al.*, 2000; Sovik, 2000.

61 The studies, in the order in which τ estimates are reported below, are Malcolm and Lieberman (1965), O'Neill (1987), Rapoport and Boebel (1992), Bloomfield (1994), Ochs (1995, games with 9 and 4 payoffs), Mookerjhee and Sopher (1997), Rapoport and Almadoss (2000, $r = 8, 20$), Binmore *et al.* (2001), Tang (2001, games 3 and 1), games 1, 3, 4. Readers please let us know of published studies we have overlooked; we plan to include them in a later draft.

62 In each game, data from those $n - 1$ out of the n possible strategies with the most extreme predicted equilibrium probabilities are used to fit the models. Excluding the n th strategy reduces the dependence among data points, since all n frequencies (and predictions) obviously add up to one.

References

Anderson, C. and Camerer, C. F. (2001) 'Experience-weighted Attraction Learning in Sender – Receiver Signaling Games', *Economic Theory*, 16, 3, 689–718.

Andreoni, J. and Miller, J. (1993) 'Rational Cooperation in the Finitely Repeated Prisoner's Dilemma: Experimental Evidence', *Economic Journal*, 103, 570–85.

Armantier, Olivier (2004) 'Does Observation Influence Learning?', *Games and Economic Behavior*, 46, February, 221–39.

Arthur, B. (1991) 'Designing Economic Agents That Act Like Human Agents: A Behavioral Approach to Bounded Rationality', *American Economic Review*, 81, 2, 353–9.

Babcock, Linda and Loewenstein, George (1997) 'Explaining Bargaining Impasses: The Role of Self-serving Biases', *Journal of Economic Perspectives*, 11, 109–26.

Babcock, L., Loewenstein, G., Issacharoff, S. and Camerer, C. F. (1995) 'Biased Judgments of Fairness in Bargaining', *American Economic Review*, 85, 1337–43.

Bajari, P. and Hortacsu, A. (2003) 'Winner's Curse, Reserve Prices, and Endogeneous Entry: Empirical Insights from eBay', *RAND Journal of Economics*, Summer, 329–55.

Binmore, K., Swierzbinski, J. and Proulx, C. (2001) 'Does Maximin Work? An Experimental Study', *Economic Journal*, 111, 445–64.

Bloomfield, R. (1994) 'Learning a Mixed Strategy Equilibrium in the Laboratory', *Journal of Economic Behavior and Organization*, 25, 411–36.

Blume, A., DeJong, D., Neumann, G. and Savin, N. (1999) 'Learning in Sender–Receiver Games', Working paper, University of Iowa.

Brown, G. (1951) 'Iterative Solution of Games by Fictitious Play', in *Activity Analysis of Production and Allocation* T. C. Uoopmans (ed.), New York, John Wiley.

Bush, R. and Mosteller, F. (1955) *Stochastic Models for Learning*, New York, John Wiley.

Cabrales, A. and Garcia-Fontes, W. (2000) 'Estimating Learning Models with Experimental Data', Working paper, University of Pompeu Febra, Barcelona 501.

Cachon, G. P. and Camerer, C. F. (1996) 'Loss-avoidance and Forward Induction in Experimental Coordination Games', *Quarterly Journal of Economics*, 111, 165–94.

Camerer, C. F. (1987) 'Do Biases in Probability Judgment Matter in Markets? Experimental Evidence', *American Economic Review*, 77, 981–97.

Camerer, C. F. (1997) 'Progress in Behavioral Game Theory', *Journal of Economic Perspectives*, 11, 167–88.

Camerer, C. F. (2003) *Behavioral Game Theory: Experiments on Strategic Interaction*, Princeton, NJ: Princeton University Press.

Camerer, C. F. and Ho, T.-H. (1998) 'EWA Learning in Normal-form Games: Probability Rules, Heterogeneity and Time Variation', *Journal of Mathematical Psychology*, 42, 305–26.

Camerer, C. F. and Ho, T.-H. (1999) 'Experience-weighted Attraction Learning in Normal-form Games', *Econometrica*, 67, 827–74.

Camerer, C. F. and Ho, T.-H. (2001) 'Strategic Learning and Teaching in Games', in S. Hoch and H. Kunreuther (eds), *Wharton on Decision Making*, New York: Wiley.

Camerer, C. F. and Lovallo, D. (1999) 'Overconfidence and Excess Entry: An Experimental Approach', *American Economic Review*, 89, 306–18.

Camerer, C. F. and Weigelt, K. (1988) 'Experimental Tests of a Sequential Equilibrium Reputation Model', *Econometrica*, 56, 1–36.

Camerer, C. F. and Weigelt, K. (1993) 'Convergence in Experimental Double Auctions for Stochastically Lived Assets', in D. Friedman and J. Rust (eds), *The Double Auction Market: Theories, Institutions and Experimental Evaluations*, Redwood City, Calif.: Addison-Wesley, 355–96.

Camerer, C. F., Ho, T.-H. and Chong, J. K. (2002a) 'Sophisticated EWA Learning and Strategic Teaching in Repeated Games', *Journal of Economic Theory*, 104, 137–88.

Camerer, C. F., Ho, T.-H. and Chong, J. K. (2002b) 'A Cognitive Hierarchy Theory of One-shot Games: Some Preliminary Results'. See http://www.hss.caltech.edu/camerer/thinking2002.pdf , accessed December.

Camerer, C. F. Ho, T.-H. and Chong, J. K. (2003) 'Models of Thinking, Learning and Teaching in Games', *American Economic Review*, 93, May, 192–5.

Camerer, C. F., Hsia, D. and Ho, T.-H. (2002) 'EWA Learning in Bilateral Call Markets', in A. Rapoport and R. Zwick (eds), *Experimental Business Research*, Dordrecht: Kluwer, 255–85.

Camerer, C. F., Johnson, E., Sen, S. and Rymon, T. (1993) 'Cognition and Framing in Sequential Bargaining for Gains and Losses', in K. Binmore, A. Kirman and P. Tani (eds), *Frontiers of Game Theory*, Cambridge, Mass., MIT Press, 27–48.

Capra, M. (1999) 'Noisy Expectation Formation in One-shot Games', Unpublished dissertation, University of Virginia.

Capra, M., Goeree, J., Gomez, R. and Holt, C. (1999) 'Anomalous Behavior in a Traveler's Dilemma', *American Economic Review*, 89, 678–90.

Cheung, Y.-W. and Friedman, D. (1997) 'Individual Learning in Normal Form Games: Some Laboratory Results', *Games and Economic Behavior*, 19, 46–76.

Christiansen, M. H. and Chater, N. (1999) 'Toward a Connectionist Model of Recursion in Human Linguistic Performance', *Cognitive Science*, 23, 157–205.

Clark, K. and Sefton, M. (1999) 'Matching Protocols in Experimental Games', Working paper, University of Manchester. See http://nt2.ec.man.ac.uk/ses/discpap/downloads/9917.pdf.

Cooper, D. and Kagel, John (2003a) 'Lessons Learned: Generalizing Learning Across Games', *American Economic Review*, 93, May, 202–7.

Cooper, D. and Kagel, John (2003b) 'Are Two Heads Better Than One? Team versus Individual Play in Signaling Games', Working paper, Case Western, July. See http://www.econ.ohio-state.edu/kagel/teamsD71.1.pdf.

Cooper, D. and Van Huyck, J. (2003) 'Evidence on the Equivalence of the Strategic and Extensive Form Representation of Games', *Journal of Economic Theory*, 110, 2, June, 290–308.

Costa-Gomes, Miguel and Weizsäcker, G. (2003) 'Stated Beliefs and Play in Normal-form Games', Working paper, ISER and Harvard, November. See http://kuznets.fas.harvard.edu/weizsack/papers/CGW3.pdf.

Costa-Gomes, M., Crawford, V. and Broseta, B. (2001) 'Cognition and Behavior in Normal-form Games: An Experimental Study', *Econometrica*, 69, 1193–235.

Crawford, V. (1997) 'Theory and Experiment in the Analysis of Strategic Interactions', in D. Kreps and K. Wallis (eds), *Advances in Economics and Econometrics: Theory and Applications*, Seventh World Congress, Vol. i, Cambridge University Press.

Crick, F. (1988) *What Mad Pursuit?*, New York: Sloan Foundation.

Croson, R. T. A. (2000) 'Thinking Like a Game Theorist: Factors Affecting the Frequency of Equilibrium Play', *Journal of Economic Behavior and Organization*, 41, 3, 299–314.

Cross, J. (1983) *A Theory of Adaptive Learning Economic Behavior*, New York, Cambridge University Press.

Duffy, J. and Feltovich, N. (1999) 'Does Observation of Others Affect Learning in Strategic Environments? An Experimental Study', *International Journal of Game Theory*, 28, 131–52.

Erev, I. and Roth, A. (1998) 'Predicting How People Play Games: Reinforcement Learning in Experimental Games with Unique, Mixed-strategy Equilibria', *American Economic Review*, 88, 848–81.

Erev, I., Bereby-Meyer, Y. and Roth, A. (1999) 'The Effect of Adding a Constant to All Payoffs: Experimental Investigation, and a Reinforcement Learning Model with Self-Adjusting Speed of Learning', *Journal of Economic Behavior and Organization*, 39, 111–28.

Fudenberg, D. and Levine, D. (1989) 'Reputation and Equilibrium Selection in Games with a Patient Player', *Econometrica*, 57, 759–78.

Fudenberg, D. and Levine, D. (1998) *The Theory of Learning in Games*, Boston, Mass., MIT Press.

Goeree, J. K. and Holt, C. A. (1999) 'Stochastic Game Theory: For Playing Games, Not Just for Doing Theory', *Proceedings of the National Academy of Sciences*, 96, 10564–7.

Goeree, J. K. and Holt, C. A. (2001) 'Ten Little Treasures of Game Theory, and Ten Intuitive Contradictions', *American Economic Review*, 91, December, 1402–22.

Goeree, J. K. and Holt, C. A. (2004) 'A Theory of Noisy Introspection', *Games and Economic Behavior*, 46, February, 365–82.

Güth, Werner (2000) 'Boundedly Rational Decision Emergence – A General Perspective and Some Selective Illustrations', *Journal of Economic Psychology*, 21, 433–58.

Harley, C. (1981) 'Learning the Evolutionary Stable Strategies', *Journal of Theoretical Biology*, 89, 611–33.

Heller, D. and Sarin, R. (2000) 'Parametric Adaptive Learning', Working paper, University of Chicago.

Ho, T.-H. and Chong, J. K. (2003) 'A Parsimonious Model of SKU Choice', *Journal of Marketing Research*, 40, 351–65.

Ho, T.-H. and Weigelt, K. (1996) 'Task Complexity, Equilibrium Selection, and Learning: An Experimental Study', *Management Science*, 42, 659–79.

Ho, T.-H., Camerer, C. F. and Chong, J. K. (2001) 'Economic Value of EWA Lite: A Functional Theory of Learning in Games', Working paper, University of Pennsylvania. See http://www.hss.caltech.edu/camerer/camerer.html.

Ho, T.-H., Camerer, C. F. and Weigelt, K. (1998) 'Iterated Dominance and Iterated Best Response in Experimental 'p-Beauty Contests', *American Economic Review*, 88, 947–69.

Ho, T.-H., Wang, X. and Camerer, C. F. (1999) 'Individual Differences and Payoff Learning in Games', Working paper, University of Pennsylvania.

Holt, Debra (1999) 'An Empirical Model of Strategic Choice with an Application to Coordination', *Games and Economic Behavior*, 27, 86–105.

Hopkins, E. (2002) 'Two Competing Models of How People Learn in Games', *Econometrica*, 70, 2141–66.

Huberman, G. and Rubinstein, A. (2000) 'Correct Belief, Wrong Action and a Puzzling Gender Difference', Working paper, Tel Aviv University.

Jehiel, P. (2002) 'Predicting by Analogy and Limited Foresight in Games', Working paper, University College London.

Jehiel, P. (forthcoming) Bounded Rationality and Imperfect Learning: Game Theory versus AI', *Greek Economic Review*.

Johnson, E. J., Camerer, C. F., Sen, S. and Rymon, T. (2002) 'Detecting Backward Induction in Sequential Bargaining Experiments', *Journal of Economic Theory*, 104, May, 16–47.

Jung, Y. J., Kagel, J. H. and Levin, D. (1994) 'On the Existence of Predatory Pricing: An Experimental Study of Reputation and Entry Deterrence in the Chain-store Game', *RAND Journal of Economics*, 25, 72–93.

Kahneman, D. (1988) 'Experimental Economics: A Psychological Perspective', in R. Tietz, W. Albers and R. Selten (eds), *Bounded Rational Behavior in Experimental Games and Markets*, New York, Springer Verlag, 11–18.

Kahneman, Daniel (2003) 'Maps of Bounded Rationality: Psychology for Behavioral Economics', *American Economic Review*, 93, December, 1449–75.

Kaufman, H. and Becker, G. M. (1961) 'The Empirical Determination of Game-theoretical Strategies', *Journal of Experimental Psychology*, 61, 462–8.

Keynes, J.M. (1936) *The General Theory of Interest, Employment and Money*, London, Macmillan.

Kocher, M. G. and Sutter, M. (forthcoming) 'When the 'Decision Maker' Matters: Individual versus Team Behavior in Experimental "Beauty-contest" Games', *Economic Journal*.

Kreps, D. and Wilson, R. (1982) 'Reputation and Imperfect Information', *Journal of Economic Theory*, 27, 253–79.

Lei, V., Noussair, C. and Plott, C. (2001) 'Non-speculative Bubbles in Experimental Asset Markets: Lack of Common Knowledge of Rationality', *Econometrica*, 69, 813–59.

Lucas, R. G. (1986) 'Adaptive Behavior and Economic Theory', *Journal of Business*, 59, October, S401–S426.

Marcet, Albert and Nicolini, Juan P. (2003) 'Recurrent Hyperinflations and Learning', *American Economic Review*, 93, December, 1476–98.

Mailath, G. (1998) 'Do People Play Nash Equilibrium? Lessons from Evolutionary Game Theory', *Journal of Economic Literature*, 36, 1347–74.

Malcolm, D. and Lieberman, B. (1965) 'The Behavior of Responsive Individuals Playing a Two-person Zero-sum Game Requiring the Use of Mixed Strategies', *Psychonomic Science*, 2, 373–4.

McAllister, P. H. (1991) 'Adaptive Approaches to Stochastic Programming', *Annals of Operations Research*, 30, 45–62.

McKelvey, R. D. and Palfrey, T. R. (1992) 'An Experimental Study of the Centipede Game', *Econometrica*, 60, 803–36.

McKelvey, R. D. and Palfrey, T. R. (1995) 'Quantal Response Equilibria for Normal-form Games', *Games and Economic Behavior*, 7, 6–38.

McKelvey, R. D. and Palfrey, T. R. (1998) 'Quantal Response Equilibria for Extensive-form Games', *Experimental Economics*, 1, 9–41.

Mookerjee, D. and Sopher, B. (1994) 'Learning Behavior in an Experimental Matching Pennies Game', *Games and Economic Behavior*, 7, 62–91.

Mookerjee, D. and Sopher, B. (1997) 'Learning and Decision Costs in Experimental Constant-sum Games', *Games and Economic Behavior*, 19, 97–132.

Nagel, R. (1995) 'Experimental Results on Interactive Competitive Guessing', *American Economic Review*, 85, 1313–26.

Nagel, R. (1999) 'A Review of Beauty Contest Games', in D. Budescu, I. Erev and R. Zwick (eds), *Games and Human Behavior: Essays in honor of Amnon Rapoport*, Lawrence Erlbaum Associates, Mahwah NJ, 105–42.

Nagel, R., Bosch-Domenech, A., Satorra, A. and Garcia-Montalvo, J. (2002) 'One, Two, (Three), Infinity: Newspaper and Lab Beauty-contest Experiments', *American Economic Review* 1, 92, 1687–1701.

Neral, J. and Ochs, J. (1992) 'The Sequential Equilibrium Theory of Reputation Building: A Further Test', *Econometrica*, 60, 1151–69.

Nyarko, Y. and Schotter, A. (2002) 'An Experimental Study of Belief Learning Using Elicited Beliefs', *Econometrica*, 70, May, 971–1005.

Ochs, J. (1995) 'Games with Unique, Mixed Strategy Equilibria: An Experimental Study', *Games and Economic Behavior*, 10, 202–17.

Ochs, J. (1999) 'Entry in Experimental Market Games', in D. Budescu, I. Erev and R. Zwick (eds), *Games and Human Behavior: Essays in Honor of Amnon Rapoport*, Mahwah, NJ, Lawrence Erlbaum Associates.

O'Neill, B. (1987) 'Nonmetric Test of the Minimax Theory of Two-person Zero-sum Games', *Proceedings of the National Academy of Sciences*, 84, 2106–9.

Palfrey, T. R. and Rosenthal, H. (1988) 'Private Incentives in Social Dilemmas: The Effects of Incomplete Information and Altruism', *Journal of Public Economics*, 35, 309–32.

Partow, J. and Schotter, A. (1993) 'Does Game Theory Predict Well for the Wrong Reasons? An Experimental Investigation', Working paper No. 93–46, C. V. Starr Center for Applied Economics, New York University.

Plott, C. R. (1986) 'Dimensions of Parallelism: Some Policy Applications of Experimental Methods', in A. E. Roth (ed.), *Laboratory Experimentation in Economics: Six Points of View*, Cambridge University Press.

Plott, C. R. (1997) 'Laboratory Experimental Testbeds: Applications to the PCS Auctions', *Journal of Economics and Management Strategy*, 6, 605–38.

Prendergast, C. (1999) 'The Provision of Incentives in Firms', *Journal of Economic Literature*, 37, March, 7–63.

Rapoport, A. and Amaldoss, W. (2000) 'Mixed Strategies and Iterative Elimination of Strongly Dominated Strategies: An Experimental Investigation of States of Knowledge', *Journal of Economic Behavior and Organization*, 42, 483–521.

Rapoport, A. and Boebel, R. B. (1992) 'Mixed Strategies in Strictly Competitive Games: A Further Test of the Minimax Hypothesis', *Games and Economic Behavior*, 4, 261–83.

Rapoport, A. and Erev, I. (1998) 'Coordination, "Magic", and Reinforcement Learning in a Market Entry Game', *Games and Economic Behavior*, 23, 146–75.

Rapoport, A., Lo, K.-C. and Zwick, R. (1999) 'Choice of Prizes Allocated by Multiple Lotteries with Endogenously Determined Probabilities', Working paper, University of Arizona, Department of Management and Policy.

Rassenti, S. J., Smith, V. L. and Wilson, B. J. (2001) 'Turning Off the Lights', *Regulation*, Fall, 70–6.

Roth, A. and Erev, I. (1995) 'Learning in Extensive-form Games: Experimental Data and Simple Dynamic Models in the Intermediate Term', *Games and Economic Behavior*, 8, 164–212.

Roth, A. and Xing, X. (1994) 'Jumping the Gun: Imperfections and Institutions Related to the Timing of Market Transactions', *American Economic Review*, 84, 992–1044.

Roth, A., Barron, G., Erev, I. and Slonim, R. (2000) 'Equilibrium and Learning in Economic Environments: The Predictive Value of Approximations', Working paper, Harvard University.

Salmon, T. (2001) 'An Evaluation of Econometric Models of Adaptive Learning', *Econometrica*, 69, 1597–628.

Salmon, T. (2003) 'Evidence for "Learning to Learn" Behavior in Normal-form Games', Paper, Florida State University, August. See http://garnet.acns.fsu.edu/tsalmon/EvidV2.pdf.

Sargent, T. (1999) *The Conquest of American Inflation*, Princeton, NJ, Princeton University Press.

Schelling, T. (1960) *The Strategy of Conflict*, Cambridge Mass Harvard University Press.

Schotter, A., Weigelt, K. and Wilson, C. (1994) 'A Laboratory Investigation of Multiperson Rationality and Presentation Effects', *Games and Economic Behavior*, 6, 445–68.

Seale, D. A. and Rapoport, A. (2000) 'Elicitation of Strategy Profiles in Large Group Coordination Games', *Experimental Economics*, 3, 153–79.

Selten, R. (1978) 'The Chain Store Paradox', *Theory and Decision*, 9, 127–59.

Selten, R. (1998) 'Features of Experimentally Observed Bounded Rationality', *European Economic Review*, 42, 413–36.

Selten, R. and Stoecker, R. (1986) 'End Behavior in Sequences of Finite Prisoner's Dilemma Supergames: A Learning Theory Approach', *Journal of Economic Behavior and Organization*, 7, 47–70.

Smith, V. L., Suchanek, G. and Williams, A. (1988) 'Bubbles, Crashes and Endogeneous Expectations in Experimental Spot Asset Markets', *Econometrica*, 56, 1119–51.

Sonsino, D., Erev, I. and Gilat, S. (2000) 'On the Likelihood of Repeated Zero-sum Betting by Adaptive (Human) Agents', Technion, Israel Institute of Technology.

Sovik, Y. (2000) 'Impossible Bets: An Experimental Study', Mimeo University of Oslo Department of Economics.

Stahl, D. O. (1996) 'Boundedly Rational Rule Learning in a Guessing Game', *Games and Economic Behavior*, 16, 303–30.

Stahl, Dale O. (1999) 'Sophisticated Learning and Learning Sophistication', Working paper, University of Texas at Austin.

Stahl, D. O. (2000) 'Local Rule Learning in Symmetric Normal-form Games: Theory and Evidence', *Games and Economic Behavior*, 32, 105–38.

Stahl, D. O. and Wilson, P. (1995) 'On Players Models of Other Players: Theory and Experimental Evidence', *Games and Economic Behavior*, 10, 213–54.

Sundali, J. A., Rapoport, A. and Seale, D. A. (1995) 'Coordination in Market Entry Games with Symmetric Players', *Organizational Behavior and Human Decision Processes*, 64, 203–18.

Sutton, R. and Barto, A. (1998) *Reinforcement Learning: An Introduction*, Boston, Mass., MIT Press.

Svensson, L. E. O. (2001) 'Comments on Michael Woodford Paper', presented at Knowledge, Information and Expectations in Modern Macroeconomics: In Honor of Edmund S. Phelps', Columbia University, 5–6 October.

Tang, F.-F. (2001) 'Anticipatory Learning in Two-person Games: Some Experimental Results', *Journal of Economic Behavior and Organization*, 44, 221–32.

Timmerman, A. G. (1993) 'How Learning in Financial Markets Generates Excess Volatility and Predictability in Stock Prices', *Quarterly Journal of Economics*, 108, November, 1135–45.

Tirole, J. (1985) 'Asset Bubbles and Overlapping Generations', *Econometrica*, 53, 1071–100 (reprinted 1499–528).

Van Damme, E. (1999) 'Game Theory: The Next Stage', in L. A. Gerard-Varet, Alan P. Kirman and M. Ruggiero (eds), *Economics Beyond the Millenium*, Oxford University Press, 184–214.

Van Huyck, J., Battalio, R. and Beil, R. (1990) 'Tacit Cooperation Games, Strategic Uncertainty, and Coordination Failure', *The American Economic Review*, 80, 234–48.

Van Huyck, J., Battalio, R. and Rankin, F. W. (2001) 'Selection Dynamics and Adaptive Behavior Without Much Information', Working paper, Texas A & M Department of Economics.

Van Huyck, J., Cook, J. and Battalio, R. (1997) 'Adaptive Behavior and Coordination Failure', *Journal of Economic Behavior and Organization*, 32, 483–503.

Von Neumann, J. and Morgenstern, O. (1944) *The Theory of Games and Economic Behavior*, Princeton, NJ, Princeton University Press.

Warglien, M., Devetag, M. G. and Legrenzi, P. (1998) 'Mental Models and Naive Play in Normal Form Games', Working paper, Universita Ca' Foscari di Venezia.

Watson, J. (1993) 'A "Reputation" Refinement without Equilibrium', *Econometrica*, 61, 199–205.

Watson, J. and Battigali, P. (1997) 'On "Reputation" Refinements with Heterogeneous Beliefs', *Econometrica*, 65, 363–74.

Weber, Roberto (forthcoming) ' "Learning" with No Feedback in a Competitive Guessing Game', *Games and Economic Behavior*.

Weibull, J. (1995) *Evolutionary Game Theory*, Cambridge, Mass., MIT Press.

Weizsäcker, G. (2003) 'Ignoring the Rationality of Others: Evidence from Experimental Normal-form Games', *Games and Economic Behavior*, 44, 145–71.

Wilcox, Nat (2003) 'Heterogeneity and Learning Models', Working paper, University of Houston. See http://www.uh.edu/ñwilcox/papers/hetero_and_learning_W.pdf.

Woodford, M. (2001) 'Imperfect Common Knowledge and the Effects of Monetary Policy', Paper presented at 'Knowledge, Information and Expectations in Modern Macroeconomics: In Honor of Edmund S. Phelps', Columbia University, 5–6 October.

9
Double Auction Markets with Stochastic Supply and Demand Schedules: Call Markets and Continuous Auction Trading Mechanisms

John H. Kagel *

Introduction

Performance under two different double auction trading mechanisms is investigated: a call market and a continuous double auction trading mechanism. Both auctions are two-sided, with several buyers and several sellers. A call market is a discrete trading mechanism in which buyers (sellers) submit a single bid (offer) in each trading period and the market clears according to well-defined rules of who trades and at what prices. In a continuous double auction, trades can occur at any time in the trading period, with buyers and sellers free to update unaccepted bids and offers continuously. Both trading mechanisms have wide applications in field settings, and have also been the subject of intense experimental study (see Holt, 1995, for a survey of experimental work).

This chapter adds to the literature through an experimental investigation of double auction (DA) markets, in which buyers' valuations and sellers' costs are randomly drawn in each trading period. Although this procedure is common practice in experimental studies of one-sided auctions, it is rarely employed in experimental studies of DA markets, as these typically involve stationary supply and demand schedules. A random value environment is natural for investigating Bayesian–Nash equilibrium theories of price formation in DA markets (Wilson, 1987; Satterthwaite and Williams, 1989a, 1989b; Friedman, 1991) since traders are assumed to have incomplete information regarding each other's valuations. In contrast, with stationary supply and

* Research was supported in part by grants from the National Science Foundation. Chi-Ren Dow, Susan Garvin, Wei Lo, Chris Jacobs, Juping Jin and William Vogt provided valuable research assistance. The research has benefited from discussions with Dan Friedman and Mark Satterthwaite, and the comments of a referee. I alone am responsible for any errors and omissions.

demand schedules, traders effectively acquire complete information regarding market clearing price and quantity after several trading periods. Arguably, random valuation procedures are also (i) more representative of field settings, since supply and demand schedules are rarely, if ever, stationary from one period to the next; and (ii) provide the appropriate vehicle for examining the Hayek (1945) hypothesis – that markets are capable of achieving (close to) the competitive equilibrium (CE) price and quantity resulting from truthful revelation – as this hypothesis is intended to apply to markets where traders have incomplete information regarding the CE.

Behaviour is studied in markets with two buyers and two sellers ($m = 2$) and in markets with eight buyers and eight sellers ($m = 8$). The call market studied is the buyer's bid double auction (BBDA) (Satterthwaite and Williams, 1989a, 1989b) in which fully rational traders achieve near 100 per cent efficiency with $m = 8$ (ibid., 1989a). The continuous double auction (CDA) trading mechanism studied uses New York stock exchange rules but no specialists book. The analysis proceeds on two levels: (i) comparisons of efficiency and price convergence between the two mechanisms and compared to outcomes in markets with stationary supply and demand schedules; and (ii) comparisons of behaviour with theoretical predictions for the two mechanisms.

Both trading mechanisms achieve relatively high efficiency levels (75 per cent or higher) even in thin markets with $m = 2$. The CDA achieves substantially higher average efficiency levels than the BBDA both in markets with $m = 2$ (87 per cent versus 77.1 per cent average efficiency) and with $m = 8$ (95.1 per cent versus 88.9 per cent). There are no significant differences in price levels, relative to the CE norm, between the two institutions, with both large and small numbers of traders. Thus, consistent with received wisdom from experiments with stationary supply and demand schedules (Holt, 1995), CDA outcomes are close to the CE level in markets with stochastic supply and demand schedules, and achieve higher efficiency levels than a sealed-bid trading mechanism.[1] However, unlike markets with stationary supply and demand schedules, there is no tendency for efficiency or prices to converge to the CE level with increased trader experience *within* an experimental session. Rather, if anything, the data suggest greater deviations from the CE norm in the small markets with two buyers and two sellers.

The key theoretical prediction for the BBDA, higher efficiency with increased numbers of traders, is satisfied, with remarkably high efficiency levels (94 per cent) observed for experienced traders with $m = 8$. However, buyers tend to bid more than predicted, particularly with $m = 2$, and sellers do not follow consistently the dominant strategy of offering at cost (typically offering at above cost). The first result is consistent with bidding above the risk-neutral Nash equilibrium (RNNE) in one-sided, first-price, private-value auctions (behaviour that has sometimes been attributed to risk aversion). The second result is consistent with deviations from the dominant bidding strategy in one-sided, second-price, private-value auctions and in uniform price, multiple unit auctions. Deviations from the dominant bidding strategy are attributed to (i) the non-transparency of the strategy; and (ii) the relatively small costs associated with the deviations. Experimental sessions in which computerized sellers follow the dominant bidding strategy of offering at cost are used to test if the limited degree of strategic buyer misrepresentation observed might be in response to sellers' errors at offering above cost. There is no evidence to support this conjecture.

Contrasting theoretical predictions of the Wilson (1987) and Friedman (1991) models of the CDA price formation process are compared with those of zero intelligence (ZI) traders (Gode and Sunder, 1993). Consistent with both the Wilson and Friedman models, there is a strong tendency for higher-valued buyers and lower-cost sellers to trade first. Although these propensities are significantly less than the Wilson and Friedman model predictions, they are greater than in the ZI simulations. Further, there are fewer units traded in the experiments for all market sizes than in the ZI simulations. The net result is that much of the inefficiency found in markets with $m = 2$ results from fewer units traded than the CE level predicts, as both the Wilson and Friedman models imply. Finally, there is a clear tendency for price changes to be correlated negatively within an auction period, which is inconsistent with both the Friedman and Wilson specifications, but is similar to what is found in ZI simulations. However, the average absolute price changes are substantially smaller than in the ZI simulations, and the opportunity to arbitrage prices is quite limited.

There have been a handful of earlier studies of DA markets with random supply and demand schedules, all of which have focused on testing Bayesian–Nash equilibrium models of price formation in DA markets (Cason and Friedman (1993, 1996) for continuous double auctions; Kagel and Voght, (1993) and Cason and Friedman (1997) for call markets).[2] In addition to reporting results from a new data set, this chapter differs from these earlier reports by comparing performance explicitly across the BBDA and the CDA.[3] I also employ more extreme variation in the number of traders within the CDA.[4] One important result of this manipulation is that it reveals that inefficiencies in small markets result from *too few* trades occurring relative to the CE model prediction, as the Bayesian–Nash equilibrium trading models predict. In contrast, in larger markets inefficiencies result from *too many* units trading. I offer some conjectures as to the basis for these differences resulting from market size.

The plan of this chapter is as follows: the next section characterizes the theoretical implications of the Bayesian–Nash equilibrium models for the two trading mechanisms studied. These are contrasted with predictions for 'zero intelligence' traders (Gode and Sunder, 1993), as this provides a useful benchmark for completely non-strategic behaviour. The third section outlines the experimental procedures. The results of the experiment are reported in the fourth section, and a brief concluding section summarizes our main results.

Theoretical implications

The buyer's bid double auction mechanism

The BBDA (Satterthwaite and Williams, 1989a, 1989b) is a call market where bids and offers are collected from traders, supply and demand schedules are constructed, a market clearing price is established, and trades are executed at the market clearing price. In the BBDA, each buyer (seller) draws a single redemption value, $x_{2i}(x_{1i})$, from a known probability distribution $F_2(F_1)$ defined on the interval $[\underline{x}, \bar{x}]$. In the BBDA, buyers who get to trade earn profits equal to $(x_{2i} - p)$, where p is the market price. Sellers who get to trade earn profits equal to $(p - x_{1i})$. Buyers and sellers who do not trade earn zero profits.

In the BBDA, price is selected at the upper endpoint of the closed interval determining the market clearing-price, with all trade occurring at this price. This is determined as follows: all bids and offers are arranged in non-decreasing order $s_1 \leq s_2 \leq \ldots \leq s_{2m}$, where m is the number of buyers and sellers in the market (the number of buyers is assumed to equal the number of sellers, as in the experiment). Price is set at $p = s_{m+1}$. In the case where a single bid/offer uniquely determines s_{m+1}, supply is exactly equal to demand, and every buyer whose bid is at least p purchases an item and every seller whose offer is less than p sells the item. In a case where at least two bids/offers equal s_{m+1}, demand may exceed supply. The BBDA then prescribes that items are allocated, beginning with the buyer who bid the most and working down the list of buyers whose bids are at least p. If a point is reached where two or more buyers submitted identical bids and there is insufficient supply to serve them (excess demand), then the available supply is rationed among these bidders using a lottery that assigns each buyer an equal chance of receiving an item.[5]

Satterthwaite and Williams (1989a) demonstrate that each seller in the BBDA has a dominant strategy to offer at cost (x_{1i}). In response to this, buyers bid less than their reservation value (x_{2i}). This strategic misrepresentation causes the BBDA to be *ex post* inefficient. However, the amount of buyer misrepresentation decreases rapidly as market size increases: with uniform distributions of traders' redemption values, and risk-neutral buyers, expected efficiency increases from 92.6 per cent with $m = 2$ to 99.6 per cent with $m = 8$ (ibid., 1989a; where efficiency is defined as realized consumer and producer surplus as a percentage of the maximum possible consumer and producer surplus). The buyer's bid function underlying this rapid increase in efficiency, given a uniform distribution of redemption values, is:

$$b_i = \frac{m}{m+1} x_{2i} \tag{9.1}$$

In contrast, using a simple fixed price rule, with prices set at the midpoint of the distribution functions underlying redemption values, efficiency increases at a substantially slower rate (from 78.1 per cent with $m = 2$, to 85.4 per cent with $m = 8$; ibid., 1989a).

Figure 9.1 illustrates two cases of the BBDA with $m = 4$. In Figure 9.1(a), Buyers 4 and 3 and Sellers 1 and 2 trade at a price set by Seller 3's offer (note that Seller 3 does *not* trade here). In Figure 9.1(b), again Buyers 4 and 3 and Sellers 1 and 2 trade, but in this case price is set at Buyer 3's bid.[6]

The BBDA provides an explicit trading procedure that achieves efficiency levels quite close to those that could be achieved using an optimal revelation mechanism (Myerson and Satterthwaite, 1983). However, unlike optimal revelation mechanisms, the BBDA does not require a change in the mechanism's rules as the underlying distribution of valuations and costs change. Further, although Satterthwaite and Williams (1989a) do not analyse the effects of limitations on agents rationality and information processing on auction outcomes, they argue that 'our result that all equilibrium strategies of the BBDA in a large market are close to truthful revelation suggests that cognitive limitations are unimportant in large markets' (ibid., p. 479). This implication does not necessarily follow, however, should it be the case that

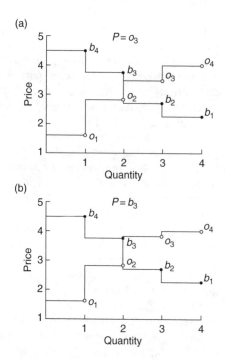

Figure 9.1 Price and quantity determination in BBDA

equilibrium strategies are hard to understand, and there are simpler strategies that appear intuitively reasonable but are far from truthful revelation.[7]

The continuous double auction mechanism

In a CDA trades can occur any time in a trading period, with prices free to vary from one trade to the next. Wilson (1987) models the CDA price formation process as a sequential equilibrium of an extensive form game in which traders privately draw a single redemption value from a commonly-known joint distribution. The model is concerned with the price formation process *within* a given trading period.[8] The basic idea is that traders play a waiting game, but they are impatient as a result of possible pre-emption of gains by other traders. At some point, a trader makes a 'serious' offer, one that has a positive probability of being accepted in a sequential equilibrium. If this offer is not accepted immediately, the trader making the offer steadily improves it until it is accepted, as in a Dutch auction (other traders remain passive during this process). One of the striking predictions of the model is that, at any point in time, transactions occur between the highest-value buyer and the lowest-cost seller remaining in the market. Further, inefficiencies result strictly from lost trading opportunities as lower-value buyers and higher-cost sellers, who would trade in the absence of strategic considerations, fail to trade in the time allotted.

In principle, the Wilson model is capable of making very precise predictions regarding who trades and when. However, as Cason and Friedman (1993) point out, it is not practical to test these predictions, as solutions to the model are defined implicitly by a nested set of partial differential equations whose boundary conditions at each stage are derived recursively from the solution to subsequent-stage partial differential equations (with some arbitrariness as to the final stage specification), and no numerical algorithms are presently available to solve the equations even for very simple value distributions and simple sets of auxiliary hypotheses. Nevertheless, the model has a number of reasonably precise qualitative implications that can readily be tested; for example, as already stated, inefficiencies result strictly from lost trading opportunities, and transactions occur between the highest-value buyer and lowest-cost seller remaining in the market.

Friedman (1991) develops a Bayesian game-against-nature model of the CDA trading process. According to Friedman, agents are boundedly rational with limited strategic capabilities. Agents carry with them reservation prices for buying and selling based on their valuations. Sellers are willing to undercut the standing market price as long as they can do so without selling below their reservation price, and they accept the market bid whenever it is greater than their reservation price; buyers behave analogously. To complete the model, Friedman (1991) employs a drastic simplification: traders are assumed to ignore the impact of their own current bids and offers on subsequent bids and offers. This 'game-against-nature' assumption, together with Bayesian updating and auxiliary assumptions similar to Wilson's (for example, risk neutrality), give reservation prices as solutions to the optimal stopping problem associated with current estimates of 'nature's' bid and offer generating process.

If the reservation prices associated with trader valuations were known, the Friedman model could be solved for precise bids, asks and acceptances. Without this, the model still has a number of distinctive qualitative implications, several of which are quite similar to the Wilson model's predictions: (i) early transactions will be between higher-value buyers and lower-cost sellers; and (ii) efficiencies are close to 100 per cent, with inefficiencies resulting from lost trading opportunities as lower-value buyers and higher-cost sellers, who would trade in the absence of strategic considerations, fail to trade in the time allotted. Other qualitative predictions differ from the Wilson model. The one I look at is that changes in transaction prices will be positively correlated, with this effect being most pronounced for early trades. This positive correlation between price changes results from the fact that, as traders' beliefs change in response to new bids and offers, their reservation prices shift unexpectedly. These unanticipated shifts in bids and offers can be shown to result in positively correlated changes in transaction prices. In contrast, the Wilson model implies zero correlation between changes in transactions prices, since auto-correlation between prices would open up opportunities for price arbitrage that fully rational agents would exploit.[9]

Zero intelligence traders

Gode and Sunder (1993) show that for the CDA, zero intelligence (ZI) traders achieve remarkably high efficiency levels (between 95 per cent and 100 per cent), often achieving higher efficiencies than human traders in the first period of an experiment with stationary supply and demand schedules. ZI traders are completely non-strategic, with sellers offering at or above cost (but no higher than $\underline{x}$) and buyers bidding at or below resale values (but no lower than $\underline{x}$).[10] Bids and offers are determined randomly within these intervals, with traders churning out bids and offers repeatedly. Employing New York stock exchange rules, so that if a bid (offer) is to have standing in the market it must improve on the current market bid (offer), bids and offers are selected at random from traders. A transaction occurs whenever a new bid (offer) exceeds (is less than) the current market offer (bid), with the transaction price determined by the current market offer (bid). ZI traders are fast enough, relative to the market period time constraint, that transactions stop when there is no longer any room for mutually profitable trades to occur.

As already noted, for the CDA, ZI traders achieve remarkably high efficiency levels. However, unlike the Wilson and Friedman models, inefficiencies result from too much (rather than too little) trading. As such, extra-marginal traders (those traders whose redemption values lie just beyond the CE) get to trade, or extra-marginal traders displace infra-marginal traders who would get to trade at a CE outcome. As in the Wilson and Friedman models, with ZI traders, early transactions tend to occur between higher-value buyers and lower-cost sellers. But these effects are less pronounced. For ZI traders, transaction prices are independent draws from a distribution that changes over time as successful traders leave the market. Cason and Friedman (1993) show that this implies negative auto-correlation between changes in transaction prices, in contrast to the Friedman model's prediction of positive auto-correlation.

Kagel and Vogt (1993) show that ZI traders achieve very low efficiency levels in the BBDA, considerably lower than achieved with a fixed-price rule set at the mid-point of the interval from which valuations are drawn.[11] The reason for the poor efficiency outcomes (relative to behaviour) in the BBDA is that the static (one-shot) nature of the game does not permit ZI traders to correct for bids and offers within an auction period when they fail to find trading partners (and ZI traders do not learn across auction periods). In contrast, the dynamic CDA permits ZI traders to make virtually unlimited numbers of bids and offers within the time constraint for trading, so that 'bad' bids and offers (those that do not result in trades) are easily overcome by new bids and offers resulting in profitable (and relatively efficient) transactions. That is, the CDA permits 'bad' bids and offers to be corrected, whereas the BBDA does not, so that ZI traders perform better in the former than in the latter.

ZI traders are employed as a reference point against which to evaluate behaviour in CDAs since (i) they provide a useful benchmark for completely non-strategic behaviour; and (ii) the algorithm achieves very high efficiency levels in the CDA and, according to Cason and Friedman (1993, 1996), it predicts quite well the volume of transactions, the order of transactions relative to redemption values, and the auto-correlation in transaction prices within a given market period. In contrast, for the BBDA, we document briefly the extremely poor efficiency outcomes for ZI compared to real traders, and then drop the use of ZI traders as a reference point, since it is clear that it is hopelessly inadequate for organizing behaviour in these auctions.

Experimental procedures

Redemption values were drawn randomly from a uniform distribution [0, $5.00], with new random draws in each trading period. The parameters of the distribution were read out loud as part of the instructions and were posted publicly. Each experimental session consisted of a pre-announced number of trading periods. For inexperienced subjects there were two dry runs (with no money at stake) after the instructions were read.

The goal was to recruit sixteen subjects for each experimental session. Markets with equal numbers of buyers and sellers were used throughout. Auction markets with $m = 2$ and $m = 8$ were conducted under both mechanisms.

In markets with $m = 2$ traders were randomly assigned to a new small market group in each trading period.

Six BBDA sessions were conducted, three with live sellers and buyers, and three with live buyers and computerized sellers. In both cases, two inexperienced subject sessions were conducted first, followed by a third session recruiting subjects from the first two sessions.[12] In the sessions with computerized sellers, it was announced that 'sellers' were following the dominant strategy of offering at cost.

Five CDA sessions were conducted. The first two used inexperienced subjects. This was followed by three experienced subject sessions, recruiting subjects from the first two sessions and using subjects who had first participated in a CDA as part of a classroom teaching exercise.[13] In one of the experienced-subject sessions only twelve subjects arrived on time, and in the other only fourteen were there on time. These sessions were run as planned, with the large market auctions employing $m = 6$ and $m = 7$, respectively, and the small market auctions employing $m = 2$ (in the session with fourteen subjects, one small auction market operated with $m = 3$).

The BBDA auctions used a dual market bidding procedure. With dual market bidding, using the same redemption values, traders first participate in a market with $m = 2$, but before the market clearing price is established, they play again in a market with $m = 8$. Payment is made in only one of the two markets, which is determined randomly after both sets of bids and offers have been submitted.[14] In determining the effect of increases in the number of traders on behaviour, the dual market technique minimizes the extraneous variability resulting from changes in redemption values and the subject population. That is, it operationalizes responses to *ceteris paribus* changes in the number of traders, since the same subjects bid with exactly the same redemption values, with only m changing. The dual market technique has been employed to good effect in one-sided auctions, resulting reliably in increased bidding with increased numbers of bidders in first-price, private-value auctions (Dyer *et al.*, 1989; Battalio *et al.*, 1990). In order to simplify instructions as much as possible, auctions with inexperienced subjects began with several periods of bidding in a single market before introducing the dual market technique.

The BBDA auctions used software developed explicitly for this purpose. Although there were no restrictions on bids or offers, the instructions did point out that the only possible way to lose money was to bid (offer) above (below) valuation (cost). Following each auction period, all bids and offers were reported back to traders for the market(s) in which they participated, along with the underlying redemption values (individual subject IDs were suppressed). In addition, all traders learnt the market price, the number of transactions, and their own earnings. Finally, in the auctions with live buyers and live sellers, halfway through the session subjects switched roles, with buyers becoming sellers and vice versa.

The CDA auctions were conducted using a modified version of software developed by Shyam Sunder at Carnegie Mellon University. This software did not permit multiple market bidding (which can get rather cumbersome in a CDA), so an ABA design was employed, with agents first bidding in a single large market, then bidding in one of several small markets, followed by bidding in a single large market. Here, too, there were no restrictions that bids (offers) had to be less (greater) than or equal to valuations (costs). However, as in the BBDA, the instructions pointed out that the only possible way to lose money was to bid (offer) above (below) valuation (cost). The computer program provides a graphical representation of bids, offers and transaction prices, along with a ticker-tape reporting of the same information. There was no switching of roles between buyers and sellers in these auctions, as the software did not readily permit it, and the software and ticker-tape feedback were sufficiently unfamiliar to subjects that a number of them asked to play the same role on returning for a second session (we accommodated these requests as much as possible).[15]

For both mechanisms, subjects were provided with a starting capital balance of $5.00, in lieu of a participation fee, against which any potential net losses would be subtracted. Although buyers (sellers) should never bid above (below) their redemption value (cost) under either mechanism, there were occasional mistakes in the early auction periods in the CDA which resulted in losses, and a number of bids (offers) of this sort throughout the BBDA that resulted in occasional losses. However, no subject's cash balance ever dropped much below $5.00.

Experimental results

BBDA auctions

Table 9.1 reports measures of market performance for the BBDA auctions with live buyers and live sellers. For sessions 1 and 2, which used inexperienced bidders, we have dropped the first several (8) auction periods (which involved bidding in a single auction market). Two efficiency measures are reported in the top part of Table 9.1: the first measure computes efficiency in each auction period separately and takes the average of these efficiency measures (excluding those auctions with $m = 2$, in which the maximum possible consumer and producer surplus was zero). The second measure sums realized consumer and producer surplus across auction periods and divides by the sum of the maximum possible consumer and producer surplus.[16] The second measure is the one employed in Satterthwaite and Williams (1989a) in computing the expected increase in efficiency under the BBDA with increases in m. The first measure provides some indication of the variation in efficiency between auction periods, and accentuates the improvements in efficiency resulting from increases in m.

Table 9.1 Measures of market performance: BBDA efficiency

Efficiency

BBDA auction number (no. auction periods)[a]	Average across auction periods (standard error of mean)				Sum of surplus across auction periods			
	m = 2		m = 8		m = 2		m = 8	
	Actual	Predicted	Actual	Predicted	Actual	Predicted	Actual	Predicted
1 (16)	0.752 (0.076)	0.780 (0.056)	0.832 (0.047)	0.989 (0.005)	0.843	0.924	0.871	0.993
2 (16)	0.786 (0.052)	0.795 (0.054)	0.875 (0.047)	0.998 (0.002)	0.870	0.939	0.885	1.00
3× (25)	0.774 (0.047)	0.784 (0.046)	0.935 (0.019)	0.996 (0.002)	0.890	0.942	0.940	0.996
Average	0.771 (0.033)	0.786 (0.029)	0.889 (0.021)	0.995 (0.002)	0.871	0.936	0.906	0.996

BBDA auction number (no. auction periods)[a]	Quantity deviations from CE prediction (standard error of mean)				Price deviations from CE prediction (standard error of mean)			
	m = 2		m = 8		m = 2		m = 8	
	Actual	Predicted	Actual	Predicted	Actual	Predicted	Actual	Predicted
1 (16)	−0.132 (0.066)	−0.340 (0.066)	0.125 (0.180)	−0.438 (0.128)	0.229 (0.072)	0.031 (0.026)	0.151 (0.046)	0.015 (0.010)
2 (16)	−0.226 (0.064)	−0.264 (0.061)	−0.625 (0.155)	−0.063 (0.063)	0.079 (0.031)	0.020 (0.015)	0.136 (0.042)	0.000 (0.000)
3× (25)	−0.179 (0.051)	−0.295 (0.052)	−0.080 (0.114)	−0.240 (0.087)	0.051 (0.023)	0.020 (0.012)	0.061 (0.035)	0.006 (0.005)
Average	−0.179 (0.034)	−0.299 (0.034)	−0.175 (0.091)	−0.246 (0.058)	0.111 (0.025)	0.023 (0.010)	0.107 (0.023)	0.007 (0.004)

Notes: a With $m = 2$ there are 4 auction markets operating simultaneously in each auction period.
b Negative numbers indicate fewer trades than the CE prediction with truthful revelation.
c Positive numbers indicate prices are higher than the CE prediction with truthful revelation.

Both measures show similar results. In all cases, actual efficiency is less than predicted based on Satterthwaite and Williams' equilibrium analysis for idealized (risk-neutral) traders. These differences are small, however, particularly with $m = 2$, where the average efficiency across auction periods measure is not significantly different from the BBDA prediction. Differences between realized and idealized efficiency become larger, and achieve statistical significance with $m = 8$. Note, however, that realized efficiencies are closest to the BBDA prediction for experienced traders (session 3x), suggesting that experience may improve performance under this mechanism. Finally, in all three sessions, efficiency increases with increases in m, as the theory predicts, with the largest increase occurring in the session with experienced traders.

Realized efficiency is much higher than with ZI traders, who achieve consistently low efficiencies averaging 0.3 and 0.35 with $m = 2$ and $m = 8$, respectively (using the sum of the surplus measure).[17] Thus, efficiency is considerably closer to the level predicted for perfectly rational (risk-neutral) traders than for completely non-strategic, ZI traders. Realized efficiency is also considerably higher than with a fixed price rule of $2.50, where efficiency averages 0.79 and 0.84 with $m = 2$ and $m = 8$, respectively (again, using the sum of the surplus measure). One way to view the fixed price rule is as a clumsy instrument that is bound to achieve 100 per cent efficiency with increases in m, as the dominant strategy of bidding (offering) at redemption values is completely transparent.[18] In contrast, the BBDA opens up the possibility of achieving substantially higher efficiency with small numbers of traders, while at the same time creating the possibility that the trading rules are so complicated that traders fail to respond to the strategic possibilities, or respond incorrectly to them. Consequently, what this comparison shows is that, while human traders do not respond in an idealized fashion to the BBDA mechanism, 'mistakes' are not large enough, or frequent enough, to offset the promised improvements in efficiency relative to a fixed price rule.

The lower part of Table 9.1 shows the price and quantity data underlying these efficiency results. With the exception of session 1 with $m = 8$, the quantity traded is always below the CE prediction, and prices are always above the CE prediction, so that price and quantity deviations are consistent, directionally, with the BBDA predictions. However, quantity traded is usually a little greater than the BBDA's predictions, and prices are consistently, and substantially, greater than the BBDA's predictions. As shown below, the latter results from sellers' tendencies to offer at above-cost and buyers' tendencies to bid closer to valuations than the BBDA predicts.

Table 9.2 shows individual trader's bids and offers relative to the BBDA's predictions. With $m = 2$, buyers bid significantly more than the risk-neutral BBDA predicts, as they bid an average of $0.28 below their valuations compared to the BBDA prediction of bidding $0.81 below valuations. Not shown in Table 9.2 is that with $m = 8$ there was a small, but statistically significant, increase in bids relative to valuations averaging $0.08 (compared to

Table 9.2 Individual subject behaviour in the BBDA

BBD auction number	Buyers' misrepresentation:[a] mean values across subjects (standard error of the mean)				Sellers' misrepresentation:[b] mean values across subjects (standard error of the mean)	
	$m = 2$		$m = 8$		$m = 2$	$m = 8$
	Actual	Predicted	Actual	Predicted	Actual	Actual
1	0.242	0.800	0.099	0.267	0.222	0.144
	(0.199)	(0.039)	(0.197)	(0.013)	(0.113)	(0.120)
2	0.326	0.765	0.342	0.255	0.351	0.391
	(0.045)	(0.034)	(0.124)	(0.011)	(0.079)	(0.083)
3×	0.271	0.853	0.184	0.284	0.164	0.073
	(0.052)	(0.029)	(0.064)	(0.010)	(0.098)	(0.121)
Average	0.280	0.806	0.208	0.269	0.245	0.203
	(0.078)	(0.028)	(0.080)	(0.007)	(0.056)	(0.065)

Notes: [a] Buyers' misrepresentation = (resale value − bid).
[b] Sellers' misrepresentation = (offer − cost); truthful revelation is a dominant strategy for sellers.

a predicted increase of $0.54).[19] This small increase in realized bids, along with the sharp increase in predicted bids, resulted in bids being much closer to the risk-neutral BBDA prediction with $m = 8$ (average reduction in bids relative to valuations of $0.21 versus the BBDA prediction of $0.27 shown in Table 9.2).

Bidder behaviour in the BBDA has a number of similarities to behaviour in one-sided private-value auctions. The overbidding relative to the risk-neutral BBDA prediction with $m = 2$ is not unlike the overbidding relative to the RNNE found in one-sided, first-price sealed-bid auctions. The overbidding in one-sided, private-value auctions is most extreme with small numbers of bidders, in which case it is not uncommon for bidders to take home 50 per cent or less of predicted RNNE profits (see Kagel and Roth (1992), table 4). Some of the overbidding in one-sided, first-price auctions may be a result of risk aversion, and some may be the result of buyers' misperceptions and/or inexperience (see Kagel (1995) for a review of the experimental literature on this point). Some of these same forces are likely to be at work in the BBDA as well. Also, increases in bidding with increased numbers of rivals is reliably reported in one-sided, first-price private-value auctions. This result is replicated in the BBDA as well.

Table 9.2 shows that sellers failed to follow the dominant strategy of offering at cost, usually offering at above cost. Not reported in Table 9.2 is that several (6) sellers offered within $0.10 of cost in 90 per cent or more of the auctions. In contrast, none of the buyers ever bid within $0.10 of their valuations 90 per cent of the time or more. This suggests that the dominant bidding strategy had some (weak) drawing power in these auctions.

Deviations from the dominant offer strategy are not unexpected, as deviations from dominant bidding strategies are reported in one-sided, second-price private-value auctions (Kagel *et al.*, 1987; Kagel and Levin, 1993) and in one-sided, multiple-unit, uniform-price auctions (Cox *et al.*, 1985). As with the one-sided auctions, the dominant offer strategy is far from transparent in the BBDA. Further, the expected cost of deviating from the dominant strategy was relatively small, averaging $0.05 per auction period ($0.11 conditional on selling), so that sellers were close to playing best responses.[20] As such, any trial and error search process that might be expected to help subjects find the dominant offer strategy would generate relatively weak feedback effects.[21]

Given the extent of seller misrepresentation reported, it is not clear that buyers should bid according to (1). Further, it is possible that the smaller than predicted under-revelation on the part of buyers constitutes a (possibly mistaken) strategic response to sellers offering at above cost. The BBDAs with computerized sellers were designed to investigate the latter question. The answer, in short, is that bidder behaviour is quite similar to auctions with live sellers. With $m = 2$, buyers bid significantly more than the risk-neutral BBDA predicts (average buyer misrepresentation of $0.37 compared to the BBDA prediction of $0.82). In contrast, with $m = 8$, there were minimal differences between predicted and actual misrepresentation ($0.266 actual versus $0.272 predicted). Both patterns are quite similar to those reported in Table 9.2 for BBDA auctions with live sellers.

CDA auctions

Table 9.3 reports measures of market performance for the CDAs. For Sessions 1 and 2, with inexperienced bidders, the first several (5) auction periods have been dropped as there is some tendency for mistakes (bidding above valuation or offering at below cost) to occur in these periods as a result of unfamiliarity with the software. As already noted, we also use ZI traders as the reference point since (i) ZI traders do such a good job of organizing much CDA trading data; and (ii) because of the absence of point predictions for both the Wilson and Friedman models.

Efficiency measures are reported in the top part of Table 9.3. In auctions with $m = 2$, average efficiency across auction periods is quite high and significantly *above* the level reported under the BBDA mechanism ($t = 2.11, p < 0.05$, two-tailed t-test). However, efficiency is significantly *below* the ZI benchmark. With $m = 8$, efficiency tends to increase relative to markets with $m = 2$ (the one exception is Session 7 using the sum of the surplus measure). Using the average across auction periods measure, with $m = 8$, average efficiency is higher than in the BBDA auctions with live sellers ($t = 1.82, p < 0.10$, two-tailed t-test), while being essentially the same as in the ZI simulations.[22]

Table 9.3 Measures of market performance: CDA

CDA auction number (no. auction periods)[a]	Efficiency							
	Average across auction periods (standard error of mean)				Sum of surplus across auction periods			
	m = 2		m = 8		m = 2		m = 8	
	Actual	ZI	Actual	ZI	Actual	ZI	Actual	ZI
7 (32/7)	0.825 (0.077)	0.962 (0.009)	0.939 (0.042)	0.955 (0.006)	0.963	0.957	0.942	0.956
8 (32/11)	0.923 (0.047)	0.968 (0.009)	0.977 (0.016)	0.951 (0.004)	0.941	0.965	0.978	0.951
9× (24/0)	0.899 (0.060)	0.946 (0.015)	–	–	0.939	0.936	–	–
10× (16/0)	0.927 (0.040)	0.967 (0.013)	–	–	0.902	0.959	–	–
11× (32/17)	0.804 (0.068)	0.965 (0.010)	0.938 (0.005)	0.954 (0.004)	0.898	0.954	0.959	0.954
Average	0.870 (0.028)	0.962 (0.005)	0.951 (0.026)	0.953 (0.025)	0.932	0.954	0.961	0.953

Table 9.3 (Continued)

CDA auction number (no. auction periods)	Quantity deviations from CE prediction[b] (standard error of mean)				Average absolute price deviation from CE prediction (standard error of mean)							
					Average based on individual transactions (standard error of mean)				Average based on average auction period price (standard error of mean)			
	m = 2		m = 8		m = 2		m = 8		m = 2		m = 8	
	Actual	ZI	Actual	ZI	Actual	ZI	Actual	ZI	Actual	ZI	Actual	ZI
7 (32/7)	−0.125 (0.110)	0.142 (0.032)	0.143 (0.340)	0.343 (0.043)	0.106 (0.070)	0.065 (0.023)	0.097 (0.046)	0.300 (0.021)	0.113 (0.079)	0.054 (0.020)	0.089 (0.043)	0.091 (0.014)
8 (32/11)	0.0 (0.053)	0.081 (0.023)	0.363 (0.152)	0.436 (0.038)	0.073 (0.048)	0.024 (0.007)	0.131 (0.045)	0.297 (0.017)	0.075 (0.050)	0.010 (0.004)	0.071 (0.037)	0.086 (0.011)
9× (24/0)	−0.130 (0.072)	0.087 (0.026)	–	–	0.018 (0.013)	0.041 (0.015)	–	–	0.004 (0.004)	0.020 (0.009)	–	–
10× (16/0)	0.0 (0.0)	0.013 (0.013)	–	–	0.0 (0.0)	0.032 (0.016)	–	–	0.0 (0.0)	0.015 (0.010)	–	–
11× (32/17)	−0.143 (0.085)	0.086 (0.024)	0.118 (0.146)	0.426 (0.031)	0.049 (0.033)	0.060 (0.017)	0.127 (0.037)	0.226 (0.011)	0.053 (0.037)	0.047 (0.015)	0.062 (0.033)	0.051 (0.006)
Average	−0.085 (0.036)	0.087 (0.012)	0.200 (0.107)	0.413 (0.021)	0.054 (0.020)	0.046 (0.077)	0.122 (0.024)	0.263 (0.009)	0.054 (0.021)	0.031 (0.060)	0.070 (0.021)	0.070 (0.006)

Notes: [a] Number of auction periods with m = 2 followed by auction periods with m = 8.
[b] Negative numbers indicate fewer trades than the CE prediction with truthful revelation.

Although efficiency levels are quite high, unlike CDAs with stationary supply and demand schedules, there is no tendency for market efficiency to improve with replication within a given experimental session (and even some tendency for efficiency to decrease with experience in markets with $m = 2$). The evidence for this is twofold. First, we ran a number of different regression specifications with auction period as a right-hand side variable, finding no systematic patterns in the data (no continuous improvements or drop-offs in efficiency over time). Second, for each experimental market we identified the median efficiency level for that market (100 per cent in all cases) and counted the frequency with which efficiency deviated from the median in the first half compared to the second half of the auction within a given experimental session. We then pooled the data for the large markets ($m \geq 6$) and for the small markets ($m = 2$), separately, and conducted a simple binomial test for significant differences in deviations from the median. There were no significant differences between halves in the large markets (44 per cent – 11 out of 25 – of the deviations occurred in the second half of the sessions, $p > 0.10$, two-tailed t-test). However, in the markets with $m = 2$, 66.7 per cent of the deviations (18 out of 27) occurred in the second half of the sessions, which is significantly different from the 50 per cent reference point at the 10 per cent level (two-tailed-test).

Unlike auctions with stationary supply and demand schedules, there is no reason to expect efficiency to increase over time within a given experimental session, as traders must search to find mutually beneficial trades in each auction period with past realizations being no help in the search process. From this perspective, the apparent decrease in efficiency in the small markets is somewhat puzzling. Perhaps it is a statistical aberration. Alternatively, it could be that traders are learning to behave more strategically over time, which implies missed trading opportunities and reduced efficiency. This would be consistent with the overall pattern of more strategic play observed in markets with fewer numbers of traders (discussed in some detail below).

The lower part of Table 9.3 shows the price and quantity data underlying these efficiency results. With $m = 2$, average quantity traded is significantly below the CE model's prediction, consistent with the Wilson and Friedman models' predictions. In contrast, with $m = 2$, ZI traders *always* trade more than the CE model predicts.[23] With $m = 8$, people trade a little above the CE model's prediction but still below the ZI level. That is, with the increase in the number of traders, valuations are congested enough in the neighbourhood of the CE that excess numbers of mutually profitable trades reliably occur. But the number of such trades is still less than in the ZI simulations.[24]

One possible explanation for why there are fewer trades than the CE level with $m = 2$ and excess numbers of trades with $m = 8$ is that agents play the game the way Wilson and Friedman suggest, but there are errors in implementing these strategies. With the increase in m there is increased congestion which, in conjunction with a constant error rate, is enough to result in more trades than the CE level or the Wilson and Friedman models (without errors) predict. This is supported by the fact that, in experimental session 9 x, with $m = 6$, in one-third of all auctions (5 out of 15) there were fewer units traded than the CE model predicts, and there were no auctions with more units traded. In contrast, in the ZI simulations all deviations from the CE output level involve more units traded. This conjecture is also supported by the ZI simulations themselves. In going from ZI simulations with $m = 2$ to $m = 8$, the underlying stochastic bidding strategies have been held constant (bid a random number within one's budget constraint), but there is increased bidder congestion. The net result is that in 8.7 per cent of the simulated auction markets with $m = 2$ there are units traded in excess of the CE, compared to 39.4 per cent with $m = 8$.

With $m = 2$, average absolute price deviations do not differ significantly from the ZI benchmark, and average less than $0.06 per auction from the CE prediction. Although average absolute price deviations were $0.11 from the CE in the corresponding BBDAs, these differences between institutions are not significant at conventional levels ($t = 0.89$). With $m = 8$, average absolute price deviations were just over $0.12 per auction, well below the ZI benchmark and only marginally higher than the price deviations observed in the BBDA ($0.11).[25] Overall, price deviations from the CE level are relatively small, as judged by the ZI benchmark and in comparison to the BBDA auctions.[26] This is somewhat surprising, to this investigator at least, given the search process underlying the price formation process in these markets. Finally, employing the same techniques used to evaluate changes in efficiency over time to evaluate price convergence, we conclude that, unlike markets with stationary supply and demand schedules, there is no tendency for prices to converge to the CE outcome across market periods within an experimental session.[27] This is hardly surprising, since each auction period essentially sets off a new search for mutually acceptable trading prices, with minimal hints as to what is the relevant equilibrium price interval based on past auction outcomes.

Tables 9.4 and 9.5 examine the pattern of transactions in CDAs. Table 9.4 reports rank order correlations between the order in which transactions occurred with the ranking of traders' redemption values. In both large and small markets there are statistically significant correlations indicating that the highest-value buyers and lowest cost sellers tend to trade first. In markets with $m = 2$, the realized correlations are significantly stronger on the seller's side of the market than implied by the ZI simulations (80 per cent of the ZI correlations fall between 0.16 and 0.82).[28] In markets with $m > 2$, there is essentially no difference between the strength of the experimental correlations and the ZI simulations (pooled values are virtually the same). With the notable exception of sellers' correlations in sessions 8 and 10 x with $m = 2$, rank order correlations are well below 1.0, the predicted value of both the Wilson and Friedman models.

Table 9.4 Rank order correlations: transaction number and ranking of buyer valuations and seller costs

| Experimental session | Small markets (prob=0) | | | | Large markets (prob=0) | | | |
| | Buyers[b] | | Sellers[b] | | Buyers[b] | | Sellers[b] | |
	Actual	ZI[c]	Actual	ZI[c]	Actual	ZI[c]	Actual	ZI[c]
7 (m=8)[a]	0.25 (0.23)	0.36 (0.06)	0.50 (0.01)	0.59 (0.05)	0.51 (<0.01)	0.33 (0.05)	0.37 (0.06)	0.28 (0.04)
8 (m=8)[a]	0.76 (<0.01)	0.56 (0.04)	1.00 (<0.01)	0.59 (0.05)	0.40 (<0.01)	0.33 (0.02)	0.18 (0.24)	0.33 (0.03)
9× (m=6)[a]	0.24 (0.26)	0.46 (0.07)	0.62 (<0.01)	0.53 (0.05)	0.34 (0.04)	0.44 (0.04)	0.33 (0.04)	0.42 (0.03)
10× (m=7)[a]	0.42 (0.02)	0.64 (0.05)	1.00 (<0.01)	0.49 (0.05)	0.26 (0.04)	0.39 (0.03)	0.31 (0.01)	0.33 (0.02)
11× (m=8)[a]	0.52 (<0.01)	0.56 (0.05)	0.76 (<0.01)	0.56 (0.05)	0.24 (0.05)	0.25 (0.03)	0.44 (<0.01)	0.37 (0.03)
Pooled	0.46 (<0.01)	0.52 (0.03)	0.81 (<0.01)	0.55 (0.02)	0.34 (<0.01)	0.35 (0.02)	0.36 (<0.01)	0.35 (0.01)

Notes: [a] Number of buyers and sellers in large market. Small markets $m = 2$.
[b] Buyers' valuations ranked from 1 to m, starting with highest valuation; sellers' costs ranked from 1 to m starting with lowest cost.
[c] Mean of ZI simulations with standard error of mean in parentheses.

Table 9.5 Consumer and producer surplus as a function of transaction rank[a] (values in cents)

Experimental session	Small markets: transaction order				Large markets: transaction order							
	1		2		1		2		3		4	
	Actual	ZI	Actual	ZI	Actual	ZI	Actual	ZI	Actual	ZI	Actual	ZI
7 (m=8)[b]	261.7	205.9	151.6	137.2	259.3	255.8	284.1	239.0	239.1	216.8	148.0	165.6
	t=1.63 (0.10)		t=0.22 (0.83)		t=0.09 (0.93)		t=1.10 (0.27)		t=0.53 (0.60)		t=-0.37 (0.71)	
8 (m=8)[b]	234.2	219.0	48.0	93.2	268.9	251.2	208.5	232.6	226.6	197.7	167.6	180.1
	t=0.66 (0.51)		t=-1.38 (0.18)		t=0.54 (0.59)		t=-0.74 (0.46)		t=0.88 (0.38)		t=-0.32 (0.75)	
9× (m=6)[b]	247.8	208.7	142.7	135.7	278.5	258.6	211.0	207.4	201.4	156.1	ND	
	t=1.63 (0.10)		t=0.13 (0.89)		t=0.73 (0.47)		t=0.11 (0.92)		t=1.03 (0.30)			
10× (m=7)[b]	168.7	171.4	79.0	170.8	254.8	261.6	220.9	221.8	170.3	196.4	204.6	160.3
	t=-0.08 (0.93)		t=-0.88 (0.42)		t=-0.27 (0.79)		t=0.04 (0.97)		t=-1.01 (0.31)		t=1.60 (0.11)	
11× (m=8)[b]	201.4	172.7	97.6	110.2	307.3	264.7	197.7	240.0	237.3	201.7	187.0	165.9
	t=1.14 (0.25)		t=-0.26 (0.79)		t=1.61 (0.11)		t=-1.55 (0.12)		t=1.25 (0.21)		t=0.67 (0.50)	
Average	225.4	197.1	104.9	121.6	275.9	256.3	217.4	223.7	212.6	192.7	178.4	159.5
	t=2.28 (0.03)		t=-0.67 (0.51)		t=1.52 (0.13)		t=-0.45 (0.66)		t=1.33 (0.18)		t=1.13 (0.26)	
Maximum possible[c]	240.5 (12.8)		65.8 (20.6)		388.0 (7.6)		252.7 (12.4)		154.5 (11.9)		94.1 (17.6)	

Notes: ND – no data
[a] T-tests for difference between actual and ZI surplus. Probability $t = 0$ in parentheses (2-tailed, t-test).
[b] Number of buyers and sellers in large market. Small markets $m = 2$.
[c] Maximum possible is conditional on the number of units actually traded. Standard errors of the mean reported in parentheses.

Table 9.5 reports total consumer plus producer surplus by the order in which trades occurred. In markets with $m = 2$, the first transaction generated an average surplus of $2.25 compared to an average of $1.97 in the ZI simulations ($t = 2.28, p < 0.03$, two-tailed t-test) and a maximum possible surplus for the units traded of $2.41 per auction. Surplus generated is below the ZI simulations for the second unit traded, but this is to be expected, since with $m = 2$ at most two units can be traded. In the large markets, more surplus is also generated on the first unit traded than in the ZI simulations, although the difference in this case is not statistically significant at conventional levels, and is well below the maximum possible surplus. Further, average surplus decreases monotonically for later transactions and is higher than in the ZI simulations for transactions 3 and 4 (but not for 2), which is weakly consistent with the Wilson and Friedman models' predictions.

Given the data in Tables 9.4 and 9.5, I conclude that there is a clear tendency for higher-value buyers and lower-cost sellers to trade first, and for more surplus to be generated in early transactions. Both of these tendencies are stronger than in the ZI simulations, although the difference between experimental and ZI data is only statistically significant with $m = 2$. As such, I conclude that the pattern of transaction partners is closer to the Wilson and Friedman models than to a completely random trading process, although there are still substantial deviations from the idealized pattern that both models suggest.[29]

One aspect of these results worth speculating about is why the trading partner pattern moves closer to the ZI prediction as m increases. One factor, already mentioned, is that as m increases the average distance between traders' redemption values becomes smaller. As a result of this congestion, given a constant noise level in actual transactions, relative to the trading processes that Friedman or Wilson formulate, one would expect to see greater deviations from the trading pattern predicted. Further, systematic increases or decreases in traders' impatience to transact can exaggerate or dampen this effect. From the data in Tables 9.4 and 9.5 it appears that whatever tendencies there are in this direction involve greater impatience to trade, since in going from small to large markets, (i) reductions in the realized correlation coefficients in Table 9.4 are more extreme, particularly for sellers, than the changes in the ZI correlations (by definition ZI traders' impatience does not change as m increases); and (ii) the increase in consumer and producer surplus on the first unit traded is somewhat larger for the ZI simulations than for the experimental data (59.2 versus 50.5).

Table 9.6 reports the covariance of price changes within each auction period for the large markets, and the average absolute size of these price changes.[30] As noted, the ZI model implies a negative covariance for price changes of around -0.50, which is very close to what the ZI simulations yield for the sample data. For sessions 8, 10x and 11x, for which we have a reasonable number of observations, the covariance in the actual price

Table 9.6 Within auction period price changes: CDA, large markets

Experimental session[a]	Covariance of price changes		Average absolute price changes[c] (standard error price change)	
	Actual (prob=0)	ZI[b] (standard error of mean)	Actual	ZI[b]
7 (26)	−0.287 (0.34)	−0.445 (0.04)	25.9 (19.2)	91.8 (73.9)
8 (46)	−0.623 (< 0.01)	−0.506 (0.03)	35.2 (36.0)	92.4 (72.8)
9× (16)	−0.014 (0.97)	−0.532 (0.03)	16.1 (51.4)	93.4 (75.6)
10× (84)	−0.370 (0.02)	−0.487 (0.03)	36.0 (14.5)	86.4 (71.0)
11× (60)	−0.525 (< 0.01)	−0.502 (0.03)	40.4 (52.3)	87.9 (69.9)

Notes: [a] Number of price change observations in parentheses (actual data).
   [b] Average over 20 simulations for each experimental session. Covariance of price changes = $Cov[U_t, U_{t-1}]$ where $U_t = P_t - P_{t-1}$ and P_t is the price of transaction t for a given action period.
   [c] Values are in cents.

changes is statistically significant at better than the 5 per cent level and close to the level found in the ZI simulations. However, as the last two columns in Table 9.6 show, average absolute price changes are substantially larger in the ZI simulations than in the experimental data for each and every experimental session. Thus, while the negative covariances reported are consistent with the ZI benchmark (and earlier reports of negative price change covariances in Cason and Friedman, 1993),[31] average absolute price changes are sharply lower than the ZI simulations suggest.

The negative covariances imply that price increases are followed consistently by price decreases, which is clearly inconsistent with Friedman's model. To the extent that traders can profitably arbitrage these price changes, they are also inconsistent with the Wilson specification. However, arbitrage is difficult in these markets, since traders each have a single unit to trade and no re-sales were permitted. Thus what room for arbitrage there is requires agents to hold off a transaction until after the second unit has traded, since prior to this traders do not have the information required to determine if they could profit by holding out and getting a more favourable (expected) price. But there is a positive probability of not being able to trade after two units have been transacted – with $m = 8$, 6 per cent of the time there is no third unit traded and 34 per cent of the time there is no fourth unit traded (the corresponding values are 19 per cent and 46 per cent if we pool the data from $m = 6$ and 7 with the $m = 8$ sessions). The net expected profit for a trader waiting to cash in on these arbitrage opportunities is approximately $0.11 for the third unit traded and −$0.12 for the fourth unit traded.[32] Thus the arbitrage opportunities are limited and the incentives relatively small – substantially smaller than if one were competing against true ZI traders. As such, I count these results as slightly favouring the Wilson price formation process compared to the Friedman model.

Summary and conclusions

This chapter experimentally investigates a call market trading mechanism (the BBDA) and a CDA trading mechanism for two-sided markets. The primary procedural innovation involves employing fully stochastic supply and demand schedules so that traders' redemption values, along with the CE price and quantity, vary stochastically between auction periods. Fully stochastic supply and demand schedules are necessary to test recent Bayesian–Nash equilibrium models of price formation developed for these trading mechanisms and, arguably, provide the appropriate vehicle for testing the Hayek (1945) hypothesis – that markets are able to achieve close to CE price and quantity resulting from truthful revelation of redemption values.

Both auction mechanisms achieve reasonably high efficiency levels (75 per cent or higher) even in very thin markets with $m = 2$. Consistent with received wisdom from experiments with stationary supply and demand schedules, prices and efficiency in the CDA are close to the CE level, and efficiency is higher than in the sealed bid trading mechanism (the BBDA). However, unlike markets with stationary supply and demand schedules, there is no tendency in the CDA for prices or efficiency to converge to the CE level with increased trader experience within an auction session, and some evidence suggesting reduced efficiency over time in markets with only two buyers and two sellers. The latter may be attributed to increased strategic play with increased experience in the small markets, which would result in reduced efficiency relative to the CE level.

In the BBDA, efficiency levels are consistently closer to those implied by idealized (risk neutral) traders (Satterthwaite and Williams, 1989a, 1989b) than to non-strategic, ZI traders or a fixed price rule. Further, as predicted, efficiency increases with increased numbers of traders. However, in small markets, buyers consistently bid well above the risk-neutral prediction, even with computerized sellers who followed the dominant strategy of offering at cost. Further, real sellers tend to offer at above cost. The latter is largely attributed to (i) the fact that the dominant bidding strategy is far from transparent; and (ii) the relatively small cost associated with deviating from it.

In the CDA with small markets, inefficiencies consistently resulted from too few rather than too many trades relative to the CE level. This is consistent with strategic bidding models of the price formation process (Wilson, 1987; Friedman, 1991) and contrary to the implications of the ZI (Gode and Sunder, 1993) trading algorithm. Markets with larger numbers of buyers and sellers ($m = 8$) consistently produce more trades than the CE level, but fewer trades than implied by the ZI algorithm. Early transactions consistently produced more total surplus than later transactions, as the Wilson and Friedman models suggest. Further, the amount of surplus generated in early trades is consistently greater than implied by the random trading process underlying the ZI algorithm, with these differences being most pronounced in markets with fewer traders. I conjecture that the closer conformity of behaviour to non-strategic trading models with larger numbers of traders is caused by congestion effects in conjunction with random variation in bidding strategies that are not accounted for in the Wilson and Friedman models.

Notes

1 As Holt (1995) notes, these comparisons ignore costs associated with gathering buyers and sellers and the greater time cost of conducting a CDA.

2 There have also been a number of hybrid implementations of random and stationary environments with buyers' and sellers' individual valuations changing from period to period, but with aggregate supply and demand schedules remaining stationary (see Aliprantis and Plott (1991) for example), or with stationary supply and demand schedules whose intercepts shift randomly so that equilibrium price varies while equilibrium quantity is stationary (McCabe *et al.*, 1993).

3 Since this chapter was first written, Cason and Friedman (1998) report an experiment comparing a CDA mechanism and call-market mechanisms similar to the BBDA, in markets with four buyers and four sellers.

4 Variation in number of traders is, among other things, essential to examining the hypothesis that thicker markets promote economic efficiency.

5 The BBDA is a special case (where $k = 1$) of the more general k–DA mechanism described in Rustichini *et al.* (1994) and Satterthwaite and Williams (1993).

6 Smith *et al.* (1982), using stationary supply and demand schedules, implement a sealed-bid-offer trading mechanism that is close in spirit to the k–DA mechanism with $k = 0.5$ (their *PQ* mechanism). However, in this experiment, buyers and sellers had multiple units for sale and had to submit a *single* bid or offer for *all* units, which results in important strategic differences between their game and the single unit k – DA game.

7 I am grateful to my referee for this important counterpoint to Satterthwaite and Williams' argument.

8 Easley and Ledyard (1993) were the first to offer a theoretical model designed to explain price formation in laboratory auction markets. Their model is concerned primarily with how prices and quantities settle down to competitive equilibrium levels *across* trading periods in markets with *stationary* supply and demand schedules.

9 Cason and Friedman (1993) offer an additional qualitative difference between the Friedman and Wilson models: this concerns which traders are involved

in successive improvements of offers prior to the completion of a transaction. I do not pursue this qualitative difference here, in part because I know of no way of determining reliably when 'serious' bids and offers begin, which is what the Wilson model requires.

10 These are budget-constrained ZI traders. Gode and Sunder also simulate performance of unconstrained ZI traders. Without budget constraints, buyers can and do bid above value and sellers can and do offer at below cost, so that a number of inefficient trades occur, resulting in sharply lower efficiency levels than in the budget-constrained case.

11 Under the fixed price rule, traders are assumed to follow the dominant strategy of truthful revelation. In cases where the market does not clear, the computer randomly rations trade on the long side of the market.

12 Kagel and Vogt (1993) report an earlier set of BBDA auctions with live buyers and sellers. The sessions here differ from these earlier ones through an enhanced set of instructions that included diagrams and problems to help subjects to better understand the BBDA rules. There are no major differences between behaviour under the two sets of instructions.

13 The teaching exercise never discussed bidding strategies or models, but was used to illustrate how markets operate and to operationalize the concepts of consumer and producer surplus.

14 This eliminates any possibility of portfolio effects, with subjects hedging their bets between the two auctions.

15 This is not to criticize the software, which is relatively user-friendly. Williams (1980) reports that inexperienced subjects in computerized CDAs make more mistakes and take longer to converge to the CE outcome than in auctions done by hand. This is attributed to subjects needing to gain familiarity with the software. In contrast, the mechanics underlying the BBDA are much easier for subjects to deal with.

16 With $m = 2$ there were occasional trades even when maximum possible consumer and producer surplus was zero (this results from buyers bidding above value or sellers offering below cost). These efficiency losses are not captured in the first measure, but are incorporated into the second.

17 All ZI calculations consist of twenty simulations for each experimental auction period.

18 This is the assumption underlying the calculations reported. This prediction should, of course, be tested experimentally.

19 Treating average differences in individual subject bids with $m = 2$ and $m = 8$ as the unit of observation, there was an increase of $0.14 in session 1 ($t = 1.99, df = 15$, $p < 0.05$, 1-tailed paired t-test), a reduction of $0.02 in session 2 ($t = -0.13$), an increase of $0.09 in market $3 \times$ ($t = 2.18, df = 15, p < 0.025$, 1-tailed paired t-test), and an overall increase in bids of $0.08 ($t = 1.42, df = 47$, $p < 0.10$, 1-tailed, paired t-test).

20 Bidding is above the dominant strategy in one-sided buyers auctions. With symmetry this should translate into bidding below cost for sellers in the BBDA, as in both cases it increases the chances of winning an item (when you do not want to!). Similar to the results reported here, Cason and Friedman (1997) observe bidding above cost for sellers in call markets where they have a dominant strategy to bid their value. But they also find buyers bidding below value when they have a dominant strategy to bid their value, unlike the pattern generally reported in one-sided auctions.

21   The basic reason for the low cost of deviating from the dominant offer strategy is that adoption of the strategy would have no effect on who sells for 91 per cent of all offers. These calculations are for $m = 8$. With $m = 2$, the corresponding costs are $0.07 per auction ($0.16 conditional on trading), and adopting the dominant strategy would have no effect on who sells for 89 per cent of all offers. Calculations are based on each seller, in turn, adopting the dominant strategy, with all others trading as they did.

22   Pooling the data from sessions 9x ($m = 6$) and 10x ($m = 7$) with the $m = 8$ data does not change this conclusion as efficiency averages 95.86 per cent with an even smaller standard error of the mean.

23   With $m = 2$, in 18 out of 136 auctions the number of units traded deviated from the CE quantity. In 14 of these 18 cases *fewer* units were traded than the CE prediction. In contrast, ZI traders failed to trade at the CE quantity in 51 out of 680 simulations. *All* of these involved too many units traded. The Z statistic for the frequency of over-trading in the ZI simulations versus the realized data is 7.05 ($p < 0.01$).

24   Pooling the data from sessions 9x ($m = 6$) and 10x ($m = 7$) with the $m = 8$ data results in even fewer trades relative to the ZI benchmark as 9x generates fewer trades than even the CE benchmark; that is, it is closer to the $m = 2$ pattern than the $m = 8$ pattern.

25   Pooling data from 9x and 10x with the $m = 8$ data leaves these conclusions unaffected as prices averaged $0.125 versus $0.122 for the $m = 8$ sessions alone.

26   Friedman (1992) advocates averaging transactions prices within an auction period and then computing average absolute price differences across auction periods to put the data on a more comparable footing relative to call markets, which impose a single price for all trades. This was not done here. Doing so makes very little difference when $m = 2$, since there is usually only one unit traded. For the larger markets this results in eliminating the difference between ZI price deviations and realized price deviations, and results in marginally lower prices than observed in the BBDAs.

27   Once again, the regression analysis shows no consistent time trends. The non-parametric price deviation analysis yields median price deviations of zero in all the small markets, and zero (or pennies above it) in the large markets. Of all price deviations, 52.9 per cent (18/34) occur in the second-half sessions with $m \geq 6$, and 63.6 per cent (7/11) in the $m = 2$ sessions. Neither proportion is significant at the 10 per cent (two-tailed) level.

28   This is based on twenty simulations for each small market. For sessions 8 and 10x, additional simulations were run to determine the likelihood that the sellers' rank order correlation of 1.0 resulted from an unusual distribution of costs. In Session 8, 6 simulations out of 100 resulted in a rank order correlation of 1.0, and in session 10x, 2 simulations out of 100 produced this result. Thus, it is quite unlikely that the correlations for 8 and 10x were a result of chance factors alone.

29   Our conclusion in this respect differs from Cason and Friedman (1993 and 1996). They report weak evidence that higher-value buyers and lower-cost sellers tend to trade first, that gains from trade decrease as more units are transacted, and they conclude that this pattern is consistent with the one implied by the ZI algorithm. One reason our results differ from Cason and Friedman is that they did not have any data for small markets ($m = 2$) where the experimental data conform more closely to the Friedman and Wilson models' predictions.

30 Cov (u_t, u_{t-1}) where $u_t = P_t - P_{t-1}$ and P_t is the price of transaction t within a given auction period. Note, under the CDA rules there was no specialists book, so that after each transaction all existing bids and offers were cancelled, leaving no room for the specialists' book to contribute to the negative price change correlations reported.

31 Although the data here shows no evidence for reductions in these negative price correlations with experience, Cason and Friedman (1996) argue that a meta-analysis, using their data, ours and some additional data, shows clear evidence of such an effect.

32 As a buyer or seller there is only a 50 per cent chance of profiting on the third unit as prices may increase or decrease. This yields an expected gain of $0.175 (50 per cent of the average absolute price difference). From this I have subtracted half of the expected total surplus realized on the third unit traded (Table 9.5) after multiplying by the probability of not trading with $m = 8$. Similar calculations are made for the fourth unit. These are approximate, 'back of an envelope' calculations.

References

Aliprantis, C. D. and Plott, C. R. (1991) 'Experiments with Competitive Equilibrium in Overlapping Generations Economies', Mimeo, California Institute of Technology, Pasadena, Calif.

Battalio, R. C., Kogut, C. A. and Meyer, D. J. (1990) 'Individual and Market Bidding in a Vickrey First-price Auction: Varying Market Size and Information', in L. Green and J. H. Kagel (eds), *Advances in Behavioural Economics*, Vol. 2, Norwood, NJ, Ablex.

Cason, T. N. and Friedman, D. (1993) 'Empirical Analysis of Price Formation in Double Auction Markets', in D. Friedman and J. Rust (eds), *The Double Auction Market: Institutions, Theory and Evidence*, SFI Studies in Sciences of Complexity, Redwood City, Calif., Addison-Wesley.

Cason, T. N. and Friedman, D. (1996) 'Price Formation in Double Auction Markets', *Journal of Economic Dynamics and Control*, 20, 1307–37.

Cason, T. N. and Friedman, D. (1997) 'Price Formation in Single Call Markets', *Econometrica*, 65, 311–46.

Cason, T. N. and Friedman, D. (1998) 'Price Formation and Exchange in Thin Markets: A Laboratory Comparison of Institutions', in P. Howitt, E. de Antoni and E. Liezonhufvud (eds), *Money, Markets and Method: Essays in Honor of Robert W. Clower* Northhmpton, Mass., Edward Elgar.

Cox, J. C., Smith, V. L. and Walker, J. M. (1985) 'Expected Revenue in Discriminative and Uniform Price Sealed Bid Auctions', in V. L. Smith (ed.), *Research in Experimental Economics*, Vol. 3, Greenwich, Conn., JAI Press.

Dyer, D., Kagel, J. H. and Levin, D. (1989) 'Resolving Uncertainty About the Number of Bidders in Independent Private-Value Auctions: An Experimental Analysis', *Rand Journal of Economics*, 20, 268–79.

Easly, D. and Ledyard, J. O. (1993) 'Theories of Price Formation and Exchange in Oral Auctions', in D. Friedman and J. Rust (eds), *The Double Auction Market: Institutions, Theory and Evidence*, SFI Studies in Sciences of Complexity, Redwood City, Calif., Addison-Wesley.

Friedman, D. (1991) 'A Simple Testable Model of Double Auction Markets', *Journal of Economic Behaviour and Organizations*, 16, 47–70.

Friedman, D. (1993) 'How Trading Institutions Affect Financial Market Performance: Some Laboratory Evidence', *Economic Inquiry*, 31, 410–35.

Gode, D. and Sunder, S. (1993) 'Allocative Efficiency of Markets with Zero-Intelligence Traders: Market as a Partial Substitute for Individual Rationality', *Journal of Political Economy*, 101, 119–37.

Hayek, A. (1945) 'The Use of Knowledge in Society', *American Economic Review*, 35, 519–30.

Holt, Charles A. (1995) 'Industrial Organization: A Survey of Laboratory Research', in J. H. Kagel and A. E. Roth (eds), *The Handbook of Experimental Economics*, Princeton, NJ, Princeton University Press.

Kagel, J. H. (1995) 'Auctions: A Survey of Experimental Research', in J. H. Kagel and A. E. Roth (eds), *The Handbook of Experimental Economics*, Princeton, NJ, Princeton University Press.

Kagel, J. H. and Levin, D. (1993) 'Independent Private Value Auctions: Bidder Behavior in First, Second and Third-Price Auctions with Varying Numbers of Bidders', *Economic Journal*, 103, 868–79.

Kagel, J. H. and Roth, A. E. (1992) 'Theory and Misbehavior in First-Price Auctions: Comment', *American Economic Review*, 82, 1379–91.

Kagel, J. H. and Vogt, W. (1993) 'Buyers' Bid Double Auctions: Preliminary Experimental Results', in D. Friedman and J. Rust (eds), *The Double Auction Market: Institutions, Theory and Evidence*, SFI Studies in Sciences of Complexity, Redwood City, Calif., Addison-Wesley.

Kagel, J. H., Harstad, R. M. and Levin, D. (1987) 'Information Impact and Allocation Rules in Auctions with Affiliated Private Values: A Laboratory Study', *Econometrica*, 55, 1275–304.

McCabe, K., Rassenti, S. and Smith, V. (1993) 'Designing a Uniform Price Double Auction: An Experimental Evaluation', in D. Friedman and J. Rust (eds), *The Double Auction Market: Institutions, Theory and Evidence*, SFI Studies in Sciences of Complexity, Redwood City, Calif., Addison-Wesley.

Myerson, R. and Satterthwaite, M. (1983) 'Efficient Mechanisms for Bilateral Trade', *Journal of Economic Theory*, 29, 265–81.

Rustichini, A., Satterthwaite, M. and Williams, S. (1994) 'Convergence to Efficiency in a Simple Market with Incomplete Information', *Econometrica*, 62, 1041–64.

Satterthwaite, M. A. and Williams, S. R. (1989a) 'The Rate of Convergence to Efficiency in Buyer's Bid Double Auction as the Market Becomes Large', *Review of Economic Studies*, 56, 477–98.

Satterthwaite, M. A. and Williams, S. R. (1989b) 'Bilateral Trade with the Sealed Bid k-Double Auction: Existence and Efficiency', *Journal of Economic Theory*, 48, 107–33.

Satterthwaite, M. A. and Williams, S. R. (1993) 'Theories of Price Formation and Exchange in k-Double Auctions', in D. Friedman and J. Rust (eds), *The Double Auction Market: Institutions, Theory and Evidence*, SFI Studies in Sciences of Complexity, Redwood City, Calif., Addison-Wesley.

Smith, V. L., Williams, A. W., Bratton, W. K. and Vannoni, M. G. (1982) 'Competitive Market Institutions: Double Auction vs. Sealed Bid-Offer Auctions', *American Economic Review*, March, 72, 58–77.

Williams, A. W. (1980) 'Computerized Double Auction Markets: Some Initial Experimental Results', *Journal of Business*, 53, 235–8.

Wilson, R. (1987) 'On Equilibria of Bid-Ask Markets', in G. Feiwel (ed.), *Arrow and the Ascent of Modern Economic Theory*, London, Macmillan, 375–414.

10
Multistage Sealed-bid k-Double-Auctions: An Experimental Study of Bilateral Bargaining

*James E. Parco, Amnon Rapoport, Darryl A. Seale, William E. Stein and Rami Zwick**

Introduction

Mechanisms that structure bargaining between a potential buyer and potential seller are of perpetual interest to both academics and practitioners. An 'ideal bargaining mechanism' would enable bargaining parties to reach an agreement 100 per cent of the time when a deal is possible and collectively realize all possible gains from trade. However, the only individually rational way to achieve maximum efficiency would be for the bargaining parties truthfully to reveal their private valuation, which is typically not in the players' unilateral best interests. The result: bargaining parties end up 'walking away' from a potentially profitable deal, especially when there is a narrow trading range. The single-stage sealed-bid k-double-auction mechanism is no exception. This particular trading mechanism requires that both buyer and seller simultaneously and independently submit an offer to buy/sell (the seller submits an ask s and the buyer submits a bid b). Each player has an independent predetermined reservation value for the indivisible object of potential trade denoted by v_s and v_b, for the seller and buyer, respectively. Only when both parties have made their offers, are the offers revealed. If the buyer is willing to pay at least as much as the seller is asking, trade occurs at price p and gains from trade for the seller and buyer are $p - v_s$ and $v_b - p$, respectively. If there is no trade, then neither trader incurs any cost and the gain for each is taken to be zero. Trade occurs at the price $p = kb + (1 - k)s$, if $b \geqslant s$. If $b < s$, then no trade takes place and the buyer pays the seller nothing. $k \, (0 \leqslant k \leqslant 1)$ is simply a parameter of the mechanism that determines the trade price p. In this chapter we consider the case of $k = 1/2$, commonly referred to as the 'midpoint trading rule' and implying that $p = (b + s)/2$

* Our research was supported by a grant from the Hong Kong Grants Committee (Project HKUST6225/99H). Special thanks to Maya Rosenblatt for her outstanding assistance with data collection.

(see Satterthwaite and Williams, (1989) and Parco (2003) for discussion of the cases $k = 0$ and $k = 1$.)

Chatterjee and Samuelson (1983) constructed an equilibrium for this trading mechanism in which each player's ask or bid is a (possibly piecewise) linear function of his/her reservation value. At about the same time, Myerson and Satterthwaite (1983) showed that this linear equilibrium has *ex ante* the highest expected gain from trade of any equilibrium of any (suitably restricted) bargaining mechanism (Leininger *et al.*, 1989). These two seminal papers were the first to explore the nature of equilibria in the double-auction and establish properties that all equilibria must satisfy. They have had a strong impact on the study of two-person bargaining under two-sided incomplete information, and have given rise to considerable theoretical research (for example, Leininger *et al.*, 1989; Satterthwaite and Williams, 1989, 1993; Brams and Kilgour, 1996; Ausubel *et al.*, 2002). In addition, the 1983 paper by Chatterjee and Samuelson has stimulated several experimental studies of the double-auction mechanism (for example, Radner and Schotter, 1989; Daniel *et al.*, 1998; Rapoport *et al.*, 1998; Seale *et al.*, 2001; Parco, 2002; Parco and Rapoport, 2003) that have provided qualified support to the linear equilibrium.

The information assumptions of the double-auction are as follows. Each trader's reservation value is assumed to be a random variable whose value is included in some interval. The reservation values of the seller and buyer, v_s and v_b, are assumed to be drawn independently from the distribution functions F and G defined over the respective intervals $[\alpha_s, \beta_s]$ and $[\alpha_b, \beta_b]$. These two distributions are common knowledge to the traders. However, the realization of v_s is private information to the seller, and that of v_b private information to the buyer. Hence, when trade commences, each trader knows his/her own reservation value with certainty but only the distribution of the reservation values of the other trader.

Assume that both traders are risk-neutral.[1] A pure strategy for the seller is a function $S(\bullet)$ that specifies an ask $s = S(v_s)$ for each of his/her possible reservation values. Similarly, $B(\bullet)$ is a pure strategy for the buyer that specifies a bid $b = B(v_b)$ for each of his/her possible reservation values. A pair of strategies (S, B) is a Bayesian–Nash equilibrium if B is a best reply to S and, simultaneously, S is a best reply to B. S is a best reply to B if, conditional on the seller's knowledge of the buyer's distribution G, the buyer's bid strategy B, and the realization of v_s, $S(v_s)$ maximizes his/her expected gain from trade for each value of v_s. The buyer's best-response function is similarly defined.

It is known that the double-auction has multiple equilibria (Leininger *et al.*, 1989). Chatterjee and Samuelson constructed a piecewise linear equilibrium solution (LES) for this two-sided incomplete information bargaining game. It is known to be *ex ante* efficient when F and G are each uniform on the interval [0, 1]. Experimental evidence from several bargaining experiments with uniform distributions sharing the same support (Radner and Schotter, 1989;

Valley *et al.*, 2002) or with uniform distributions with unequal and overlapping supports (Daniel *et al.*, 1998, Rapoport *et al.*, 1998; Seale *et al.*, 2001) have provided qualified support for the LES.

The sealed-bid *k*-double-auction is a very simple mechanism for single-round, two-person bargaining under the simultaneous protocol of play. In commenting on it, Chatterjee and Samuelson (1983) conceded that this mechanism 'fails to capture the pattern of reciprocal concessions observed in everyday life' (p. 849). Nevertheless, they justified it as a useful idealization and a starting point for additional investigations, and then proceeded to write: 'A more general model would allow the bargainers multiple rounds in which to exchange offers (potentially incurring "transaction" costs in the process)' (ibid., p. 849). This chapter reports the results of extending the double-auction from one to two rounds, with simultaneous play on each round. We focus on the asymmetric case – most common in practice – where the distributions of the reservation values of the buyer and seller do not share the same support and, consequently, one trader may have an information advantage (in the sense described below) over the other.

The next section presents and discusses the extension of the double-auction to two-round bargaining, where traders make offers simultaneously in each round. Our main hypothesis is that, if payoffs are not discounted, then in the first round of the bargaining process the traders will reveal limited information about their reservation values and will not strike a deal, whereas in the second and final stage they will play the Chatterjee–Samuelson (C–S) linear equilibrium. We propose to test this hypothesis in an iterated two-stage double-auction played by several groups of subjects who differ from one another in their strategic sophistication. Our second hypothesis is that strategically sophisticated bargainers will learn to reveal less information in Round 1 faster than inexperienced bargainers. The third section describes the experimental procedure and the different experimental conditions, and the fourth presents the results on both the aggregate and individual levels. The fifth section concludes.

The two-stage sealed-bid *k*-double auction

Multistage bargaining exhibits sequential rationality in a way that single-stage mechanisms such as the double-auction does not. To illustrate this, consider the case where F and G are each distributed uniformly on the interval [0, 100]. Here, the LES specifies the following strategies for the buyer and seller:[2]

$$b = \begin{cases} \text{at most } v_b & \text{if } 0 \leqslant v_b \leqslant 25 \\ 25/3 + 2v_b/3 & \text{if } 25 < v_b \leqslant 100 \end{cases} \tag{10.1}$$

and

$$s = \begin{cases} 25 + 2v_s/3 & \text{if } 0 \leqslant v_s < 75 \\ \text{at least } v_s & \text{if } 75 \leqslant v_s \leqslant 100 \end{cases} \tag{10.2}$$

Now, consider the case where $v_s = 37.5$ and $v_b = 49$. Since $v_b \geqslant v_s$, there is a potential gain to be realized if trade occurs. However, Equation (10.1) and (10.2) imply that in equilibrium $s = 50$ and $b = 41$, and therefore trade will not occur. Because the C–S strategies are invertible, if the buyer and seller learn each other's bid it is common knowledge to them after they fail to trade that some unrealized gain from trade exists. Then, contrary to the rules of the single-stage double-auction, they have an incentive to participate in at least another round of bargaining. Linhart *et al.* (1992) conclude, 'Consequently, if bargaining is regarded as a voluntary activity, the double-auction rule that bargaining is restricted to a single ask/bid pair lacks credibility' (p. 7). We therefore consider the case of a multistage double-auction with N rounds, where the bids/asks on each round are made simultaneously. The value of N is assumed to be common knowledge. This multistage *simultaneous* double-auction (MSDA) differs from *sequential* bargaining with alternating offers, as studied by Rubinstein (1982, 1985).

The MSDA asymmetric case

The MSDA experiment (described in the following section) was limited to two rounds of offers and designed as an asymmetric double-auction, where the buyer knew more about the reservation values of the seller than the seller knew about the reservation values of the buyer. F was uniform over [0, 100] and G uniform over [0, 200]. Because the support of F was included in the wider support of G (and both were uniform), the buyer is considered to have had an 'information advantage' (Rapoport *et al.*, 1998) over the seller.

The LES for the seller in the *single-stage* double-auction (see Daniel *et al.*, (1998) and Rapoport and Fuller (1995), who used the same distributions F and G) is linear in v_s:

$$s = S(v_s) = 50 + 2v_s/3 \qquad \text{if } 0 \leqslant v_s \leqslant 100 \tag{10.3}$$

The LES for the buyer is piecewise linear in v_b:

$$b = B(v_b) = \begin{cases} \text{at most } v_b & \text{if } 0 \leqslant v_b \leqslant 50 \\ 50 + 2(v_b - 50)/3 & \text{if } 50 \leqslant v_b \leqslant 150 \\ 350/3 & \text{if } 150 \leqslant v_b \leqslant 200 \end{cases} \tag{10.4}$$

Because the seller never asks below 50, it does not matter what the buyer does if $0 \leqslant v_b < 50$. Figure 10.1 displays the LES functions in Equations (10.3) and (10.4) for the buyer and seller.

Now we write the general equations for computing the LES for uniform F and G. Assume that F is distributed uniformly on the interval $[0, \beta_s]$, G is distributed uniformly on the interval $[\alpha_b, 200]$, and $0 \leqslant \alpha_b \leqslant \beta_s \leqslant 200$. Assume that $k = 1/2$, and denote the equilibrium solutions for the seller and buyer by S^* and B^*, respectively. Then (see Stein and Parco, 2001):

$$S^*(v_s) = 50 + 2(\alpha_b - 50)/3 \quad \text{if } 0 \leqslant v_s < \max(50, \alpha_b) - 50 \qquad (10.5)$$

$$S^*(v_s) = 50 + 2v_s/3 \quad \text{if } \max(50, \alpha_b) - 50 \leqslant v_s \leqslant \min(150, \beta_s) \quad (10.6)$$

$$B^*(v_b) = 50 + 2(v_b - 50)/3 \quad \text{if } \max(50, \alpha_b) \leqslant v_b \leqslant \min(150, \beta_s) + 50 \quad (10.7)$$

$$B^*(v_b) = 50 + 2\beta_s/3 \quad \text{if } \min(150, \beta_s) + 50 < v_b \leqslant 200 \qquad (10.8)$$

It is easy to verify that if $\alpha_b = 0$ and $\beta_s = 100$, as in our experiment, then Equations (10.5)–(10.8) reduce to Equations (10.3) and (10.4).

Consider a two-stage asymmetric MSDA with $F = \text{Uniform } [0, 100]$ and $G = \text{Uniform } [0, 200]$. Assume that players restrict themselves to Bayesian updating that yields a uniform distribution: if the buyer bids $b^* \geqslant 0$ and the seller asks s^*, then the seller revises the prior distribution of v_b to $G' = \text{Uniform} [b^*, 200]$ and the buyer revises the prior of v_s to $F' = \text{Uniform} [0, \min (100, s^*)]$.

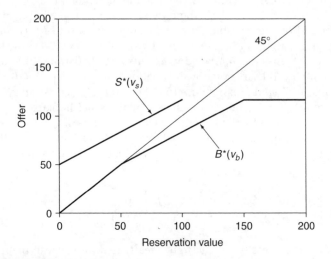

Figure 10.1 LES bid and ask functions

The following Proposition characterizes one possible equilibrium under these assumptions.

PROPOSITION 1:
A subgame-perfect equilibrium is:

Stage 1: *The buyer bids* $0 < b^* \leqslant 50$ *and the seller asks* $s^* \geqslant 350/3 = 116.67$
   (b^* can vary with v_b and s^* with v_s).
Stage 2: *The players use the LES strategy given in Equations (10.3)–(10.4).*

PROOF:
First we consider the buyer's response to the seller's strategy as stated in the Proposition. We already know that the Stage 2 strategy is an equilibrium for that subgame. In Stage 1 the buyer bids some value b^*. The seller assumes that the buyer is individually rational so that $b^* \leqslant v_b$. We assume the seller will update the information about the buyer's v_b to the new interval $[b^*, 200]$. Using the stated strategy, the seller assumes that the distribution G' is still uniform over this interval. (The precise distribution depends on the method the buyer uses to select his/her bid b^*.) The buyer knows that the seller is using a uniform distribution for G'.

We now want to see if there is an incentive for the buyer to deviate from his/her strategy. For any $0 < b^* \leqslant 50$, there is no chance for a Stage 1 agreement. However, we also need to determine if the revised information in G' will cause the seller to play a Stage 2 strategy different from the LES in Equations (10.3)–(10.4). Given the distributions F and G', the solution of Equations (10.5) to (10.8) is the same as Equations (10.3) and (10.4). To verify this fact, notice that the only change is that the value of α_b, the lower bound of the distribution of v_b, changes from 0 to b^*. However, α_b appears on the right side of Equations (10.5), (10.6) and (10.7) only through the expression $\max(50, \alpha_b)$. For any $\alpha_b = b^* \leqslant 50$, Equation (10.5) does not hold for any v_s and can be ignored. Equation (10.6) reduces to (10.3), while Equations (10.7) and (10.8) reduce to (10.4). So the seller's strategy will not change in Stage 2. Therefore, no change by the seller in Stage 2 and so no incentive to change by the buyer in Stage 1.

On the other hand, if the buyer bids $50 < b^* \leqslant 350/3$ in Stage 1, then the seller's strategy in Stage 2 will be given by Equations (10.5) and (10.6), with $\alpha_b = b^*$:

$$s = \begin{cases} 50 + 2(b^* - 50)/3 & \text{if } 0 \leqslant v_s < b^* - 50 \\ 50 + 2v_s/3 & \text{if } b^* - 50 \leqslant v_s \leqslant 100 \end{cases} \tag{10.9}$$

This function can easily be seen to be greater than or equal to the function in Equation (10.3) for each v_s. Thus the seller will never decrease, and will possibly *increase*, the asked value in Stage 2 compared to Equation (10.3). Furthermore, Equation (10.9) yields a maximum asking price of 350/3. Therefore the buyer concludes that (i) trying to make an agreement in Stage 1 (by bidding at least 350/3) is not advantageous since the seller will ask a lower price in Stage 2; and (ii) bidding $50 < b^* < 350/3$ as a 'signal' to the seller, knowing that an agreement will not be made in Stage 1, is also not advantageous because it can cause the seller to increase the asking price in Stage 2. Therefore the buyer is better off with $b^* \leqslant 50$ in Stage 1 and waiting until Stage 2 before making a serious bid.

Similar logic holds for the seller. Could the seller be better off asking less than 350/3 in Stage 1? In order to try making a Stage 1 deal, the seller must ask 50 or less. Let us assume the most favourable scenario for the seller, consistent with the buyer's strategy: the buyer bids $b^* = 50$ for all v_b and the seller asks 50 for some v_s. (We are even assuming that the buyer violates individual rationality to make the most favourable case for the seller.) The profit to the seller will then be $50 - v_s$. The seller certainly will not want to make a deal if $v_s > 50$. If $v_s = 0$, then the seller gets a profit of 50. We need to compare this with the expected profit from a Stage 2 deal. From Equations (10.3) and (10.4), and integrating the profit function over the region where a deal will take place, after some algebra, it can be shown that a seller with $v_s = 0$ will expect to make a profit of 325/6. Since this exceeds the (optimistic) profit of 50 from a Stage 1 deal, this seller should wait. For any other $v_s < 50$ a similar result can be shown: the optimistic Stage 1 profit $50 - v_s$ is less than the expected profit in Stage 2. So no Stage 1 deal should be attempted.

Now we need to check the effect of an unsuccessful attempt by the seller in Stage 1 on the buyer's Stage 2 bid. If the seller asks $s^* \geqslant 100$ in Stage 1, then the buyer gains no information to update his/her prior over v_s from $[0, 100]$ and thus the buyer's Stage 2 bid is still given by Equation (10.4). Instead, if the seller asks for $s^* < 100$ then, by assumption, the buyer will update F to be uniform on $[0, s^*]$. Then the buyer's Stage 2 strategy is given by Equations (10.7) and (10.8), with $\beta_s = s^* < 100$:

$$b = \begin{cases} 50 + 2(v_b - 50)/3 & \text{if } 50 \leqslant v_b < s^* + 50 \\ 50 + 2s^*/3 & \text{if } s^* + 50 \leqslant v_b \leqslant 200 \end{cases} \qquad (10.10)$$

It is easy to see that Equation (10.10) will yield a smaller bid than Equation (10.4) for each value of v_b. This is not to the seller's advantage, so $s^* < 100$ will be avoided. (While any $s^* \geqslant 100$ will give a best response to the buyer's bid we saw previously, $s^* \geqslant 350/3$ was used to make sure the buyer did not change his/her Stage 1 strategy.) Thus, the seller cannot profitably deviate from the proposed strategy, concluding the proof.

Method

Design considerations

Because our major purpose was to determine whether the addition of a second round of bargaining alters the results of the single-stage double-auction, we included a control condition in which subjects played the single-stage double-auction game. Table 10.1 describes the experimental design. In total, five groups of twenty subjects each (a total of 100 subjects) participated in three different conditions: two groups of subjects in a Baseline (control) condition; two in the Inexperienced condition; and one in the Sophisticated condition. The Baseline and Inexperienced conditions each included two separate groups of twenty subjects each. The Sophisticated condition included only a single group.[3]

The question is often asked regarding how well the results of experiments that almost always recruit college undergraduates to play for a couple of hours for small or modest financial stakes generalize to workers in firms, corporations forming alliances, firms bargaining with each other and so on. As Camerer (2003) noted, these doubts about the generalizability and relevance of the experimental findings should be taken as a demand for more elaborate experiments, with subjects drawn from other populations and games played for higher stakes, rather than a wholesale dismissal of the experimental method. This criticism applies directly to two-person bargaining. Casual observation suggests that people differ from one another in the way they conduct trade and the strategies they employ in bargaining. Sophisticated bargainers in the open markets in Hong Kong and other countries in South-East Asia typically behave quite differently from inexperienced ones and are easy to spot in the marketplace. Indeed, a major purpose of university courses on bargaining and negotiation is to inform students of various bargaining mechanisms and tactics, teach them what to expect in a range of bargaining situations, and instruct them how to negotiate effectively. Responding to this challenge, we have made an attempt to recruit subjects from two different populations. One population includes undergraduate students, most commonly used in economic experiments, who are not familiar with the double-auction mechanism and play for small financial gains. The other population includes strategically sophisticated Ph.D.

Table 10.1 Experimental design for the multistage bargaining game

Condition*	No. of stages	No. of groups	No. of trials
Baseline	1	2	50
Inexperienced	2	2	50
Sophisticated	2	1	25

Notes: *In all three conditions, $F \sim$ uniform $[0, 100]$, $G \sim$ uniform $[0, 200]$, and $k = 1/2$. Each group consists of ten buyers and ten sellers.

students in economics who are familiar with the double-auction and play for higher stakes.

Subjects

The Baseline condition included two groups of undergraduate students, each comprising twenty subjects. All the students volunteered to participate in the experiment for payoff contingent on performance. In addition, they received partial class credit.

The two Inexperienced groups included subjects who were recruited from the same undergraduate population through the same standard recruiting procedures. None of the subjects had participated previously in bargaining experiments using the double-auction mechanism. These subjects were recruited through advertisements placed on public boards on campus promising payoff contingent on performance. They also received $5.00 show-up fee for their participation.

The Sophisticated condition included twenty participants in a summer workshop on experimental economics sponsored by the International Foundation of Experimental Economics and conducted at the Economic Science Laboratory at the University of Arizona. The workshop lasted for a week. The participants either already had a Ph.D. in economics or were working towards it. All of them were knowledgeable in game theory, experienced with strategic thinking, and familiar with games under incomplete information. Most of them had participated or supervised various laboratory experiments in economics. Participants in the workshop were not reimbursed for their travel expenses. Rather, they were afforded the opportunity to recover them during the workshop through participation in a sequence of experiments, typically two a day, with payoffs contingent on performance. Consequently, subjects in this condition were paid 2.5 times higher than the Inexperienced subjects, with a mean payoff of $61.00. It is interesting to note that the Sophisticated subjects took more than twice the time to register their bids and asks, and therefore completed only 25 (rather than 50) trials during a slightly longer session. There is no doubt that they were highly motivated, approached the game seriously, deliberated on their decisions for much longer than did the Inexperienced subjects, and attempted to make rational decisions.

Procedure

The same procedure was used in all three conditions. On arrival, each subject was asked to draw a chip from a bag containing chips numbered 1–20. Subjects who drew a number between 1 and 10 were assigned the roles of buyer, and those drawing a chip between 11 and 20 were assigned the roles of seller. Trader roles were fixed during the entire duration of the experiment. Communication between the subjects was strictly forbidden.

The instructions (see Appendix on page 231) explained and illustrated the MSDA. The subjects were instructed that they would participate in 50 trials, and that their bargaining partners would be changed randomly from one trial to another. The random assignment design was used in order to minimize reputation effects. Knowing the number of trials (50) and number of traders of the opposite role (10), a subject could figure out that s/he would be paired with the same subject about five times. However, the identity of the bargaining partner in any particular trial was not disclosed. Rapoport and Fuller (1995), Daniel *et al.* (1998), Rapoport *et al.* (1998) and Seale *et al.* (2001) implemented this type of procedure in previous studies.

All 50 trials (25 for the Sophisticated condition) were structured in exactly the same way. At the beginning of Round 1, the computer assigned each subject his or her reservation value. Then, the buyer (seller) was prompted to state his/her bid (ask). Subjects were required to confirm their offers. They could review their previous responses and outcomes in earlier trials by viewing a separate screen. Once all the twenty subjects had stated their offers in Round 1, the computer compared them and determined for each pair separately whether an agreement had been reached. At the end of Round 1, each subject was informed of his/her decision, the decision of his/her bargaining partner in that round, and, if an agreement had been reached, the trade price p and the earnings for the trial. In all three conditions, if an agreement was reached in Round 1, then the trial terminated with traders realizing their profit. In the two-stage conditions, if no agreement was reached in Round 1, then the subjects moved to the second and final (costless) round (with the same reservation values), which was structured in exactly the same way. No public information about the decisions and outcomes of the other eighteen subjects in the session was ever disclosed. The experiment was self-paced, with the slowest subject in each trial dictating the pace. Each session lasted two hours (2.5 hours for the Sophisticated condition). Once all the trials were completed, each subject was paid separately, contingent on his/her performance, and dismissed.

Results

This section is organized as follows. First, we compare groups *within condition*. Finding no differences, we aggregate the results over groups and compare the conditions to one another. Our results show that strategic sophistication matters, and therefore that aggregation across conditions is not justified. Second, for each round separately, we report individual bid and ask functions, and then present the aggregate results. Our focus is on individual behaviour. A major finding is that, to varying degrees, different populations reveal their reservation values in Stage 1 during the early trials of the session. However, in support of the main hypothesis of the chapter, over the course of the session most subjects learn not to disclose information in Round 1.

Within and between treatment comparisons

For each buyer in the Baseline condition we computed his/her mean bid over the 50 trials. Comparison of Groups 1 and 2 failed to reject the null hypothesis of equality in mean bids ($t = 1.23, p > 0.1$). Similarly, for each seller in the Baseline condition we computed the mean ask across trials. Comparison of Groups 1 and 2 also failed to reject the null hypothesis of equality in mean offers ($t = 0.55, p > 0.2$). The Baseline condition replicates Daniel *et al.*'s Experiment 1 (1998), where the same procedure and same parameter values were used, and the subjects were drawn from a similar population of inexperienced undergraduates. Comparison of the Baseline condition with Daniel *et al.*'s Experiment 1 in terms of number of deals, deviations from the LES, and achieved surplus also yielded no significant differences (see Parco, 2002).

As mentioned earlier, given that $F \sim$ Uniform [0, 100] and $G \sim$ Uniform [0, 200], the LES implies an intercept $y = 50$ and slope $m = 2/3$ for the seller's ask function across all his/her reservation values. The buyer's bid function is piecewise linear with slopes $m = 1.0, m = 2/3$, and $m = 0$ for the ranges $0 \leqslant v_b < 50, 50 \leqslant v_b < 150$, and $150 \leqslant v_b < 200$, respectively, and corresponding intercepts $y = 0, y = 50$, and $y = 116.67$. Simple linear regression was used to test these hypotheses for the sellers, and a spline regression was used for the buyers. The spline regression method uses ordinary least squares to find the best-fitting piecewise linear function by adjusting the 'knots' joining the segments at the predicted reservation values of $v_b = 50$ and $v_b = 150$ while simultaneously adjusting the slopes of the three line segments.

Row 1 of Table 10.2 presents the results of these analyses (slope and intercept) for the Baseline condition. Also presented in the table are the predicted LES slopes and intercepts (last row). Inspection of the table shows that the LES accounts for the buyers' mean bids in the Baseline condition remarkably well (compare, for example, the observed slope $m = 0.96$ to the predicted slope $m = 1.00$ for the range $0 \leqslant v_b < 50$, the observed slope $y = 0.58$ to the

Table 10.2 Spline regression coefficients for buyers and sellers on Round 2 by experimental condition for the present study and for Experiment 1 of Daniel *et al.* (1998) (DSR)

| | Buyers | | | | | | | Sellers | | |
| | $0 \leqslant v_b < 50$ | | $50 \leqslant v_b < 150$ | | $150 \leqslant v_b \leqslant 200$ | | | $0 \leqslant v_s \leqslant 100$ | | |
Condition	Slope	Interc.	Slope	Interc.	Slope	Interc.	R^2	Slope	Interc.	R^2
Baseline	0.96	1.00	0.58	48.8	0.17	106.6	0.76	0.72	35.2	0.32
Sophisticated	1.09	−3.4	0.44	51.2	−0.03	95.6	0.90	0.59	40.1	0.58
Inexperienced	0.99	−0.6	0.53	48.8	0.09	102.2	0.76	0.60	44.0	0.39
LES	**1.00**	**0**	**0.67**	**50.0**	**0**	**116.7**		**0.67**	**50.0**	

predicted slope $y = 2/3$ for the range $50 \leqslant \nu_b < 150$, and the observed slope $y = 0.17$ to the predicted slope $y = 0.$). In contrast, the sellers were more timid than predicted, as judged from the intercept ($y = 35.2$ rather than the predicted $y = 50$) of their ask function. Figure 10.2 displays the best fitting

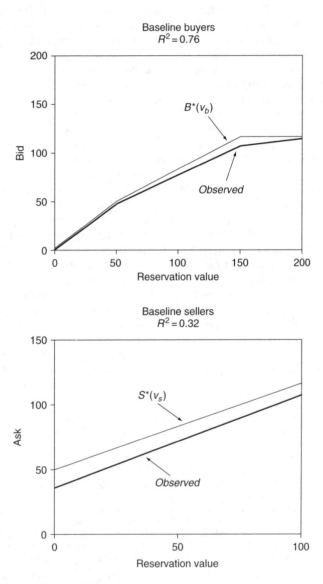

Figure 10.2 Best fitting OLS regression lines

OLS regression functions for the buyers and sellers in the Baseline condition. For each trader class, the regression lines are displayed separately for trials 1–25 and trials 26–50. It is evident that there is not much difference between the two blocks of twenty-five trials; that the LES accounts for the buyers' bids very well; and that the sellers' asks fall closer to the LES line than the 45-degree line that describes truth-telling behaviour. The latter result replicates the findings of DSR.

We also observe no difference between the two groups in the Inexperienced condition. For the two groups of buyers there was no significant difference between the mean number of trades in Round 1 ($p > 0.5$) and Round 2 ($p > 0.2$). Similarly for the two groups of sellers, there was no significant difference in the number of agreements reached in Round 1 ($p > 0.3$) and Round 2 ($p > 0.2$). A comparison of the two groups in terms of mean payoffs for the entire session also showed no significant difference for the buyers ($p > 0.4$) and sellers ($p > 0.1$). Consequently, the data from the two Inexperienced groups were combined.

Comparisons of the Inexperienced and Sophisticated conditions revealed highly significant differences ($p < 0.001$) between them in the number of deals made in either Round 1 or Round 2, and in the mean individual payoffs (stated in terms of the experimental currency). We shall describe and remark on these differences after first examining the individual data.

Individual bids and asks: Round 1

Figure 10.3 exhibits the fifty Round 1 bids for each of the twenty Inexperienced buyers. The reservation values (the same for all buyers but presented in a different random order for each buyer) are displayed on the horizontal axis, and the actual bids on the vertical axis. Superimposed on the individual bids is a horizontal line at $b = 50$. Bids falling at or below this line do not support the hypothesis of no revelation. Also recorded in each plot is the number of deals (that could range from 0 to 50) struck by each buyer. Figure 10.3 shows considerable between-subjects variation in the number of Round 1 deals ranging from 0 (Buyers 11, 16, 18 and 20) to 22 (Buyer 6).

There are considerable individual differences in the individual bids in Round 1. We hypothesized that Round 1 bids would not exceed 50. It is clear from Figure 10.3 that, with six exceptions (Buyers 3, 4, 11, 16, 18 and 20) this hypothesis was not supported. Mean individual bids ranged from 12 (Buyer 18) to 87 (Buyer 6), with an overall mean of 42. There is no way of telling from the individual plots which bids were made earlier and which made later in the 50-trial sequence. Analysis of trial-to-trial mean bids, that we report later, shows that bidding behaviour changed systematically – in the direction of no revelation – as the subjects gained more experience with the MSDA mechanism and more information about the sellers' behaviour.

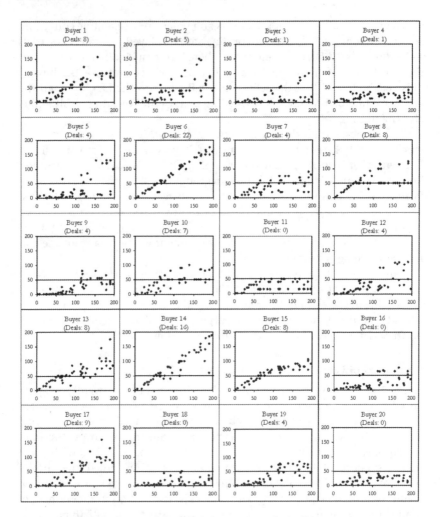

Figure 10.3   Bids by buyers in Round 1: inexperienced condition

Of particular interest are the individual bids of the ten Sophisticated buyers. Using the same format as in Figure 10.3, the individual bids for Round 1 are exhibited in Figure 10.4. With no more than three exceptions, nine of the ten buyers did not disclose their reservation values. The only exception is Buyer 29. Only 16.4 per cent of the Round 1 bids exceeded 50, and all but nine of these bids occurred in the first 10 trials. As a result of these bidding patterns (and the asks of the sellers), 5 of the ten buyers (Buyers 21, 22, 24, 27, 28) never reached an agreement on Round 1, and five more (Buyers 23, 25, 26, 29, 30) reached only a single agreement (see Figure 10.4). These results contrast sharply with those displayed in Figure 10.3 for the buyers

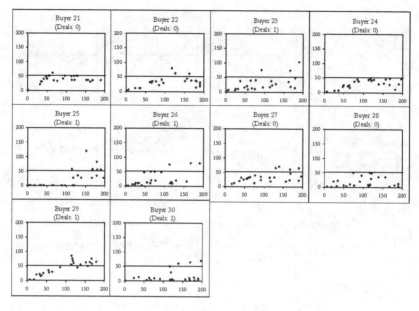

Figure 10.4 Bids by buyers in Round 1: sophisticated condition

in the Inexperienced group. Clearly, the strategically sophisticated buyers realized not to reveal information about their reservation values in Round 1.

Turning next to the sellers' decisions, Figure 10.5 exhibits the fifty Round 1 asks for each of the twenty Inexperienced sellers. Note that the sellers' reservation values (horizontal axis) range between 0 and 100, whereas their asks (vertical axis) range between 0 and 200. On only 62.6 per cent of all trials (20 × 50 = 1000) were the Round 1 asks equal to or higher than 100. Strong support for the main hypothesis is provided by Sellers 1, 3, 4, 13, 14 and 17, who made only a few offers below 100. But the majority of the sellers violated the hypothesis on at least 50 per cent of the trials. Mean Round 1 asks ranged between 60 (Seller 9) and 159 (Seller 14), with an overall mean of 114.

Most of the sellers in the Sophisticated group submitted Round 1 asks of between 90 and 120, with 59.2 per cent of all Round 1 offers being equal to or higher than 100. However, we do not see as many low asks as in Figure 10.5. Seller 27 consistently asked 160 on trials 1–25, and Seller 23 – possibly in attempt to reveal her identity – made four outrageous offers (234, 911, 911, 911) on trials 14–17. But all other sellers exhibited more or less the same offer patterns. Six of them (Sellers 21, 22, 24, 25, 26 and 30) consistently made offers that varied from one another by no more than a few points, and all around the 100 mark. With one exception (Seller 24), no seller reached

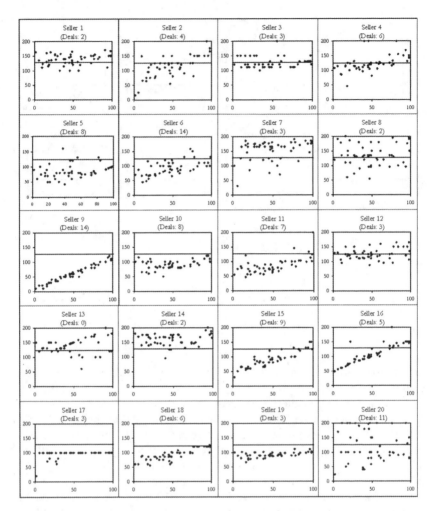

Figure 10.5   Asks by sellers in Round 1: inexperienced condition

more than a single agreement in Round 1. We only observed similar patterns with two of the twenty Inexperienced sellers (Sellers 3 and 17 in Figure 10.5).

Aggregate bids and asks: Round 1

Figure 10.7 displays separately for each of the two-stage conditions the running mean (in steps of 5) of Round 1 asks and bids over the course of the experiment. Recall that the Sophisticated subjects completed only twenty-five trials compared to fifty trials by the Inexperienced subjects. Two findings stand out. First, across all trials, Sophisticated subjects were the most aggressive, with buyers in this condition bidding about the same as the buyers in

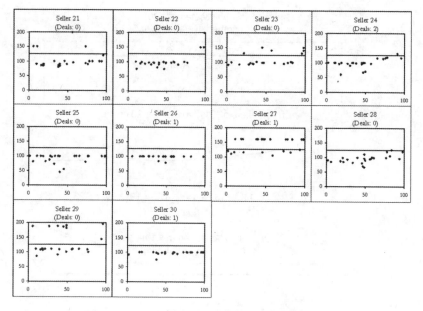

Figure 10.6   Asks by sellers in Round 1: sophisticated condition

the Inexperienced condition, whereas sellers in this condition asked significantly more than did sellers in the Inexperienced condition. Second, there is clear evidence of learning in both conditions. Sophisticated buyers started out immediately in Round 1 with mean bids about 50 that decreased steadily across the course of the experiment, stabilizing at about 23. Mean asks of the Sophisticated sellers quickly increased to well above 100 by trial 7 and above 117 by trial 15. Inexperienced buyers also recognized the value of not revealing information, exhibiting behaviour very similar to that of the Sophisticated buyers. In contrast, Inexperienced sellers made significantly lower, less aggressive Round 1 asks, and continued to reveal information about their reservation values throughout the entire course of the experiment.

The effects of learning and strategic sophistication manifested themselves in the number of deals struck in Round 1. Table 10.3 reports the number of deals made in Rounds 1 and 2 together (row 1), the number of deals made in Round 1 only (row 2), and the percentage of Round 1 deals out of the total number of deals (row 3). The results are presented separately for all the trials (either 50 for the Inexperienced condition or 25 for the Sophisticated condition) in the left-hand panel of Table 10.3, and for the last 20 trials (in both conditions) in the right-hand panel. In Round 1, Sophisticated subjects only made 5 agreements, all occurring in the first two trials. In contrast, 20.1 per cent of all deals made by the Inexperienced subjects were results of Round 1 agreements. Consistent with Figure 10.7, these percentages decreased with

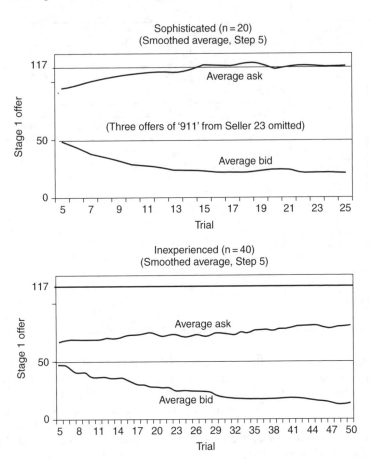

Figure 10.7   Running means of Round 1 asks and bids

Table 10.3   Number of deals by condition and round of play

No. of deals	Across all trials		Last 20 trials	
	Sophisticated	*Inexperienced*	*Sophisticated*	*Inexperienced*
Rounds 1, 2	131	561	100	227
Round 1	5	113	0	26
Percentage	3.8%	20.1%	0%	11.5%

experience from 3.8 per cent to 0.0 per cent in the Sophisticated group, and 20.1 per cent to 11.5 per cent in the Inexperienced group in the last twenty trials. Although the changes in the percentage of Round 1 deals across trials are significant (Parco, 2002), nearly one in six Round 1 deals persisted, a result mainly of a handful of subjects, almost all of them Inexperienced subjects, who continually made truthful or near truthful offers.

Individual bids and asks: Round 2

If the subjects reveal information about their reservation values in Round 1, playing LES on Round 2 is no longer optimal. Therefore the hypothesis about LES play in Round 2 is only testable with the Sophisticated subjects, who mainly revealed no information in Round 1. Focusing again on individual bids and asks, Figure 10.8 portrays the individual Round 2 bids for all ten of the Sophisticated buyers. The similarity of the bid functions between the ten buyers is striking. In nearly all cases, the buyers bid at or slightly below the LES. In only two cases did a single buyer's bid exceed the maximum LES bid of 116.67. Buyer 25 bid 120 and 123 in Rounds 1 and 2, respectively. We observe no cases where buyers' bids exceeded their reservation values.

Compared to the buyers, Sophisticated sellers exhibited more between-subject variability (see Figure 10.9). Like the buyers, there are no cases of sellers submitting offers smaller than their reservation values. And with the exception of Seller 27, who attempted to stand firm and resist being pushed down by the anonymous, information-advantaged buyers, the vast majority of the offers lie between the LES and truth-telling functions. The

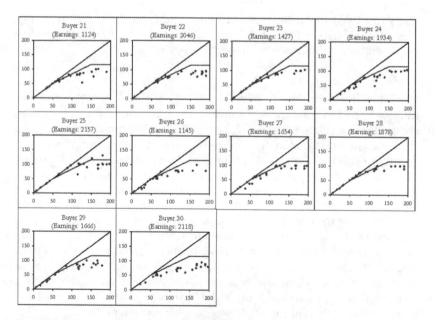

Figure 10.8 Bids by buyers in Round 2: sophisticated condition

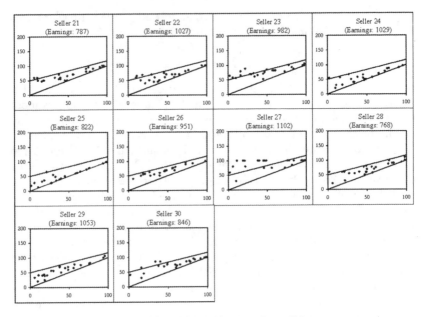

Figure 10.9   Asks by sellers in Round 2: sophisticated condition

offers displayed in Figure 10.9 are indistinguishable from the individual offers of the sellers in the Baseline condition (see Parco, 2002 for individual plots), and the individual offers reported by DSR.

Aggregate bids and asks: Round 2

Although the LES is no longer optimal for players who reveal information about their reservation values in Round 1, we present aggregate results for both two-stage conditions. Table 10.2 reported the spline regression coefficients for the Sophisticated (row 2) and Inexperienced (row 3) groups. Consider first the Sophisticated buyers. Two bids of 1 (not displayed on the individual plots in Figure 10.8) were made during Round 2 for reservation values $v_b = 18$ and $v_b = 32$, resulting in a pushing down of the intercept of the first segment of the regression line to $y = -3.4$. The slope value of this line ($m = 1.09$) is therefore an artefact because of these two outliers, as in no case did any buyer bid above his/her reservation value. For the mid-range bids ($50 < v_b \leqslant 150$), the slope ($m = 0.44$), was smaller than the predicted value $m = 2/3$, and smaller than that observed in the Baseline condition. This indicates that either the buyer's behaviour in Round 2 was more aggressive than in the single-stage double-auction mechanism, or that the Sophisticated buyers differ from the Baseline subjects. For the upper range of bids ($150 \leqslant \nu_b \leqslant 200$), the slope of the regression line ($m = -0.03$) did not differ from 0, but the intercept ($y = 95.6$) was significantly lower ($p < 0.0001$) than the LES predicted value of $y = 116.7$, testifying again to the observation made earlier of aggressive buyer behaviour in comparison to the LES.

The bids of the Inexperienced buyers, as a whole, did not differ from the bids of the Sophisticated buyers. Only the percentage of variance accounted for by the spline regression line decreased significantly, from 90 per cent in the Sophisticated condition to 76 per cent. This reflects the larger between-subjects variability in the behaviour of the Inexperienced buyers observed in Figure 10.4.

Support for the LES prediction for Round 2 asks by the sellers was considerably weaker. As shown by Table 10.2, mean asks did increase in the reservation values across the range $0 \leqslant \nu_s \leqslant 100$, but the adjusted R^2 values were considerably smaller than those for the buyers, ranging between $R^2 = 0.39$ for the Inexperienced sellers to $R^2 = 0.58$ for the Sophisticated sellers. In both groups, the slopes of the linear regression lines were significantly smaller than predicted, as were the intercepts. The information disparity hypothesis (Rapoport *et al.*, 1998) that the information-advantaged traders (buyers in our case) will 'push down' the information-disadvantaged traders (sellers in our case) was supported in both conditions.

Discussion and conclusions

The MSDA allows for several rounds of bargaining, in which asks and bids are made simultaneously on each round. It does not capture the pattern of reciprocal concessions so often observed in economic transactions, where traders respond to each other's offers. At best, it captures elements of the pattern of diplomatic or military bargaining, where the parties come to the negotiation table with pre-prepared proposals, learn about each other's proposals while attempting to limit the revelation of the concessions they are willing to make in the future, and then depart to meet again for a subsequent round of bargaining with possibly new proposals. The protracted bargaining between the USA and North Korea following the Korean War in the early 1950s, between the USA and North Vietnam in the early 1970s, and between Israel and the Palestinian authorities that took place in Camp David during the late 1990s all bear a resemblance to the MSDA. In many of such bargaining situations, the cost of delay in reaching an agreement is small in comparison to the magnitude of the potential gains with a deadline on the length of the bargaining.

The traditional way of accounting for multistage bargaining is by introducing a fixed discount factor $\delta \leqslant 1$, where δ is interpreted as being either the devaluation of profit from one round to another, or a fixed probability of breakdown in negotiations. This chapter focuses on a special case of $N = 2$ and $\delta = 1$, and then constructs an equilibrium solution in which players do not bargain seriously in Round 1, and then play the LES in Round 2. Our results provide moderate support for equilibrium play. Although players disclosed information about their reservation values in Round 1, the rate of disclosure decreased steadily with experience. Similar to the previous experimental studies of the single-stage double-auction, the LES approximated the bid and ask functions in Round 2, with the qualification that the traders who had the information advantage (buyers in our case) played more aggressively than the LES, and consequently succeeded in 'pushing down' the information-disadvantaged traders (sellers in our case) below their LES functions.

These patterns of behaviour were more pronounced when the traders were strategically sophisticated. Compare, for example, the inexperienced subjects, who in the last twenty trials reached an agreement in Round 1 in approximately 1 out of 9 trials, to the strategically sophisticated students who reached no agreement after the second trial. Our results also show that strategically sophisticated bargainers are considerably more homogenous than naïve bargainers, and therefore their offers are more predictable. Note that in our experiment the effects of strategic sophistication and magnitude of payoff (2.5 higher for the Sophisticated condition) were confounded. We cannot assess the differential effects of these two variables on the significant differences between the two conditions in length of deliberation before reaching a decision (more than twice as long in the Sophisticated condition), mean payoff per trial, and different patterns of asks and bids.

Additional limitations of the experimental design and data analysis, some of them for practical reasons, should be noted. We have made no attempt to account for the learning process (see, for example, Figure 10.7) by testing alternative adaptive learning models competitively (see, for example, Camerer (2003) for a brief review), or extending the adaptive learning model for the single-stage double-auction, that was proposed, refined and tested successfully by Daniel *et al.*, Rapoport *et al.* and Seale *et al.*, to the MSDA. We have made no attempt to relate the bids/asks in Round 1 to the ask/bids in Round 2, particularly in the case where subjects revealed information about their reservation values in Round 1 and, not reaching an agreement, proceeded to Round 2. We have made no attempt to compare behaviour in the MSDA between fixed-pair and random-pair designs where, under the fixed-pair design, reputation effects are possible as the same traders play the stage game repeatedly. And we have made no attempt to study the MSDA with $N > 2$ and $\delta < 1$ (for example, by introducing a small and fixed probability of breakdown in the negotiations at the end of each trial). We take these limitations as a challenge for extending theory construction and experimental design, and improving data analysis in future studies.

Appendix: instructions for the two-stage bilateral bargaining study

The present experiment is designed to study two-person bargaining between a buyer and seller. If you make good decisions, you may earn a considerable amount of money. The money you earn will be paid to you in cash at the end of the session.

In case you have any questions while reading the instructions, please raise your hand and the supervisor will come to help you.

Description of the task

Before the session begins, the subjects in the Laboratory will be divided randomly into two equal size groups of Buyers and Sellers.

You will participate in 50 trials. On each trial, a Buyer and Seller will be randomly paired and then bargain on the price of an unspecified object. Since you will communicate with each other via the computer, you will not know your co-bargainer's identity nor will he/she know yours. You will play the same role (either a Buyer or Seller) on all trials. However, the identity of your co-bargainer will be changed randomly from trial to trial.

At the beginning of each trial the computer will display your *reservation value* for the object. The reservation value represents how much the object is worth to you on this trial. It will change from trial to trial.

Reservation values (stated in a fictitious currency called 'francs') are determined randomly before each trial. For Buyers, reservation prices will range from 0 to 200 francs, with each value in this range equally likely. For Sellers they will range from 0 to 100 francs, with each value in this range equally likely. The ranges will be shown graphically on the computer screen before each bargain begins (see the display below). On each trial, you will know your own reservation value (assigned to you by the computer) but not the exact reservation value of your co-bargainer (you will only know that it is equally likely to be within a certain range).

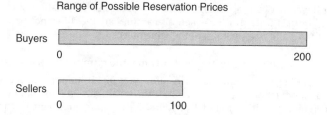

Range of Possible Reservation Prices

How do you bargain on the price?

Each trial includes at most two rounds of play.

Round 1: On round 1, after the computer displays your reservation value, you will have an opportunity to make a bid price (Buyer) or ask price (Seller) for the object. If you are the Buyer, your *bid price* represents the price you propose to pay for the object. If you are the Seller, your *ask price* represents the price you propose to accept for the object.

- If the Seller's ask price is higher than the Buyer's bid price, then no deal will be struck on round 1 and both you and your co-bargainer will move to the second round of the same trial.
- If the Seller's ask price is equal to or lower than the Buyer's bid price, then a deal will be struck and you will end this trial in an agreement. The *contract price* in this case is computed to be halfway between the buyer's bid and the seller's ask prices:

  contract price = (buyer's bid price + seller's ask price)/2

Round 2: Round 2 has the same structure as round 1 with the only exception that if no deal is struck, the trial ends in disagreement (and zero payoff to both traders).

In summary, on each trial, the buyer and seller make at most two decisions (bid price for Buyer or ask price for Seller). These decisions determine whether an agreement is reached, and if so at what contract price. An agreement may be reached on round 1. If no agreement is reached on round 1, another opportunity to reach an agreement is provided on round 2. If round 2 is reached, it may be concluded with either an agreement or disagreement.

How are your earnings determined on each trial?

- If the trial ends in disagreement (because the Seller's ask price exceeds the Buyer's bid price on both rounds of play), then you will earn nothing for this trial.
- If the trial ends (on either round 1 or 2) in agreement (because the Seller's ask price is equal to or lower than the Buyer's bid price), then your earnings will be determined by the following formulas:

  Buyer's earnings = (Buyer's reservation price − contract price)

  Seller's earnings = (contract price − Seller's reservation price)

For the Buyer, her earnings are the difference between her valuation of the object and the contract price. For the Seller, his earnings are the difference between the contract price and his valuation of the same object.

Example: The following example illustrates the computations:

Suppose the Buyer is assigned a reservation price of 110 francs, and the Seller is assigned a reservation price of 65 francs If the Buyer bids 90 francs and the seller asks 80 francs (on either round 1 or round 2), then an agreement is reached at a contract price of 85 francs ((90 + 80)/2). Using the formulas given above, the earnings are calculated to be:

  Buyer's earnings = (110 − 85) = 25

  Seller's earnings = (85 − 65) = 20

Please note the following. If the Buyer (in an effort to increase her payoff) decides to lower her bid price from 90 to 80, while the Seller (with a similar motivation to increase his payoff) changes his ask price from 80 to 85, then no deal is struck (because the Buyer's bid price is less than the Seller's ask price). In this case, both traders will earn nothing on this trial. Hence, a tradeoff exists for both the Buyer and the Seller. The more money they try to earn by decreasing their bid price (Buyer) or increasing their ask price (Seller), the more likely it is that no agreement will be reached. The key uncertainty is that each player does not know the reservation price of the other. The traders only know the range from which these prices are randomly selected.

Procedure

You will play a total of 50 trials. Each trial follows the same sequence: First, the computer will randomly match you with another trader of the opposite type, and will display your *reservation value* for the object (you will not know your co-bargainer's reservation price, only that it is equally likely to be included in a certain range). Next, you will be asked to submit your bid price (Buyer) or ask price (Seller). After both bargainers submit their offers, the computer will inform you of your co-bargainer's offer, and calculate your payoff if an agreement is reached. If an agreement is not reached, you will have a second (and last) opportunity to strike a deal on the second round of the same trial. If round 2 ends with disagreement, your payoff for the trial is zero. After you review your payoffs, you will move to the next trial, if it is not the last one.

Payment at the end of the session

At the end of the session, the computer will sum up all your earnings for the 50 trials. The supervisor will then pay you your earnings at the rate of 80 francs = $1.00. Please raise your hand to indicate to the supervisor that you have completed reading the instructions. The supervisor will then set your computer for the game. Please be patient; the game will start when everyone is ready.

Notes

1 The assumption of risk-neutrality was relaxed by Chatterjee and Samuelson (1983), who investigated the nature of the equilibria when both traders' utility functions display constant risk-aversion. Our justification for assuming risk-neutrality is based in part on the theoretical argument of Rabin (2000) and in part on the finding that previous experimental tests provided strong support to the joint hypothesis of a Chatterjee – Samuelson linear equilibrium *and* risk-neutrality.

2 Individual rationality requires $b \leqslant v_b$ and $s \geqslant v_s$, which are the only requirements if $0 \leqslant v_b \leqslant 25$ in Equation (10.1) or $75 \leqslant v_s \leqslant 100$ in Equation (10.2).

3 It was not possible to have another group of 'sophisticated' players because of the special circumstances under which they were recruited.

References

Ausubel, L., Cramton, P., and Deneckre, R. J. (2002) 'Bargaining with Incomplete Information', in R. J. Aumann and S. Hart (eds), *Handbook of Game Theory*, Vol. 3, Amsterdam, Elsevier.

Brams, S. J. and Kilgour, D. M. (1996) 'Bargaining Procedures that Induce Honesty', *Group Decision and Negotiation*, 5, 239–62.

Camerer, C. F. (2003) *Behavioral Game Theory: Experiments in Strategic Interaction*, Princeton; NJ, Princeton University Press.

Chatterjee, K. and Samuelson, W. (1983) 'Bargaining Under Incomplete Information', *Operations Research*, 31, 835–51.

Daniel, T. E., Seale, D. A. and Rapoport, A. (1998) 'Strategic Play and Adaptive Learning in the Sealed Bid Bargaining Mechanism', *Journal of Mathematical Psychology*, 42, 133–66.

Leininger, W., Linhart, P. B. and Radner, R. (1989) 'Equilibria of the Sealed-Bid Mechanism for Bargaining with Incomplete Information', *Journal of Economic Theory*, 48, 63–106.

Linhart, P. B., Radner, R. and Satterthwaite. M. A. (eds) (1992) *Bargaining with Incomplete Information*, San Diego, Calif., Academic Press.

Myerson, R. B. and Satterthwaite, M. A. (1983) 'Efficient Mechanisms for Bilateral Trading, *Journal of Economic Theory*, 29, 265–81.

Parco, J. E. (2002) 'Two-person Bargaining Under Incomplete Information: An Experimental Study of New Mechanisms', UMI published Ph.D. dissertation, Eller College of Business, Department of Management and Policy, University of Arizona.

Parco, J. E. (2003) 'Price-setting Power and Information Asymmetry in Sealed Bidding', Unpublished MS, Department of Management, United States Air Force Academy.

Parco J. E. and Rapoport, A. (2003) 'Enhancing Honesty in Bargaining: An Experimental Test of the Bonus Procedure', Unpublished MS, United States Air Force Academy.

Rabin, M. (2000) 'Risk Aversion and Expected-utility Theory: A Calibration Theorem', *Econometrica*, 68, 1281–90.

Radner, R. and Schotter, A. (1989) 'The Sealed-Bid Mechanism: An Experimental Study, *Journal of Economic Theory*, 48, 179–220.

Rapoport, A. and Fuller, M. (1995) 'Bidding Strategies in a Bilateral Monopoly with Two-sided Incomplete Information', *Journal of Mathematical Psychology*, 39, 179–96.

Rapoport, A., Daniel, T. E. and Seale, D. A. (1998) 'Reinforcement-based Adaptive Learning in Asymmetric Two-person Bargaining with Incomplete Information', *Experimental Economics*, 1, 221–53.

Rubinstein, A. (1982) 'Perfect Equilibrium in a Bargaining Model', *Econometrica*, 50, 97–109.

Rubinstein, A. (1985) 'A Bargaining Model with Incomplete Information about Time Preferences', *Econometrica*, 53, 97–109.

Satterthwaite, M. A. and Williams, S. R. (1989) 'Bilateral Trade with the Sealed-Bid *k*-Double-Auction: Existence and Efficiency', *Journal of Economic Theory*, 48, 107–33.

Satterthwaite, M. A. and Williams, S. R. (1993) 'The Bayesian Theory of *k*-Double-Auctions', in D. Friedman and J. Rust (eds), *The Double-auction Market: Institutions, Theories, and Evidence*, Reading, Mass., Addison-Wesley.

Seale, D. A., Daniel, T. E. and Rapoport, A. (2001) 'The Information Advantage in Two-person Bargaining with Incomplete Information', *Journal of Economic Behavior and Organization*, 44, 177–200.

Stein, W. E. and Parco, J. E. (2001) 'A Note on Bilateral Trading with Bonus', Unpublished MS, Department of Management and Policy, University of Arizona.

Valley, K. L., Moag, J. and Bazerman, M. H. (1998) 'A Matter of Trust: Effects of Communication on the Efficiency and Distribution of Outcomes', *Journal of Economic Behavior and Organization*, 34, 211–38.

11
The Role of Learning in Arbitration: An Exploratory Experiment

*Gary E. Bolton and Elena Katok**

Introduction

Having an arbitrator decide the outcome of a negotiation avoids the losses associated with a bargaining impasse. For example, in the public sector of the USA, arbitration is commonly mandated for failed labour negotiations to save the public the costs associated with, say, a police or firefighters' strike.[1] By definition, arbitration involves an outside party, the arbitrator, with the authority to impose a binding settlement on the disputing parties. (In contrast, a mediator is someone who can suggest, but not impose, a settlement.) We say that there is a 'dispute' when bargaining ends without a voluntary agreement. When we say 'bargaining with arbitration', we mean a negotiation in which arbitration is required in case of dispute. We shall focus here on repeated bargaining relationships, which is a common context for binding arbitration.

Field evidence finds that having arbitration as a fallback lessens the likelihood that bargainers will reach a voluntary settlement (see, for example, Currie and McConnell, 1991; Ashenfelter *et al.*, 1992). The 'overuse' of arbitration can have undesirable consequences. For example, since arbitrators are often at an informational disadvantage, the imposed agreement may be less efficient than a voluntary one (Crawford, 1982). An efficient multilevel wage agreement, for example, may require detailed knowledge of both management and labour preferences. Also, having a third party settle a dispute plausibly deteriorates the working relationship of the bargaining parties, planting the seeds for additional problems and misunderstandings.

A critical question, then, is whether arbitration need necessarily impose these drawbacks; that is, whether we can hope to avoid decreasing voluntary agreements through the clever design or implementation of the arbitration process. The field evidence provides little hope that such can

* Both authors gratefully acknowledge the financial support of the National Science Foundation.

be accomplished, but of course this evidence focuses on existing mechanisms and practices. Theory provides no unequivocal answer either, in large part because the reason bargainers have disputes in the first place is not well understood. Complete information models of bargaining typically do not predict dispute occurrence (Nash, 1950; Rubenstein, 1982). Incomplete information models account for disputes, but in a way that has been called into question by the data.[2] Even if we simply assume that disputes happen, however, it is not obvious why arbitration should aggravate the problem, since using arbitration typically imposes both financial costs and risk costs (in the form of uncertainty about what the arbitrator will do) on the negotiators, giving them an incentive to avoid its use.

This chapter reports on an exploratory experiment in the spirit of Güth, Ivanova Stenzel, Königstein, and Strobel (2003). The working hypothesis of the investigation is that arbitration may in fact reduce the rate of dispute if the process is targeted at simplifying the bargaining problem. We judge the complexity of the bargaining problem by a criterion implicit in Schelling's (1963) conceptualization of bargaining as a struggle to commit to, and co-ordinate on, a commonly understood focal point. Specifically, we study a bargaining environment known to induce two focal points, each appealing to the self-interest of a different bargainer. We then introduce an arbitration process that implements a settlement from a unimodal distribution (as we shall explain, this model is thought to capture essential features of the arbitration process). The hypothesis is that arbitration can simplify the bargaining co-ordination problem if it signals the focal point for bargainer co-ordination. In two of the treatments, the mode of the arbitration settlement distribution focuses on one or the other 'natural' focal point in the bargaining environment. In the third treatment, the distribution focuses on a compromise outcome, in effect a 'synthetic' focal point, which one might think bargainers would find attractive since it is a compromise between the natural points. An interesting issue here is whether signalling a natural focal point or a synthetic compromise point is equally effective at generating co-ordination.

Two previous experiments and a new hypothesis

The inspiration for our working hypothesis is the findings of two previous experiments. Since the new experiment also builds directly on the laboratory designs of this previous work, we first describe the earlier experiments in some detail.

Two previous arbitration experiments

Ashenfelter *et al.* (1992) report an experiment aimed at determining what effect arbitration has on voluntary dispute rates. The experiment compared bargaining with no arbitration to bargaining with three different arbitration mechanisms commonly used in the field. In *conventional arbitration* the arbitrator is free to impose any settlement s/he sees fit. In *final offer arbitration* each bargainer submits a final offer to the arbitrator, who then imposes the settlement s/he thinks most reasonable. In *tri-offer arbitration* the arbitrator imposes one of three outcomes: either one of the bargainers' final offers, or the recommendation of a neutral fact-finder. Both of these latter types of arbitrators are intended to give the bargaining parties an extra incentive to find their own settlement, since both prevent the arbitrator from splitting the difference between positions of the two sides, and in this sense increase the risk of going to arbitration.

Bargaining in the Ashenfelter *et al.* experiment concerned a simple pie-splitting task, with two bargainers who negotiated repeatedly. Bargainers negotiated via a computer for up to five and a half minutes. The only moves allowed were the posting or updating of an offer; the latest offer(s) posted were visible to both bargainers. The voluntary negotiation ended when posted offers matched or time ran out. The offers were simply a number between 100 and 500. Each offer translated into a cash payoff. Bargainers knew their own payoff but were not told their partners', although, in fact, the two schedules were symmetrical. Bargainers knew they would interact with the same partner multiple times, though, to avoid end-game effects, they were not told how many rounds of play there would be.

One of the novel aspects of this experiment was the method for simulating arbitration awards, a method based on field studies of arbitrator behaviour (Ashenfelter and Bloom, 1984; Ashenfelter, 1987). These studies suggest that, because field bargainers usually have some say about who will arbitrate, arbitrators who are predictably biased towards either party *relative* to other arbitrators tend to be vetoed by the disadvantaged party. Acceptable arbitrators therefore tend to be 'statistically interchangeable': different arbitrators may make different awards, but there is little *predictable* difference. This suggests that actual arbitration award processes are well approximated by a simple stochastic draw from a fixed distribution – something that is easy to implement in the lab. In the experiment, awards were always chosen from a truncated normal distribution, with one bargainer being, on average, modestly favoured. To give their subjects a feel for likely awards, the experimenters provided all the bargainers with a list of the arbitrator's last 100 awards; for conventional arbitration, in fact a list of 100 random draws from the distribution (an analogous method was used for final offer and tri-offer arbitration). For conventional arbitration, the Ashenfelter *et al.* experiment also tested the hypothesis that greater uncertainty costs, in the sense of higher variance of arbitrator awards, diminishes the incidence of dispute.

The main finding of the experiment was that arbitration increases the dispute rate robustly. The average rate of dispute without arbitration was 11.2 per cent, and for (high variance) conventional arbitration, this increased to 28.4 per cent. For final offer arbitration, the rate of dispute was statistically significantly *higher* at 38.1 per cent. It was also higher for tri-offer offer arbitration, at 32.4 per cent, although not statistically. There was therefore no evidence that final offer or tri-offer arbitration procedures decrease disputes. The study also found that dispute rates were negatively correlated with uncertainty costs.

Bolton and Katok (1998) presented an experiment that examined aspects of bargainer behaviour where the arbitration safety net increased disputes. The experiment focused specifically on conventional arbitration. The experimental design paralleled that of Ashenfelter *et al.*, with three notable exceptions. First, while Bolton and Katok's experiment also featured a simple pie-splitting problem, the bargaining game was more structured: bargainers first made simultaneous proposals, and then either accepted the other bargainer's proposal or maintained their own (the same action space is used in the new experiment; see the third section beginning on page 240). This permitted a cleaner analysis of bargainer behaviour. Second, bargainers were given complete information about one another's payoffs, eliminating confounding with bargainer expectations of true payoffs. Third, while arbitration awards were modelled in the same way as in Ashenfelter *et al.*, as a truncated normal distribution, Bolton and Katok manipulated the location of the distribution, so that some arbitrators made symmetrical awards while others made asymmetrical ones. The experiment also manipulated the cost of dispute in the sense of the size of the pie (or arbitrator award uncertainty); that is, in some cases, going to arbitration led to a cost in the form of a somewhat smaller pie to be awarded by the arbitrator.

In spite of these differences, the baseline results of this experiment were comparable to those of Ashenfelter *et al.* Specifically, even though the bargaining game differed with respect to information completeness of action space structure, the dispute rate absent arbitration averaged 14.3 per cent. The introduction of arbitration robustly increased the rate of dispute, ranging in this experiment from 22.2 per cent to 67.3 per cent. The high end of the range was reached for the arbitrator that made asymmetrical awards, sharply favouring one bargainer. Both experiments found that higher dispute costs were correlated negatively with dispute rate (Bolton and Katok with respect to loss of bargaining pie as well as with respect to arbitrator award uncertainty). Finally, both experiments found a high degree of heterogeneity with respect to bargaining pair dispute rates.

The novel findings of the Bolton and Katok experiment began with the observation that, for both with and without an arbitrator, bargainers exhibited learning behaviour, with disputes going down as bargainers repeatedly interacted. Field investigators also find learning effects in negotiations (see, for example, Reder and Neumann (1980); Lester (1989); see Bolton and Katok (1998) for more discussion of the field evidence). The two important findings that emerged from an analysis of the learning trend were: first, the learning is attributable to the incidence of past dispute; specifically, the influence of a dispute on the probability of a future dispute was negative. The reason for this influence was that bargainers reacted to a dispute by moderating their demands in the next round of bargaining. The second important finding spoke directly to the difference between bargaining with and without arbitration. Specifically, when bargainers had arbitration as a fallback, there was less moderation in their demands after a dispute; in this sense, there was slower learning under arbitration. Hence arbitration exhibited a version of what some commentators in the field literature on arbitration call a narcotic effect; those parties who used arbitration in the past seemed to lean more heavily on it in the future than those who had not used it in the past.

A new hypothesis

The motivation for the new hypothesis begins with the observation that both of the previous experiments focused on a simple pie-splitting nego-tiation, with a natural 50–50 division focal point. Indeed, equal division was the modal voluntary settlement in all the treatments in both experi-ments – even when bargainers were not told one another's payoffs, as in the Ashenfelter experiment, and even when arbitrator awards favoured one or the other bargainer (again, regardless of whether bargainers knew this for sure). Hence the increase in disputes under arbitration might not be caused by decreased dispute costs *per se*, but rather decreased dispute costs in a negotiation environment where most people already have a strong precon-ceived notion of what the outcome from negotiating should be. The fact that Bolton and Katok observed that disputes were highest under an arbitrator known to be biased towards one bargainer could be taken as evidence of resistance to moving away from a commonly known focal point.

It seems plausible, then, that in a more complex bargaining environment, one where there are multiple focal points, an arbitrator who selects a partic-ular focal point for the bargainers might in fact decrease dispute rates, since committing to that focal point is less costly, and so more credible, than when an arbitration option is absent. The important implication is that arbitration awards that might select between existing predilections could conceivably help along the voluntary bargaining process, rather than hinder it, as observed both in field studies and in the previous experiment.

The new experiment

In the new experiment, we use a bargaining game similar to that in Bolton and Katok (1998). But we change the payoff space of the game in a way that induces two natural but competing focal points. We then manipulate arbitrator awards either to reinforce or not to reinforce one of the focal points.

The new experiment had four treatments. In all four, bargainers were paired with the same partner for twelve rounds of bargaining. In the No Arbitrator treatment (NA), bargaining was without arbitration. Bargaining in the other three treatments, A50, A63 and A75, included a conventional arbitration procedure, but the nature of arbitrator awards differed across treatments. The bargaining game and arbitration mechanisms are described in the next two segments.

Bargaining: the deadline game

In all treatments, bargaining pairs played the *deadline game*, a two-person, two-stage negotiation concerning the division of 100 chips.[3]

Stage 1

Bargainers A and B simultaneously offer v^A and v^B, respectively, where $v^A(n)$ equals the chips offered to n; $n = A, B$. Also, $v^A(A) + v^A(B) = 100$; similar for v^B. If offers are exactly compatible (that is, $v^A(A) = v^B(A)$ and $v^A(B) = v^B(B)$) then the game ends in agreement on the implied division; otherwise the game proceeds to Stage 2.

Stage 2

After reviewing both Stage 1 offers, bargainers play the bi-matrix game as in Figure 11.1, where a denotes accepting one's partner's offer (thereby dropping one's own offer); m denotes maintaining one's own offer (thereby holding out for partner to play a); and $v^{arb}(A)$ is the payoff awarded to A by the arbitrator. If both bargainers play a, then a coin flip decides which offer becomes the settlement.

The lab treatments are distinguished by what happens in the case of dispute, when both bargainers play m. In NA, both bargainers receive 0 chips. In the other three treatments, the division of the 100 chips is determined by a conventional arbitration procedure.

Arbitration

Following the procedure used in the earlier experiments, conventional arbitration awards were determined by a random draw from a truncated normal distribution. The location of the distribution differed across treatments. The

		Bargainer B	
		a	m
Bargainer A	a	v^A with prob ½ v^B with prob ½	v^B
	m	v^A	No arbitrator (NA): 0,0 Arbitration (A50, A63, A75): Arbitrator imposes a settlement

Figure 11.1   Stage 2 of the deadline game

number in the name of each treatment indicates the average award that player A could expect if bargaining ended in dispute (see Figure 11.1). For example, in A63, the average award gave 63 chips to bargainer A, and 37 chips to bargainer B. In all cases, the standard deviation was set at 2.5 chips (this is only approximate because of the truncation at 0 and 100, but the error is negligible). The particular locations of the arbitrator distributions, at 50, 63 and 75, were each intended to implement a natural or synthetic focal point, as will become clear in the next segment.

Payoffs

Payoffs for the game involved the binary lottery first introduced by Roth et al. (1981). The chips a bargainer receives in a settlement were effectively lottery tickets; the more chips were obtained, the better the bargainer's chance of winning a monetary prize in his/her own personal lottery. Bargainer A was assigned a monetary prize of $10, while bargainer B's was $30. The probability of winning a prize is equal to the proportion of tickets received. For example, if A receives 55 tickets, then s/he has a 55 per cent chance of winning $10, and a 45 per cent chance of winning nothing. A bargainer with 0 tickets receives the prize of 0 with certainty.

The unequal payoffs and the lottery procedure create a tension between two focal divisions of the chips; in this case, the 50–50 division, representing an equal chance of winning a prize, and 75–25, representing an equal division of expected value. As we shall see, this tension leads to a significantly different pattern of negotiations without arbitration than was observed in the earlier experiments. It also creates a natural test bed for the hypothesis we set out to investigate. Observe that the A50 and A75 arbitrators are located at the natural focal points, whereas the A63 arbitrator is located at a compromise, halfway between 50 and 75.

Using a binary lottery in conjunction with arbitration presents a complication (although one that can be overcome). To see this, step back for a moment to examine the costs associated with arbitration. These include the out-of-pocket costs associated with presenting the case to the arbitrator, and the implicit cost associated with the uncertainty over the final award.[4] The existence of some sort of cost is crucial. Suppose, for example, that the arbitrator always awards a 60–40 division (the larger portion going to bargainer A), meaning there is no uncertainty cost. Suppose no further cost as well. Then bargainers would be totally indifferent between settling 60–40 and having the arbitrator award 60–40. There would be no reason for bargainers to avoid arbitration or to settle on anything other than 60–40. In the Ashenfelter et al. (1992) experiment, an arbitration cost was imposed by the uncertainty associated with arbitratior awards.[5] But the binary lottery does not necessarily induce a parallel uncertainty cost: Roth et al. (1981) show that, for any given division of the binary lottery tickets, each bargainer's expected utility can be represented by the number of lottery tickets s/he receives. Hence the expected utility from going to arbitration is equal to the expected number of chips awarded by the arbitrator – independent of the amount of uncertainty associated with the award. A binary lottery alone, therefore, might induce no arbitration cost.

To avoid this difficulty, an out-of-pocket cost was imposed on bargainers for failing to reach agreement. To avoid negative cash balances, an 'agreement fee' was awarded to each bargainer if they reached agreement on how to divide the 100 chips. If bargainers did not reach agreement on their own, each received an agreement fee of 0. It is important to note that the agreement fee was also part of the negotiating pie without arbitration; that is, NA bargainers received an agreement fee of 0 if they failed to reach agreement. Hence the differences found between NA and arbitration negotiating behaviour are not attributable to the agreement fee.[6]

The agreement fee was set at a relatively high $5. We favoured a high fee because our purpose was to study learning behaviour, and there is some reason to think that the rate of learning (if there is any) will be correlated positively with the dispute cost.[7] Because time limitations restrict the number of bargaining rounds in a lab session, it is important to choose an agreement fee sufficiently high to induce learning at a statistically observable rate.[8] More generally, the results of the experiment raise some interesting issues concerning the cost of arbitration in the field, and these issues are turned to in the discussion section.

Laboratory protocol

All subjects were Penn State University students, recruited through billboards posted around the University Park campus. Participation required putting in an appearance at a special place and time, and was restricted to one session. Cash was the only incentive offered. The sample size for both the NA and A63 treatments was 24 bargaining pairs; for A50 and A75 it was 25.

Games were played through a computer interface. On arriving at the lab, participants read written instructions (see copy in Appendix 1 on page 253) and then played some practice games with the computer as bargaining partner. So as not to bias later play, the computer made decisions entirely at random (participants knew this). A brief quiz was given to check for understanding of the computer operation, and the written instructions were read aloud. Participants were then paired randomly and anonymously. To avoid end game effects, the total number of games to be played was not revealed.

The binary lottery was explained to participants as part of the directions. The computer automatically displayed both offers and their expected value to each bargainer. At the conclusion of the session, one game was selected randomly for payoff, and a lottery wheel was spun for each participant. Participants were paid their earnings for the selected game plus a $5 show-up fee (payment in cash).

For the arbitration treatments, bargainers were given a history of the arbitrator's past awards – a list of 100 random draws from the appropriate normal distribution (see Appendix 1), and were told to expect similar decisions in the future. The list included the average for the 100 draws (50 for both bargainers in A50; 63 for *A* and 37 for *B* in A63; and 75 for *A* and 25 for *B* in A75). As with previous experiments, no mention was made of the actual randomization process.

Results

We first take a descriptive tour of the data concerning dispute rates, to provide the reader with a sense of prominent features and important trends. We then apply a statistical model developed by Bolton and Katok (1998) to analyse the learning trends that are in evidence. Finally, we take a look at how arbitration affects voluntary settlements.

A preliminary analysis

The descriptive data on dispute rates are summarized in Table 11.1, and Figures 11.2 and 11.3. Each exhibit provides a somewhat different perspective on the pattern of disputes. Table 11.1 breaks out dispute rates by treatment for early, middle, late, and all rounds of negotiation. Observe first that dispute rates for NA are about 25 per cent across all rounds. This is about twice as high as that found by Bolton and Katok (1998) for their simpler, no arbitration bargaining game. Also observe that all-round dispute rates are higher for the A63 arbitrator than for NA, but are in fact lower for A50, and a bit higher for A75. Hence the results of this experiment are quite different from those of previous experiments. Also note that the rate of dispute is time-dependent. For the arbitration treatments, dispute rates are highest in the early rounds. Since they do not account for this trend, simple statistical tests on the significance of the differences across treatments tend not

Table 11.1 Dispute rate by early, middle and late rounds (standard error)

Treatment	Number of pairs	Rounds			All rounds
		1 to 4	*5 to 8*	*9 to 12*	
NA	24	0.250 (0.0602)	0.250 (0.0673)	0.240 (0.0647)	0.247 (0.1111)
A50	25	0.310 (0.0768)	0.200 (0.0456)	0.160 (0.0454)	0.223 (0.1002)
A63	24	0.427 (0.0681)	0.271 (0.0581)	0.323 (0.0745)	0.340 (0.1165)
A75	25	0.310 (0.0564)	0.250 (0.0500)	0.220 (0.0391)	0.260 (0.0849)

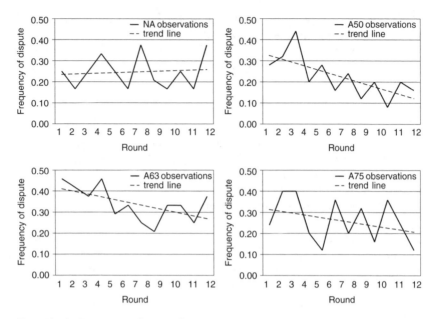

Figure 11.2 Dispute rate by round

to register significance. The formal statistical model we present in the next subsection will show that the initial propensity towards dispute is statistically higher in the arbitration treatments, but that this changes in the later rounds.

Figure 11.2 provides a round-by-round look at dispute rates, and illustrates that the dispute trend depends on the treatment. For NA, dispute rates rise modestly over time. In contrast, dispute rates clearly fall for all three arbitration treatments. The model in the following subsection will quantify these trends and confirm statistical significance. Again, this distinguishes our experiment from previous studies where either no trend was reported or bargainers were found to learn more quickly to avoid disputes without arbitration than with arbitration. In the next subsection we shall investigate the nature of these trends for the present data and present trend corrected estimates of dispute rates for the final rounds of play.

Figure 11.3 exhibits the number of disputes by bargaining pair. Consistent with the previous studies, there is a high degree of *bargainer heterogeneity* with respect to the propensity to have disputes. More formally, regressing last six round dispute rates on first six rounds rates indicates a significant positive coefficient of correlation for each of the treatments (unit = bargaining pair, $p < 0.02$ for all treatments apart from A50, where $p = 0.064$). The statistical model in the following subsection will block for heterogeneity.

245

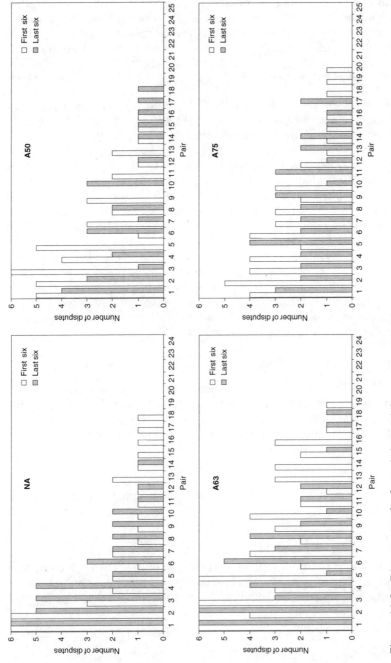

Figure 11.3 Dispute rate by bargaining pair*
Note: Some bargaining pairs never have disputes.

A learning model

To test our hypothesis, we would like to estimate formally, and then to compare across treatments, both the initial propensity to dispute and the learning rates. Following Bolton and Katok (1998), we describe two alternative models of the learning effect, and then compare their fit to the data. The better fitting model is then used to test our main hypothesis.

The first model, taken as the baseline, is a simple round-effects model:

$$d_{i,j,t} = \theta_{i,j} + \alpha_j A_j + \beta_j (t-1) + \varepsilon_{i,j,t} \tag{11.1}$$

where $d_{i,j,t}$ is the probability that pair i of treatment j will have a dispute in round t; $\theta_{i,j}$ is the pair fixed effect; A_j is the increase in the initial propensity to dispute resulting from arbitration treatment j ($A_{NA} = 0$); the α_js and the β_js are parameters to be estimated; and $\varepsilon_{i,j,t}$ is an error term.

The round-effects model attributes learning to the *amount* of experience bargainers have–that is, the model hypothesizes that simply playing the game more often influences the probability of dispute. By this model, learning may stem from something as straightforward as increased familiarity with the rules of the game, although the model is broad enough to accommodate a number of other hypotheses.

An alternative, more precise, hypothesis is that it is the history of game outcomes (dispute or settlement) that influences the probability of dispute; this is attributing learning to the *kind* of experience bargainers have, and is in keeping with field studies that emphasize the lag effect of disputes on outcomes (for example, Butler and Ehrenberg (1981) use a similar formulation). Consider, then, the outcome effects model:

$$d_{i,j,t} = \theta_{i,j} + \delta_j D_{i,j,t} + \sigma_j S_{i,j,t} + \varepsilon_{i,j,t} \tag{11.2}$$

where $D_{i,j,t}$ is the total number of disputes that (i,j) had through round $t-1$; $S_{i,j,t}$ is the total number of (non-arbitrated) settlements through $t-1$; the δ_js and the σ_js are parameters, and $\varepsilon_{i,j,t}$ is an error term. Equation (11.2) posits that each outcome has a permanent incremental impact on the probability of dispute (where 'permanent' means lasting for the duration of the experiment).[9]

The formulation in Equation (11.2) also has the advantage of affording a straightforward test of round learning against outcome learning. To see this, note that:

$$D_{i,j,t} + S_{i,j,y} \equiv t-1$$

So Equation (11.1) is equivalent to Equation (11.2) if $\delta_j = \sigma_j$. We can then test the null hypothesis of round learning against the alternative of outcome learning by estimating Equation (11.2) and checking whether the constraint $\delta_j = \sigma_j$ can be rejected.

It is convenient to define:

$$K_{i,j,t} \equiv D_{i,j,t} - S_{i,j,y}$$

Equation (11.2) can then be expressed as Equation (11.1) plus the terms $\lambda_j K_{i,j,t}$ added to the right-hand side:

$$d_{i,j,t} = \theta_{i,j} + \alpha_j A_j + \beta_j(t-1) + \lambda_j K_{i,j,t} + \varepsilon_{i,j,t} \tag{11.3}$$

where the λ_js are parameters. Testing round learning against outcome learning then reduces to testing whether $\lambda_j = 0$; rejecting is equivalent to rejecting the round-effect model in favour of the outcome model.

We present estimates obtained from least square dummy variable (LSDV) regression. The advantage of this approach is that estimates can be interpreted directly as probabilities. The drawback is that predicted probabilities are not constrained to the admissible range. Estimates for a fixed-effect logit model, however, lead to results that are comparable to those obtained with LSDV. (Random effect models are rejected by the data, or in some cases, tests indicate fixed effects are favoured.)

The estimates in column (1) of Table 11.2 are for a model including just pair fixed effects. Dropping the fixed effects and running column (1) with

Table 11.2   A comparison of round and outcome effect models: LSDV coefficient estimates (standard error)

	(1) The baseline model: pair fixed effects	(2) Rounds effects model: equations (4.1)	(3) Outcome effects model: equations (4.3)
Mean pair fixed effect (θ)	0.267*** (0.0112)	0.232*** (0.0424)	0.240*** (0.0472)
Initial effect (α_j) : j = arbitration	–	0.131*** (0.0488)	0.159*** (0.0545)
Round effect (β_j) : j = NA arbitration	–	0.002 (0.0065)	−0.007 (0.0085)
	–	−0.014*** (0.0037)	−0.035*** (0.0047)
Outcome effect (λ_j) : j = NA arbitration	–	–	−0.017 (0.0110)
	–	–	−0.048*** (0.0066)
R^2	0.306	0.315	0.348
$n = 1176$ (12 rounds for 98 pairs)			

Note: ***Two-tailed test is significant at 0.025 level.

just a constant yields an R^2 of just 0.01, indicative of the large explanatory role of bargaining pair heterogeneity. Column (2) presents estimates of the round model. The round variable for the arbitration treatments is strongly significant ($p < 0.001$), but not for NA ($p = 0.736$), all consistent with what we observed in Figure 11.2. As one would expect, the addition of round variables has a big impact on the estimate of the initial effect resulting from arbitration, increasing it from insignificantly above 0 to a very significant 13.1 per cent ($p = 0.019$). The λK terms are added to the model in column (3). The term for NA is not quite significant ($p = 0.112$), but the one for arbitration is strongly significant ($p < 0.001$). The test of the restriction that both K terms are equal to 0 is rejected ($p < 0.001$). So learning from outcomes is a better fit with the data than the baseline round-effect model. Bolton and Katok (1998) came to a similar conclusion with their data. Finally, the interested reader is referred to column (1) in Table 11.A1 in Appendix 2 on page 255–6. This version of the round-effect model breaks out initial effects and learning trends by treatment. Using this model, the hypothesis that all three arbitration treatments have similar learning trends cannot be rejected ($p = 0.610$). Similarly, the restriction that all three initial arbitration treatment effects are identical cannot be rejected ($p = 0.290$).

The full outcome model (Equation 11.2) is estimated in column (2) of Table 11.A1 Appendix 2. The coefficients for the $D_{i,j,t}$ variables are all negative, implying that the influence of a disagreement is to make future disagreement less likely; that is, bargaining pairs appear to learn from their 'mistakes'. The coefficients for the $S_{i,j,t}$ are all positive, although two of the four are not significant and all are of modest size relative to the disagreement coefficients. To economize on the exposition, Table 11.3 estimates the outcome model dropping the settlement variables. What we have to say is true for the model with settlements, estimated in Table 11.A1 in Appendix 2.

Column (1) of Table 11.3 provides a comparison of the influence of past disagreements on future disagreements for NA as well as the pooled arbitration treatments. Both coefficients are negative, although the coefficient for the arbitration variable is highly significant ($p < 0.001$), while the one for NA is smaller and not significant ($p = 0.322$). So there is more learning to avoid disputes from past mistakes in the arbitration treatments.

Column (2) tests whether learning is different between the natural focal points versus the synthetic compromise focal point. For the natural focal points, the coefficient is highly negative ($p < 0.001$). The coefficient for Δ A63 is the estimated difference the synthetic point makes; it is positive and significant ($p < 0.001$), indicating that there is less learning from mistakes for A63. Interestingly, the natural focal points are more effective at inducing learning than is the synthetic compromise focal point.

Column (3) permits an examination of whether there is any difference between the effectiveness of the two natural focal points. While the influence of A50 is somewhat larger than A75, a test of the restricted regression model

Table 11.3 Analysis of the disagreements model: LSDV coefficient estimates (standard error)

	(1)	(2)	(3)
Mean pair fixed effect (θ)	0.268*** (0.0315)	0.268*** (0.0310)	0.268*** (0.0310)
Initial effect (α_j) : j = arbitration	0.127*** (0.0373)	0.147*** (0.0369)	–
A50	–	–	0.141*** (0.0460)
A63	–	–	0.170*** (0.0472)
A75	–	–	0.135*** (0.0463)
Past disagreements (δ_j) : j = NA	−0.016 (0.0169)	−0.016 (0.0166)	−0.016 (0.0166)
Arbitration	−0.073*** (0.0094)	–	–
Natural focal point	–	−0.113*** (0.0114)	–
Δ A63	–	0.072*** (0.0118)	–
A50	–	–	−0.129*** (0.0184)
A63	–	–	−0.049*** (0.0140)
A75	–	–	−0.093*** (0.0175)
R^2	0.344	0.348	0.369
n = 1176 (12 rounds for 98 pairs)			

Note: ***Two-tailed test significant at the 0.025 level.

finds that the difference is not significant ($p = 0.451$). So there is no strong evidence that one natural focal point is more effective at inducing learning than the other.

Finally, we use Column (3) in Table 11.3 to test whether the trend-corrected dispute rates under arbitration are statistically the same in the final round of play as for NA. Predicting the 12th-round estimates for average dispute rate and associated standard errors, we find that the trend-corrected dispute rate for A50 is estimated at 8.6 per cent, and is statistically lower than the trend-corrected estimate of 22.6 per cent for NA (two-tailed $p = 0.012$). This results is robust to the inclusion of settlements in the model (that is, robust to using column (2) in Table 11.A1 in Appendix 2; two-tailed $p = 0.025$). For A75, the trend corrected estimate of the Round 12 dispute rate is 12.5 per cent, which is weakly significantly different from NA (two-tailed $p = 0.076$). The result, however, is not robust to the inclusion of settlements two-tailed $p = 0.281$). For A63, the trend-corrected estimate of the Round 12 dispute rate is 25.8 per cent, and this is not statistically different from NA, either with or without settlements; variables in the model (two-tailed $p > 0.600$ for both models). To summarize, there is robust evidence that last-round dispute rates for A50 are lower than for NA, some evidence that this is true for A75, and while trend-corrected A63 rates are estimated to be slightly higher for NA, they are not statistically so.

Settlements

What effect do the arbitrators have on settlements? Figure 11.4 displays settlements for each treatment, across all rounds. The graphs are broken out by the natural 'basin' for each of the arbitration processes. In NA, we see that equal division splits dominate equal expected value splits by about a 3-to-1 margin, while compromise solutions account for only about 10 per cent of all outcomes. This portrait changes little for A50, while for A63 and A75 we see substantial movements towards the basin implied by the arbitrator award distribution. A contingency table test of the data underlying Figure 11.3 rejects the hypothesis that all treatments exhibit the same distribution of settlements ($p < 0.001$). Note, however, that while the basin implied by the arbitrator is modal for both natural focal points, it is not for A63. There is

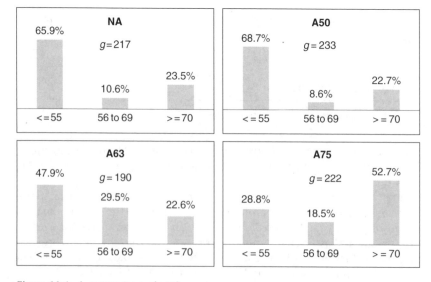

Figure 11.4 A comparison of settlements

apparently greater resistance to the decisions of the A63 arbitrator than to the others, and this may account for the slower learning we found in the models discussed on page 000.

Comparing early (first six) to late (last six) settlements finds little change in distribution for NA, A50 and A75 treatments ($p > 0.180$ in all three cases). There is a weakly significant change for A63 ($p = 0.096$). Interestingly, most of the movement in A63 is towards equal divisions; these rise from 42 per cent to 53 per cent of the total, whereas compromise solutions rise from 28 per cent to just 30 per cent.

Concluding remarks

To summarize, the analysis of the experiment produced five findings: (i) the introduction of arbitration initially increases the rate of dispute; (ii) but arbitration generally induces a learning effect not observed absent arbitration; (iii) outcome learning fits the data better than round learning; (iv) learning under either focal point arbitrator is faster than under the arbitrator positioned between the focal points – the compromise arbitrator; and (v) by the twelfth round, the dispute rate is lower under both focal point arbitrators – clearly significantly lower under one – than under no arbitration.

The explanation for these results we offer is that arbitration has two effects. First, the lower dispute costs that accompany arbitration initially induce more aggressive bargaining which leads to initially higher dispute rates. But arbitration also has a beneficial impact on disputes: in the experiment, it converts a game with two focal equilibria to a game with one, simplifying the co-ordination task, which speeds up learning. Interestingly, arbitration is more effective when the arbitrator is located at either focal point than when located at the compromise position. In terms of behaviour, this is primarily because of slower adjustment to disputes in the second stage. One potential explanation is that there is some interplay between initial bargaining positions and expectations of concession. The compromise solution requires both sides to make concessions from their 'natural' opening offers: all things being equal, I would prefer my partner make a concession first. Hence an expectation of the other side making a concession may slow down the rate of concession for both bargainers.

The major implication is that, while a short-term increase in dispute rates from the introduction of arbitration may be inevitable, a long-term increase is not. There are two important caveats: first, the laboratory experiment was done under very stable conditions. Specifically, both the bargaining problem and arbitrator behaviour were consistent across rounds. Disruption of either across time could conceivably disrupt the learning process that lowers disputes. Even if this is so, however, the experiment's results suggest a new explanation for why arbitration should increase disputes. Second, arbitrator behaviour in the experiment is transparent, in the sense that a brief history of awards conveys a pattern to bargainers. A second explanation for the difference in learning rates between focal arbitrators and compromise arbitrators suggests that this may be crucial: The arbitrator's favouritism towards one or the other bargainer's behaviour was probably easy for bargainers to comprehend (although perhaps hard for some to accept), while the reasoning for the compromise arbitrator may have been harder to grasp.

Finally, the results of the experiment are suggestive of the potential gains to be made from a more precise modelling of Schelling's concept of negotiation as a struggle to establish commitments and the role of focal points in this process. A better theoretical understanding of the basic mechanics of bargaining, particularly an understanding of why some negotiations end in an impasse, would facilitate greatly the development of more effective dispute resolution techniques, ones with fewer undesirable side effects.

Notes

1 See Lester (1984) for a description of how these systems work.
2 See, for example, the field evidence from Card (1990). Roth (1995) observes that dispute rates reported by complete and incomplete information experiments and those reported by field studies are all quite similar. He concludes that this 'raises some question about whether the incomplete information models are focusing on the underlying cause of disagreement', (p. 294).
3 This game was analysed by Harsanyi (1977). Crawford (1982) explains bargaining disputes using a modified version. Roth and Schoumaker (1983) use the game in a lab study of bargainer expectations. Bolton (1997) identifies limit-evolutionarily stable equilibria.
4 Bloom (1981) analyses the role of costs in arbitration. Some analysts identify a third cost associated with the potential damage done to a relationship when the parties are unable to settle a dispute on their own.
5 Of course, this assumes that bargainers behave in a risk-averse manner. Bolton (1995) provides evidence that this is indeed the case.
6 A somewhat more complex method of imposing an arbitration cost – one that combines elements of both out-of-pocket and uncertainty costs – would be to have the arbitrator award less than the entire 100 chips, thus effectively charging each bargainer chances in the lottery. The approach was rejected on the grounds that it would complicate the interpretation of the results significantly: if each bargainer is charged the same number of chips, then, because of the difference in the size of prizes, the cost of arbitration is effectively three times larger (in expected value terms) for bargainer B then for A. Such a sizeable difference might greatly complicate the comparison of learning rates. On the other hand, equating the cost of arbitration across bargainers would require us to charge A three times as many chips as B. A modest cost of $3 (in expected value terms) would imply a truncation of the arbitrator awards at 30 chips for bargainer A. The size of the mass point at the truncation would differ significantly across arbitration treatments, and again might greatly complicate comparison across arbitration mechanisms.
7 The adaptive learning models studied in the context of the ultimatum bargaining game by Roth and Erev (1995) and Gale *et al.* (1995) suggest that bargainers whose

payoffs vary more widely with their strategy choice learn more rapidly than those facing a smaller variance.

8 To minimize the role of boredom or fatigue in subject behaviour, lab sessions were restricted to a maximum of 90 minutes. The rationale for choosing twelve rounds of negotiation is that it was the maximum number of rounds that fitted comfortably into the maximum period.

9 Estimates of (11.2) incorporating temporary lags indicate that these variables have no substantial explanatory power. Bolton and Katok (1998) provide a detailed discussion of the formulation used and how it compares to formulations that have been applied to field data.

Appendix 1: written instructions provided to subjects

General. Please read the instructions carefully. If at any time you have questions or problems, raise your hand and the monitor will be happy to assist you. From now until the end of the session, unauthorized communication of any nature with other participants is prohibited.

At the end of the session, you will be paid a $5 cash show-up fee. During the session, you will play a series of bargaining games with another participant. Each game gives you an opportunity to earn additional cash.

Description of the Bargaining Game. The game involves two bargainers, Player A and Player B. They must decide how to divide 100 (abstract) chips. The game is played in two stages:

Stage 1. Each bargainer proposes a division of the chips. The computer will display both proposals simultaneously, meaning that neither bargainer will be able to see the other bargainer's proposal before making their own.

Stage 2. After reviewing both proposals, each bargainer decides whether to 'accept' the other bargainer's proposal or to 'maintain' their own proposal. The computer will display both decisions simultaneously, meaning that neither bargainer will be able to see the other bargainer's decision before making their own. At this point, are there any questions?

Chip Division Rules. The decisions made in Stages 1 and 2 determine how the chips are divided:

- If one bargainer 'accepts' and the other 'maintains', then the maintaining bargainer's proposal is agreed upon, and this determines the division of the chips.
- If both bargainers 'accept', then, using a process that is equivalent to a coin flip, the computer randomly chooses one bargainer's proposal as the agreed upon one.
- If both bargainers 'maintain', then the game ends in disagreement. Each bargainer receives zero (0) chips. At this point, are there any questions?

Role Assignments. You will have the same role, Player A or Player B, for all games. Your role is determined by the 'A' or 'B' that precedes the cubicle number you drew when you entered the room. Your role will also appear on your computer screen during games.

Pairing Procedure. You will be paired with the same person for all games. This person will be selected at random from the group of participants in the room who have the opposite role that you have. All pairings are anonymous: you will not know the identity of the person you are playing, nor will they know yours, nor will these identities be revealed after the session is completed. At this point, are there any questions?

Bargaining Record. Several blank 'Bargaining Records' are provided in your folder. At the conclusion of each game fill out one of these forms. Completed forms provide you with a history of your past games, and you may reference them at any time during the session.

Selection of the Payoff Game. You will be paid for one game. We will play more than one game. The one that you are paid for – the payoff game – will be selected by a lottery after all games have been completed. Each game has an equal chance of being selected as the payoff game, so it is in your interest to make as much as you can in each and every game. Immediately upon conclusion of the session, you will be paid your earnings in cash. Earnings are kept confidential.

Earning Cash from the Payoff Game. Players earn cash from the payoff game in two ways:

1. If the payoff game ended in agreement, then each bargainer automatically receives a $4 agreement fee. If the payoff game ended in disagreement (this happens only when *both* bargainers played 'maintain', and both received 0 chips), then each bargainer receives a $0 agreement fee.
2. In addition to the agreement fee, the chips you earn for the payoff game give you a chance to win an additional prize. For every game, the value of player A's prize will be $8, and the value of Player B's prize will be $32.

After the payoff game has been chosen, each player will spin the wheel behind me. The wheel is labeled with numbers 1 through 100. When you spin the wheel, each number has an equal chance of being selected. Each chip you earn in the payoff game gives you one chance of winning your prize. For example, if you earn 10 chips for the payoff game, your winning numbers on the wheel will be numbers 1 through 10. So when the wheel is spun, if it stops on a number 1 through 10, you win your prize. If it stops on any other number, you do not win. As a second example, if you earn 80 chips for the payoff game, your winning numbers on the wheel will be numbers 1 through 80. So when the wheel is spun, if it stops on a number 1 through 80, you win your prize. If it stops on any other number, you do not win. Note that if you earn 0 chips for the payoff game, then you have no chance of winning your prize.

To further explain how your cash earnings are determined, let's give the wheel a spin. *Spin wheel.* The wheel has stopped on number x. Suppose this were your spin. Had you earned x or more chips from the payoff game you would win your prize and your earning would be

Player A earnings if wins spin:	Player B earnings if wins spin:
Prize = $10	Prize = $30
Agreement Fee = 5	Agreement Fee = 5
Show-up Fee = 5	Show-up Fee = 5
TOTAL EARNINGS $20	TOTAL EARNINGS $40

Gary E. Bolton and Elena Katok   255

Had you earned less than x chips from the payoff game you would not win your prize and your earnings would be

Player A earnings if does not win spin:

Agreement Fee =   5
Show-up Fee =   5
TOTAL EARNINGS $10

Player B earnings if does not win spin:

Agreement Fee =   5
Show-up Fee =   5
TOTAL EARNINGS $10

If you did not make an agreement for the payoff game, then you would receive only the $5 show-up fee. At this point, are there any questions?

Average Value. For your convenience, the computer will automatically display the *average value* of the chips in a proposal. The average value is the *average* amount a bargainer would win on each spin if the wheel were spun many times. For example, with many spins of the wheel, a player with 10 chips wins her prize, on average, 10 per cent of the time. Hence, the average value to a Player A of a proposal giving her 10 chips would be 10 per cent (or .10) $\times$ $8 = $0.80. On the other hand, the average value of 10 chips to a Player B would be 10 per cent (or .10) $\times$ $32 = $3.20. Note that the calculation of the average value does not include the agreement fee. At this point, are there any questions?

Scratch paper and a pen have been provided (in your folder) for any further calculations you might wish to perform or if you wish to make private notes.

Consent Forms. If you wish to participate in this study, please read and sign the accompanying consent form. Please note: In order to collect your earnings from the game, you must stay until the end of the session, which will last about 90 minutes. We will now come around to collect the consent forms.

Appendix 2

Table 11.A1 Alternative approach to analysis in Table 11.2 (standard error)

	(1)	(2)
Mean pair fixed effect (θ)	0.232*** (0.0478)	0.234*** (0.0404)
Initial effect (α_j) : j = A50	0.112* (0.0689)	0.126** (0.0568)
A63	0.191*** (0.0676)	0.189*** (0.0574)
A75	0.091 (0.067)	0.093 (0.0566)
Round effect (β_j) : j = NA	0.002 (0.0065)	–
A50	−0.019*** (0.0064)	–

A63	−0.013** (0.0065)	–
A75	−0.010 (0.0064)	–
Past disagreements (δ_j) : j = NA	–	−0.024 (0.0174)
A50	–	−0.148*** (0.0200)
A63	–	−0.053*** (0.0151)
A75	–	−0.135*** (0.0206)
Past agreements (α_j): j = NA	–	0.011 (0.0081)
A50	–	0.019*** (0.0083)
A63	–	0.007 (0.0092)
A75	–	0.036*** (0.0094)
R^2	0.325	0.381

n = 1176 (12 rounds for 98 pairs)

Notes: * Two-tailed test is significant at 0.10 level; **0.05 level; ***0.025 level.

References

Ashenfelter, Orley (1987) 'A Model of Arbitrator Behavior', *American Economic Review*, 77, 342–6.

Ashenfelter, Orley and Bloom, David E. (1984) 'Models of Arbitrator Behavior: Theory and Evidence', *American Economic Review*, 74, 111–24.

Ashenfelter, Orley, Currie, Janet, Farber, Henry S. and Spiegel, Matthew (1992) 'An Experimental Comparison of Dispute Rates in Alternative Arbitration Systems', *Econometrica*, 60, 1407–33.

Bloom, David E. (1981) 'Is Arbitration *Really* Compatible with Bargaining?' *Industrial Relations*, 20, 233–44.

Bolton, Gary E. (1997) 'The Rationality of Splitting Equally', *Journal of Economic Behavior and Organization*, 32, 315–31.

Bolton, Gary E. and Katok, Elena, (1998) 'Reinterpreting Arbitration's Narcotic Effect: An Experimental Study of Learning in Repeated Bargaining', *Games and Economic Behavior*, 25, 1–23.

Bolton, Gary E. and Zwick, Rami (1995) 'Anonymity versus Punishment in Ultimatum Bargaining', *Games and Economic Behavior*, 10, 95–121.

Butler, Richard J. and Ehrenberg, Ronald G. (1981) 'Estimating the Narcotic Effect of Public Sector Impasse Procedures', *Industrial and Labour Relations Review*, 35, 3–20.

Card, David (1990) 'Strikes and Wages: A Test of an Aymmetric Information Model', *Quarterly Journal of Economics*, 105, 625–59.

Crawford, Vincent (1982) 'A Theory of Disagreement in Bargaining', *Econometrica*, 50, 607–36.

Currie, Janet and McConnell, S. (1991) 'Collective Bargaining in the Public Sector: The Effect of Legal Structure on Dispute Costs and Wages', *American Economic Review*, 81, 693–718.

Gale, J., Binmore, K. and Samuelson, L. (1995) 'Learning to Be Imperfect: The Ultimatum Game', *Games and Economic Behavior*, 8, 56–90.

Güth, Werner, Ivanova-Stenzel, Radosveta, Königstein, Manfred and Strobel, Martin (2003): 'Learning to Bid: An Experimental Study of Bid Function Adjustments in Auctions and Fair Division Games', *Economic Journal*, 113, 477–94.

Harsanyi, John C. (1977) *Rational Behavior and Bargaining Equilibrium in Games and Social Situations*, Cambridge University Press.

Lester, Richard A. (1989) 'Analysis of Experience Under New Jersey's Flexible Arbitration System', *Arbitration Journal*, 44, 14–21.

Lester, Richard A. (1984) *Labour Arbitration in State and Local Government*, Princeton, NJ, Princeton University Industrial Relations Section.

Nash, John (1950) 'The Bargaining Problem', *Econometrica*, 18, 155–62.

Reder, Melvin W. and Neumann, George R. (1980) 'Conflict and Contract: The Case of Strikes', *Journal of Political Economy*, 88, 867–86.

Roth, Alvin E. (1995) 'Bargaining Experiments', in J. Kagel and A. E. Roth (eds), *Handbook of Experimental Economics*, Princeton, NJ, Princeton University Press.

Roth, Alvin E. and Erev, Ido (1995) 'Learning in Extensive Form Games: Experimental Data and Simple Dynamic Models in the Intermediate Term', *Games and Economic Behavior*, 8, 164–212.

Roth, A. E. and Schoumaker, Françoise (1983) 'Expectations and Reputations in Bargaining: An Experimental Study', *American Economic Review*, 78, 362–72.

Roth, Alvin E., Malouf, Michael and Murnighan, J. Keith (1981) 'Sociological versus Strategic Factors in Bargaining', *Journal of Economic Behavior and Organization*, 2, 153–77.

Rubenstein, Ariel (1982) 'Perfect Equilibrium in a Bargaining Model', *Econometrica*, 51, 1, 99–109.

Schelling, Thomas C. (1963) *The Strategy of Conflict*, New York Oxford University Press (quotes excerpted from 1980 reprint).

12

Communication and Co-operation in a Common-Pool Resource Dilemma: A Field Experiment

Juan-Camilo Cardenas, T. K. Ahn and Elinor Ostrom[*]

Introduction

One tenet of classical, rational choice theory as used in non-cooperative game theory is that all players use the same model of rationality for themselves as well as for all other players. The assumption of homogeneous, self-interested actors helps theorists to model how individuals would make choices. One justification for positing homogeneous, rational, egoistic actors has been evolutionary theory (Dawkins, 1976). That is, even if individuals tried out different ways of behaving, only those who made decisions consistent with rational egoistic decisions would maximize returns. In a highly competitive environment, those who maximize returns are more likely to survive in the long run. Long ago, Armen Alchian (1950) made a cogent theoretical argument that, in a highly competitive market, selection pressure would weed out those market participants who did not maximize profits. Extensive experimental studies of behaviour in competitive market settings have supported the use of the classical, rational choice model as the only model of individual choice needed in this setting to make empirically supported predictions (Smith, 1962; Plott, 1986). Thus, continuing to use the classical model when analysing competitive markets has both strong theoretical and empirical support.

In the early 1980s, however, Werner Güth and colleagues began a series of experiments on the ultimatum game, which challenged the capability

[*] We appreciate the support of the National Science Foundation (Grant No. SES 0232072). Juan-Camilo Cardenas expresses his gratitude for a Research and Writing Grant from the John D. and Catherine T. MacArthur Foundation, and from the MacArthur Norms and Preferences Network, who funded the field experiments. The enthusiasm of Maria Claudia Lopez, Diana Maya, Ana Maria Roldan, Lilliana Mosquera and Pablo Ramos in the field made the fieldwork possible. We are appreciative of the friendly hospitality of the Colombian villagers who participated in this study. Patty Lezotte once again did a superb job of editing this manuscript for us.

of the classical model to explain behaviour in some non-market settings (see, for example, Güth *et al.*, 1982; Güth, 1995a). In an ultimatum experiment, a 'Proposer' makes a take-it-or-leave-it proposal for how to divide an amount of money, and a 'Responder' chooses to accept the offer (in which case the division is made as proposed) or reject it (in which case neither of them receives any funds). The game-theoretic prediction is that the Proposer should offer the smallest positive amount and the Responder should accept anything above zero. Nothing could be clearer in theory. Güth and his colleagues, however, found that neither prediction was substantiated in the lab (Güth, 1995a). In experiments that have been replicated many times in many countries, Proposers tend to offer between 40 per cent and 60 per cent and Responders tend to reject any offer below 20 per cent (see Camerer, 2003, tables 2.2 and 2.3, where data from fifteen experiments is summarized).[1]

Equally dramatic findings have come from experimental studies examining behaviour in social dilemmas including public goods (Isaac *et al.*, 1985; Isaac and Walker, 1988) and common-pool resources (Walker *et al.*, 1990; Ostrom *et al.*, 1992; Casari and Plott, 2003). The theoretical prediction in social dilemma games is that players will not make decisions that would lead to a group optimum. Rather, players are predicted to play strategies leading to a suboptimal Nash equilibrium. Simply allowing the players to communicate with one another without external enforcement of agreements does not change this prediction. Behaviour in public good and common-pool resource experiments, however, deviates substantially from the Nash equilibrium strategies when subjects are merely allowed to communicate with one another (Ostrom and Walker, 1991).

Given repeated findings from carefully designed and replicated experiments, multiple scholars have now concluded that the assumption of the classical model about homogeneous, own-payoff maximizing players cannot explain behaviour in a wide variety of non-market settings. A substantial number of alternative theories have been proposed to explain these findings. Many of the alternatives assume that players take the payoffs made to other players into account (either positively or negatively) in their *own* utility function (see, for example, Fischbacher *et al.*, 2001; Bolton and Ockenfels, 2000). Further, in addition to assuming that players may be other-regarding, a key assumption is that the 'nature of the other-regarding capacity could differ from person to person' (Casari and Plott, 2003, p. 243).

Güth's indirect evolutionary approach is an important theoretical step toward understanding how individuals, who do not maximize their own immediate payoffs, could survive in games with 'rational egoists' who *do* pursue the predicted strategy (Güth and Kliemt, 1998). Güth has identified a key factor that facilitates the evolution of co-operative or fair behaviour. This is the ability, on the part of the fair-minded types of players, to make contingent decisions (Güth and Yaari, 1992; Güth, 1995b; Güth *et al.*, 2000).

In other words, Güth and his colleagues have posited the existence of more than one type of player.

Brosig (2002) classified players using techniques developed by social psychologists (Liebrand, 1984; McClintock and Liebrand, 1988) into altruists (who maximized others' payoffs), individualists (who maximized their own payoffs), and co-operative types (who maximized the sum of joint payoffs). In a four-player repeated public-good game, Kurzban and Houser (2003) find three types of players: free riders (20 per cent of their sample of 84 subjects), unconditional co-operators (13 per cent), and conditional co-operators (67 per cent). They found that the strategies of these three types of players were stable across an initial round of plays as well as in an additional set of three games played after the first set was completed.

Assuming heterogeneity of types of players has recently become somewhat more accepted as scholars have tried to make coherent explanations of the extensive non-market experimental research conducted since the 1980s (Ostrom, 1998). As soon as one assumes that multiple types of players exist, however, it becomes more difficult to predict how players will in fact behave, either in the field or in an experimental laboratory. A key problem is that while one can assume that each individual knows his/her *own* type, how do participants know the types of those with whom they interact in a laboratory or field setting? How does a conditional co-operator know s/he is interacting with other conditional co-operators? Do the internal weights of a utility function remain constant over time without regard to the type of situation that players are in, or the behaviour of others in a particular situation?

A consistent finding across experimental studies in common-pool resource and public good settings is that being able to engage in face-to-face communication is a major factor enhancing the proportion of individuals who co-operate, thus producing higher group payoffs (Orbell *et al.*, 1988; Ostrom and Walker, 1991; Ledyard, 1995; Kollock, 1998; Kopelman *et al.*, 2002). In a meta-analysis of more than 100 experiments, Sally (1995) finds that face-to-face communication significantly raises the rate of co-operation in two-person games. We speculate that there is something about face-to-face communication that increases the capacity of individuals to identify the types of players with whom they are interacting. This does not imply, however, that identifying the types or intentions of others in a group guarantees co-operation. Through a group discussion, a conditional co-operator may in fact detect that there are fewer conditional co-operators than s/he originally thought, and therefore adopt a strategy of not co-operating so as to avoid the bad payoff associated with unilateral co-operation among defectors.

For some time, scholars have been trying to sort out the various aspects of the communication process within experiments to better understand its impact (Dawes *et al.*, 1977; Messick and Brewer, 1983; Dawes *et al.*, 1990). In a recent effort, Kollock (1998) summarizes the evidence for four factors that Messick and Brewer (1983) had earlier suggested as being plausible reasons for communication to enhance co-operation levels. These are: (i) communication helps players to detect the actions that others are most likely to take (or, as we develop below, detect the types of players with whom they are interacting); (ii) it allows players to make promises or commitments; (iii) it allows a process of moralization among players; and (iv) it can create or reinforce a sense of group identity. Kopelman *et al.* (2002) review these potential explanations in the psychology literature and conclude that the hypothesis suggesting that group discussions tended primarily to elicit a commitment to co-operation had the most consistent empirical support (see also Kerr and Kaufman-Gilliland, 1994; Bouas and Komorita, 1996). Frank (1988) had speculated that the more co-operative types of players were also more likely to signal their type, as well as to recognize the types of other players with whom they were paired. Brosig (2002) confirms Frank's speculations by first classifying players, then allowing them to communicate with each other, and finally asking them to predict the likely strategies of the other players.

Several scholars have stressed the importance of players being able to study each others' faces as a key part of detecting their types. Scharlemann *et al.* (2001) and Eckel and Wilson (2003), for example, explored the reaction of individuals to seeing the face with whom they were supposedly interacting in a laboratory setting. Their results support the power of smiles as a mechanism to allow players to read the intentions of others and therefore to create trustworthiness. In their design there is no face-to-face communication, but information about facial expressions are used as treatment variables among strangers. They are able to explain variations in behaviour within trust game situations, and show that when players are shown pictures of smiling individuals, they respond with more trust and co-operation. Although the pictures are not of the actual people with whom they are interacting, the facial expression does induce behavioural changes in ways that are consistent with the literature explaining how humans use such information to detect intentions by others (Frank, 1988; Schmidt and Cohn, 2001).

Bohnet and Frey (1999) explored whether it was what people said to one another or simply seeing one another that made a crucial difference to the level of co-operation in prisoner's dilemma games. They found that communication is not always required to increase levels of co-operation. Silent identification of who would be involved in an experiment was by itself sufficient to increase the level of co-operation. They found that the variance in behaviour was greater once face-to-face communication was substituted for silent identification prior to the experiment.

Other scholars who have been interested in separating out the communication aspect from the face-to-face aspect examined the impact of allowing people to communicate via computerized messages. Rocco and Warglein (1996) replicated the common-pool resource experiment of Ostrom *et al.* (1994) and obtained very similar results when they allowed subjects to communicate on a face-to-face basis. On the other hand, allowing individuals to communicate via e-mail messages was much less effective in enhancing co-operation than allowing people to engage in face-to-face communication (see also Frohlich and Oppenheimer, 1998).

In another study, Bochet *et al.* (2002) compare face-to-face communication with the exchange of messages through computer terminals. Players had the opportunity to send messages to the rest of their group members, as numerical 'possible' allocations or as verbal communication, followed by their actual decisions. The authors confirm the effectiveness of pre-play, face-to-face communication, but also find that the verbal and anonymous chat room with open verbal exchange was almost as effective as the face-to-face case. They point out that the results are potentially in contradiction with the findings by Rocco (1998) and Frolich and Oppenheimer (1998), who found that e-mail communication was not as effective as the face-to-face exchange. Bochet and co-authors suggest that e-mail creates a different environment from that of a chat room, with more difficult and slower feedback to be able to elicit intentions by other players.

Even after all the speculation and previous empirical research, the role of communication in enhancing co-operation has not yet been fully explained, especially in more complex games involving more than two players. From conducting many dilemma experiments involving more than two players (and listening to the recorded tapes and reading the transcripts of these experiments), we speculate that a great deal of the communication is devoted to two group tasks. The first task is problem clarification among the players. The second task is type-detection (see Simon and Gorgura (2003) for an analysis of the content of the common-pool resource experiments conducted at Indiana University).

The first task of problem identification is non-trivial when more than two players are involved. Group discussions allow players to teach the confused players among their group about the structure of incentives and the trade-offs between individual and group outcomes that exist in a dilemma setting. Ostrom *et al.* (1994, p. 151) suggest from the transcripts of their experiments that the discussions focused on determining the maximum possible yield and how to achieve it. Two of the main characteristics of the production function of common-pool resources, their partial excludability and partial subtractability, sometimes imply a complicated task for individual players. For example, it can be the case that no dominant strategies exist regarding the level of individual appropriation. Therefore, the same individual level could be either beneficial or harmful for group or individual outcomes. Working out the best group strategy, however, does not guarantee that players would pursue it, since the incentives to deviate from group optimum are at the core of the dilemma.

The second task, we argue, is type-detection. Once subjects have used their communication time to work out the structure of the situation and what would be the best joint outcome, they frequently turn to a discussion of what each person thinks everyone should do. They tend to make promises to one another by looking each other in the face as they are discussing their promises. Individuals begin to size up the trustworthiness and co-operativeness of the other individuals with whom they are situated and choose a best response accordingly.

McCabe and Smith (2003) suggest a cognitive model of goodwill accounting where players use a set of mental modules that allows players to gather information from the environment and from the other players to inform their decision to trust or reciprocate. One of these is closely related to the first task we suggest that groups accomplish through face-to-face communication – namely, to clarify the link between the subset of individual actions needed to obtain a group beneficial outcome. The other two modules in the McCabe and Smith cognitive model, the 'friend-or-foe' and the 'cheat detection', are associated with our argument that group discussions allow individuals to update their accounting of the goodwill they may have received from the other players with whom they interact. Through these modules, each player gathers crucial information about the others' intentions. Each player is also aware that if other players use the same modules, the player needs to send the correct signals so that the others' goodwill accounting is updated according to the best strategy that optimizes payoffs for all. Otherwise, they will be perceived as a cheat and no goodwill will be extended.

However, there are a few particular conditions specific to the problem we are studying here; that is, groups sharing a common-pool resource worth highlighting. First, group size is almost always larger than two, which increases the complexity of the information processing about the others and about type-detection. Second, as in many social dilemmas, such a process occurs within an environment of repeated rounds rather than in a one-shot game. Also, many of these situations involve a fixed group of players who interact through the game by having only partial knowledge of the individual decisions made in previous rounds, usually observing only average or aggregate outcomes given the large personal cost of observing individual actions.

In this chapter, we shall review evidence from recent experiments on common-pool resources conducted in the field to explore the existence of the mechanisms just described, and to illustrate the effectiveness of communication as an enhancing mechanism for co-operation. The features of the experimental design offer some particularities regarding the subject pool, the composition of the groups, and the institutional environments being compared.

Experimental design

The experiments reported here were conducted in villages in rural Colombia where participants were not only familiar with the use of common-pool resources – such as fisheries, water or firewood – but also knew each other and had a prior history of reputation building before the experimental sessions. Therefore we can also assume that there would be a high probability of meeting each other after the experimental session ended. This would make the use of type-detection and reading of intentions by players during the experiment more salient than when participants were total strangers, as in the case of college students, where anonymity and confidentiality of individual choices are common. Also, we expect greater heterogeneity of types in terms of rationality and familiarity with the task to be brought by participants.[2]

Our experimental design is very simple. In a session of twenty rounds, each of five players in a group has to decide a level of extraction between 1 and 8 units of a resource during two stages of ten rounds each. We framed the situation as one in which individual households have to decide about the extraction level of a resource such as fish, firewood or water. In each round, a monitor collected decisions and recorded them privately and confidentially. The monitor added the individual extraction levels and announced the total extraction for the group in that round. By knowing the group extraction and their own individual extraction, players were asked to calculate their individual earnings according to the payoff table[3] (see Appendix on page 279). However, players did not know the individual decisions of the others in the group, just their aggregate extraction. This procedure was repeated for twenty rounds. At the end of all the rounds, earnings were added for each participant, and each was privately paid in cash.

In the payoff table, one observes that increasing one's extraction yields higher individual earnings, but aggregate extraction decreases them, as is typical in any common-pool resource setting. Assuming a rational maximization of individual payoffs among material payoff-maximizers in a non-cooperative game, the Nash equilibrium is located at the bottom-right corner of the table, where each player obtains 320 points. The social optimum occurs when all players choose one unit as their level of extraction, yielding 758 points for each player.

The data that follows were gathered in a series of thirty-four sessions conducted between the years 2000 and 2002 in different villages of Colombia where people depend in part on the use of a local ecosystem. Each experimental session was conducted with five participants, all of whom lived in the same village. Groups were randomly formed, but we avoided members of the same household participating in the same session. We assume that the subjects in these field experiments would have some prior information about their neighbours in their session that they would probably use strategically in their decisions – namely, for detecting the intentions of the other four players.

Experimental treatments: baseline versus one-shot versus repeated face-to-face communication

All thirty-four sessions were run in twenty rounds, split into two stages. After the instructions were read aloud and questions answered, we started with at least one or two practice rounds. After the practice rounds, we initiated the experiment. During the first stage of ten rounds, all sessions were run under the same set of rules. The subjects were notified that the experiment would last at least ten rounds, and that during these rounds no communication among themselves would be allowed. The villagers were seated in a circle facing outwards so that the privacy of the decisions could be maintained. Once the ten rounds were over, the monitor announced that a second stage was about to start, under a new set of rules, for another ten rounds. None of the groups knew in advance during the first ten rounds the type of new rules for the second stage of the game.

We distributed the thirty-four sessions across three different treatment designs. For eight of the sessions we used a baseline treatment where the second stage was run under identical conditions – that is, decisions were made privately and no communication was allowed among the players. This is our Baseline (B) treatment. For another thirteen sessions, our One-shot communication (OSC) treatment, the subjects were allowed to have a single, face-to-face open discussion for five minutes before Round 11, but none thereafter. They were asked to turn their seats 180 degrees so that they could see each other during the discussion time.[4] Once the discussion had concluded, they had to turn their seats to face outwards again, and proceed with their individual, confidential decisions for the rest of the stage. These thirteen groups were told in advance that such a discussion would happen only once before Round 11, and that for the rest of the rounds they would make their private decisions under the same no-communication rule.

For the remaining thirteen groups, our Repeated communication (RC) treatment, we replicated the previous design, but groups were allowed to have face-to-face communication before each of the rounds from 11 to 20. Players were still asked to turn round after their face-to-face discussions and to make their decisions in private at the end of each round. Table 12.1 summarizes the experimental design, sample sizes, and treatments.

Conjectures about the effects of communication

The literature discussed in the Introduction, supporting the effectiveness of face-to-face communication in experimental studies from psychology and economics, coincides with the overall finding that humans are more likely to co-operate in social dilemmas after communication among players, even under non-binding agreements, and incomplete information about the individual choices of the others in the group. Which of the four mechanisms discussed in the introduction – detection of others' actions, making of promises, moralization, or building of group identity – induces co-operation, or whether all four are important, however, is still not agreed. If communication induces changes in individual behaviour, we should expect a significant change after Round 11 in both experiments involving communication in comparison to the Baseline treatment, where no communication was allowed for all twenty rounds.

Table 12.1 Treatments, designs and sample sizes

Treatment	Number of sessions	Number of people	Stage 1 (10 rounds)	New rule	Stage 2 (10 rounds) Round 11	Rounds 12–20
(B) Baseline	8	40	$X_1, X_2, \ldots, X_{10}$	(No change – control)	X_{11}	$X_{12}, X_{13}, \ldots, X_{20}$
(OSC) One-shot communication	13	65	$X_1, X_2, \ldots, X_{10}$	1 face-to-face group discussion ($t = 11$ only)	$(C-X)_{11}$	$X_{12}, X_{13}, \ldots, X_{20}$
(RC) Repeated communication	13	65	$X_1, X_2, \ldots, X_{10}$	Face-to-face communication before each round decision	$(C-X)_{11}$	$(C-X)_{12}, (C-X)_{13}, \ldots, (C-X)_{20}$
Total	34	170				

Regarding the OSC versus the RC treatments, it is difficult to predict with precision what differences there should be during the second stage. Some of the mechanisms through which communication may work could come into play from the very first round of communication and remain over time with no need for reinforcement. Others may require more than one group discussion. Earlier common-pool resource experiments did find one-shot communication substantially less effective than repeated communication (Ostrom *et al.*, 1994).

In our experiments in the field, we observe that, in addition to the four reasons that have already been posited for the efficacy of communication, the RC design allows participants first to clarify with each other the nature of the task and to identify collectively the socially optimum solution. Only after the structure of incentives is clarified and the optimal solution is identified do participants appear to engage in some of the other mechanisms, such as group identity, verbal commitment to the others, and ultimately the attempt to detect what they expect the others to do in the following round.

Experimental results

This section reports the experimental results. In the following discussions, B denotes the Baseline treatment in which no communication was allowed in any of the 20 rounds. OSC denotes One-shot communication treatment, in which a one-time communication was allowed between Stages 1 and 2–that is, between Rounds 10 and 11. RC denotes Repeated communication treatment, in which communication was allowed before each round of Stage 2.

Figure 12.1 plots the average levels of extraction in the three treatments with 95 per cent confidence intervals. Confidence intervals are shown to help readers see whether the differences in average levels between two treatments are significant; if two confidence intervals for a given round overlap, the difference is not significant at the 95 per cent confidence level. The average extraction levels in the B treatment are marked with circles, those in the OSC treatment with rectangles, and those in the RC treatment with triangles. As the figure shows, the average extraction levels in the first stage (Rounds 1 to 10) are very close to one another and not significantly different from each other. Thus the confidence levels are shown only for the second stage, where the differences are more pronounced.

Figure 12.1 shows that the average behaviours in different treatments are generally consistent with our conjectures outlined in the previous sections. That is, communication does help to reduce the extraction level and achieve higher levels of efficiency. Furthermore, repeated communication is more effective than one-shot communication. As the confidence intervals illustrate, however, substantial variance exists across sessions of each treatment. Appendix Figures 12.A1, 12.A2 and 12.A3

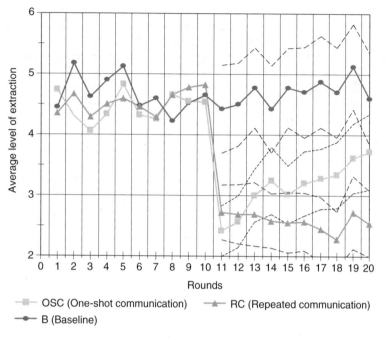

Figure 12.1 Average extraction over rounds (baseline, one-shot and repeated communication)

(see pages 280–82) present the average extraction levels in each session. Notice, from Figure 12.1, that the average extraction levels in all three treatments are above the socially efficient level of 1, but well below the equilibrium prediction of 8.

Table 12.2 provides the results of Wilcoxon rank-sum difference of means tests for selected pair-wise comparisons of average group-level extraction. We are comparing here the average group extractions rather than individual levels of extraction. For example, the first row shows that the average per-round group-level extraction in the final three rounds of Stage 1 is 22.08 in the B treatment and 23.04 in the OSC treatment. The *p*-value of 0.7720 in the last column indicates that these two averages are not significant.

We have chosen to compare the average group-level extraction in the final three rounds of Stage 1, the first three rounds of Stage 2, and the final three rounds of Stage 2. There are many other candidates for comparison, but we have chosen these three segments for the following reasons. First, the average extraction levels in the final three rounds of Stage 1 are compared to see whether the groups in different treatments are truly comparable. That is, it is possible that different treatments were assigned by accident to groups with different characteristics. If this were the case, differences across treatments

Table 12.2 Pair-wise comparisons of average group extraction levels

Rounds	Treatments compared	Average group extraction	P-values
	B versus OSC	22.08 versus 23.04	0.7720
8–10	B versus RC	22.08 versus 23.79	0.3840
	OSC versus RC	23.04 versus 23.79	0.4113
	B versus OSC	23.25 versus 13.85	0.0030
11–13	B versus RC	23.25 versus 13.51	0.0030
	OSC versus RC	13.85 versus 13.51	0.9386
	B versus OSC	24.04 versus 17.63	0.1110
18–20	B versus RC	24.04 versus 12.56	0.0007
	OSC versus RC	17.63 versus 12.56	0.0723

Note: B = baseline treatment with no communication; OSC = One-shot communication; RC = Repeated communication. Number of observations is equal to the number of groups in each treatment: 8 in B; 13 in OSC; and 13 in RC.

in Stage 2 would not be because of different levels of communication, but rather because of the differences of group characteristics that are irrelevant for our research purposes. Second, the first three rounds of Stage 2 have been chosen to see the immediate effect of communication compared to the non-communication environment in Stage 1. Third, the final three rounds of Stage 2 have been chosen to see the long-run effects of one-shot and repeated communication.

As the first set of three comparisons shows, there were no differences across treatments at the end of the first stage. We obtained the same results when we tested the significance of differences in Round 10 only, or for all ten rounds of Stage 1. The second set of tests shows that communication generated immediate positive effects. That is, both the groups in OSC and RC reduced their extraction levels significantly compared to the groups in the B treatment. The difference between OSC and RC was not significant at all ($p = 0.9386$). The third set of comparisons shows that, in the long run, only the groups in RC treatment were able to sustain the reduced levels of extraction. That is, the extraction level of the groups in OSC increased over time so that the difference in the average group-level extraction between B and OSC was no longer significant towards the end of Stage 2 ($p = 0.1110$). The groups in RC treatment, however, were able to sustain the lower levels of extraction over time. In fact, not only is the average group-level extraction of RC in the final three rounds of Stage 2 significantly lower than that in B ($p = 0.0007$), it is also significantly lower than that in OSC ($p = 0.0723$).

While Table 12.2 shows the difference across treatments, there are variations across groups in each treatment (see Appendix Table 12.A1 and Figures 12.A1, 12.A2, and 12.A3 on pages 279–82). In particular, it is worth noting that the effects of One-shot communication and Repeated communication manifest in different ways in different groups. Thus, while the average group-level extraction increased over time in OSC, some groups were able to maintain or achieve very low levels of extraction (see groups VCX13 and VCX15 in Appendix Figure 12.A2 on page 281, for example). Similarly, while the average group-level extraction across groups in RC was significantly lower than those in B or OSC, some groups in RC achieved mutual co-operation immediately after communication was allowed, and maintained the beneficial mutual co-operation successfully over time (see groups GCX5_t1 and PCCX5_t3 in Appendix Figure 12.A3 (see page 282), for example). It then took several rounds to reduce the extraction level for other groups (group CHCX5_+11 in Appendix Figure 12.A3, for example). Still other groups were not quite successful in maintaining consistently lower levels of extraction even with repeated communication (see groups GCX5_+2 and PPCX5_t2 in Appendix Figure 12.A2, for example).

On average, however, outcomes in the RC treatment groups were more efficient than in the OSC groups. Although in both treatments all groups reduced their extraction level in Round 11, groups in the OSC treatment increased the extraction levels over time after the initial communication. By the end of the second stage, the average group extraction in the OSC treatment was about four units higher than in the RC treatment, although it was still almost ten units below the group extraction in the B treatment. When comparing the extraction levels in the beginning and at the end of the second stage in each treatment, we find that only the OSC treatment shows a trend (from low to high), while the B treatment shows consistently high levels of extraction and the RC treatment shows consistently low levels of extraction over the rounds.

A more detailed look at the change in decisions by players over time provides us with the information about the underlying individual-level behaviour that resulted in the group-level patterns over time. In Figure 12.2, we present histograms of the fraction of individuals with different extraction levels, by round and by treatment for the second stage of the experiment. The first columns of the histograms show that the extraction levels in the baseline treatment are spread widely between the minimum of 1 and the maximum of 8 in each of the ten rounds.

On the other hand, the distribution of extraction levels in OSC treatment is initially skewed toward lower levels, implying that most players chose to co-operate with others immediately following the One-shot communication. Over time, however, many low extractors switch to medium- or high-level extractions. Thus, by the final rounds of the second stage, the distribution is much flatter than that in the initial rounds. The third histogram of each column shows that the distribution remains skewed over time. That is, the switch made by initial low-level extractors to high-levels of extraction did not happen in the RC treatment. The distribution pattern in OSC is closer to that in RC in the initial rounds, but becomes more similar to that of B over time.

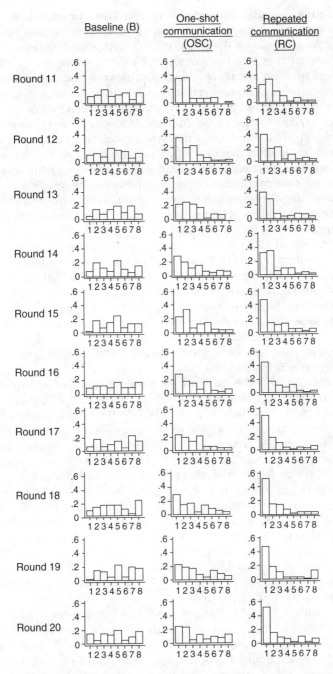

Figure 12.2 Distribution of decisions by treatment over rounds (second stage, Rounds 11–20)

The initial similarity and subsequent divergence between OSC and RC indicates that the differences after Round 11 between the two treatments are less likely because of the effects of communication on clarifying group incentives (what is best for group?), but rather because of the effects of communication on signalling and reinforcing co-operative intentions and behaviour. We have also argued that type-detection should be more effective during the RC treatment than with the OSC treatment. In the OSC treatment, a player would construct a preliminary distribution of types and would not have further information after each subsequent round to update this estimated distribution about the other four group members. This creates a higher error in detection of types than in RC, where players improve their estimate over rounds. Watching facial expressions, as well as hearing the discussions, allows players to compare commitments in the previous round with average outcomes obtained in that round and therefore update their estimates.

Our experiments involve people in villages who face the common-pool resource problem on a daily basis, providing one way of testing the external validity of earlier common-pool resource experiments utilizing undergraduate students. Running experiments in the field also presents challenges for the researcher. The fact that each of the groups is composed of five people from the same village implies that pre-play and *ex-post* incentives may come into play. Cardenas (2003a) discusses these elements in detail and suggests that there are valuable lessons that can be learnt when one takes into account the village-specific conditions.

One exercise with the current data illustrates some of these issues. As shown in Figure 12.1 and the Appendix (see pages 279–82), groups in the same treatment show large variances among them. This variation across groups persists during the second stage and shows that, even within treatments, some groups were able to achieve substantially higher levels of co-operation than others. What accounts for this within-treatment variation? We conjecture that part of the reason is the heterogeneity of types and the particular composition of each group. It is reasonable to assume that each subject in a group had prior information about the other four players in the group – a significant difference from the typical experiments with student subjects – acquired during their daily interactions with others in the village. Thus we hypothesize that the baseline composition of different types in each group was more or less known to each member of a group, and this knowledge, in turn, generated correlation between group members' behaviour in the first and second stages.

To test this, Figure 12.3 plots the correlation between the average extraction level in the final three rounds of the first stage (Rounds 8 to 10) and those in the final three rounds of the second stage (Rounds 18 to 20) for each of the thirty-four groups. Groups in the B (OSC, RC) treatment are marked by circles (rectangles, triangles) and the Pearson correlation coefficients are

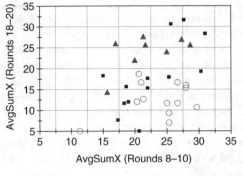

■ OSC (Pearson = 0.678) ○ RC (Pearson = 0.319)
▲ B (Pearson = 0.566)

Figure 12.3 Average group-level extraction at the end of the first and second stages, by treatment

calculated for each treatment. The figure and the correlation coefficients suggest that, particularly for the B and OSC treatments, the variation in the levels of co-operation at the end of the second stage can be explained by the level of co-operation at the end of the first stage.[5] Notice that even in the absence of communication, some groups in the B treatment maintain low levels of extraction in both first and second stages. We interpret the high correlations between the behaviour in the first and second stages as evidence that subjects in the field experiments brought knowledge into the lab about the composition of types in a group and their likely behaviour.

Though not reported in this chapter, a replication of the current experiment utilizing student subjects did not show the levels of correlation between the behaviour in the first and second stages. The experiment did, however, replicate the positive effects of communication in both the OSC and RC treatments. The fact that in experiments with villagers, group behaviour in the first stage can explain group behaviour in the second stage. That is not the case in experiments with student subjects, which indicates that our subjects in the field used information they had about the types of others in the group and brought that into the experiment.

Communication and co-operation: a look at the transcripts

We speculated earlier that at least two tasks are accomplished when players communicate with each other. First, a group discussion allows the subjects to clarify the nature of the dilemma and to make explicit agreements about group-orientated goals – in this case, that lower levels of extraction yield higher individual and group returns. The second task we posited is type-detection, which we discuss below.

With a non-linear functional form and five players, it can be a confusing task for some of the participants to know for sure what is the optimum strategy for them to pursue. In such a setting, participants need to discuss alternative strategies to determine the best one available to them. Examples from the transcripts of the audiotapes during the first round of communication for several groups illustrate how they pursued this task. In this dialogue between two players during the opening of the first round of communication, one of them explains the nature of the relationship between total extraction level and individual earnings, while the other begins to understand the dilemma:

Player 1: We run out of points, but we end up with more earnings if at least, let's say, we choose 1, it adds to 5; another example we all play 2, it adds to 10, so you think, say, about choosing 2.

Player 2: Number 2 . . .

Player 1: Then 10 minus 2 equals 8.

Player 2: 8.

Player 1: Then by choosing 2, it yields more points the lower numbers we choose.

Player 2: The lower our number, more points we get.

Player 1: Let's put that number here . . . there, more points this way.

Player 2: Yes, so that we . . .

Player 1: It is better to get more points, the more points the better, the lower the points we choose, the number right here will favour us in the table there.

In another village, where shrimp-fishing is very common, we observe a similar pattern in the opening round of group discussion. Further, notice the metaphor they use for the first round of group discussion when referring to fishing in kilos (kilograms) as the units of extraction in the experiment:

Player A: 2 kilos.

Player B: Tomorrow I am going to fish 2 kilos.

Player C: Tomorrow, 1 kilo.

Player A: Well, my thing is to be in agreement; that is, with the word of my friend here, if you know what I mean; to me, let us catch a kilo each so that it does not turn scarce later on, and then with pleasure it is good to come back the next day and catch another kilo and that way we don't run out of shrimp; if we catch 5 or 4 kilos the next day we go out there and there is no more; so, to me, I say we catch a kilo each.

Player B: Yes, that's the way, easy so that it stays good, that's the idea, the gasoline.

Player C: And then, when the kids go fishing tomorrow they will have their chance to catch a kilo each, otherwise look how we are now.

Player A: We all agree.

Several players at same time: Yes, don't worry.

In this second case, the group discussion allowed the players to relate the exercise to their own shrimp-fishing activity and easily find a strategy that will make sense regarding the need for extracting small quantities of the resource to yield higher returns to the players.

In the other sessions for which we have audio recordings, we observe a regular sequential pattern during the conversations. The pattern could be described as a sequence of the following steps for building an effective agreement for co-operation:

Step 1 Identification of the goal for the group and clarification to all group members that a lower level of aggregate extraction can increase individual earnings. In most cases, one or two players make these comments.

Step 2 An agreement or ratification of the need for every player to choose a low level of extraction. In several cases, it was as explicit as agreeing that one unit per player would achieve the maximum earnings for the group. In other cases, it was more frequent to observe agreements such as 'low numbers for extraction'.

This pattern occurred for most sessions during Round 11 in the second stage under communication. The first round of discussion usually focused primarily on these first two elements. We found no reference to the detection of types during the first round of group discussion.

However, during the next rounds (12–20) for those groups that were allowed repeated communication in the RC treatment, we observed at least two more steps or kinds of interventions by players:

Step 3 Reinforcement of an agreement that members had previously made. In several cases of repeated communication, we observed a permanent call by some in the group to maintain a low level of extraction. Others brought up the comparison of rounds with higher and lower yields in the payoffs table, or how detrimental it was for the group that one player increased his or her extraction to obtain higher payoffs.

Step 4 Discussions about types and type-detection strategies. Often, some players would argue that total extraction was high in previous rounds, and that somebody was probably increasing his or her level of extraction, causing damage to the rest of the group. For some groups, such a call was at times directed at someone in particular, usually when group members knew each other well.

For the latter case, we were able to observe examples like the following dialogue in one of the sessions under repeated communication:

Player X: The idea, the idea is that we all choose 1 in all rounds, I think that . . . We are all expecting to make some gains here.
Player Y: Of course!
Player X: Therefore everybody gets together . . . That means there is no God, as long as no one breaks the trust.
Player Z: The thing is trusting the other.
Player Y: No, I'm all set.
Player X: That is, we all know each other, no not knowing others, no one wants to partner with no one, so that everyone chose 1 and that's it, that's it.
Player Y: Ha, ha, yes.
Player Z: It is like a mirror work.
Player Y: Beware! It makes me giggle, I know that after a while . . . I'm going to be one of those . . .
Player X: Look, if anyone cheats now that we start adding the group total . . .
Player Y: We will end up knowing . . .
Player X: It is going to be right here, at the moment when the total adds to more than five, at that moment someone cheated.
Player Z: Someone . . .
Player X: At that moment we will know, and well, we will understand.
Player Y: We will know that . . .
Player X: We will know that . . . anyway, well, let's choose 1 and that's it, and we all gain and ready.
Player Y: Listen . . .
Player X: The sum will always have to be five.

In this case, they gave each other several warnings on how easy it would be to detect if someone had cheated from the initial agreement of each choosing one unit for a group total of five units. By stressing an ethical responsibility for each player, the absence of a God, and the importance of trust, they were stressing their expectations that others would keep to the agreement and, if not, that they would know who had defected. In fact, days after the conclusion of the sessions, and because the experimenter usually holds workshops with the community members to show preliminary results and discuss parallels with their actual common-pool resources, we observed that many participants had already identified who from the different groups acted more selfishly during the experiments even though care was taken to make payments in private.

Conclusion

We have approached the problem of co-operation in common-pool resource dilemmas by first exploring the role of face-to-face communication in a repeated game setting. We posit that communication helps players to choose a strategy that improves their payoffs above those predicted by the conventional model of a Nash equilibrium. The consistent finding is that face-to-face communication has a powerful effect on increasing trust and co-operation in experimental settings. As we discuss in the Introduction, a complex set of factors affects likely outcomes. Group identity, reputation building, creation of normative feelings, fear of social ostracism, and the emergence of commitments are examples of factors explaining why communication works. We suggest that communication also helps players to improve their capacity to detect the types of other group members.

In the experiments presented here, we allowed villagers from the same community to engage in a face-to-face discussion during the second stage of a common-pool resource situation. This gave the players, we argue, an opportunity to update their priors about the types of players within their groups and choose their extraction level accordingly. As discussed in our model, better type-detection does not guarantee greater levels of co-operation. Detecting a sufficiently large number of rational egoists in a group will induce conditional co-operators to act as egoists to reduce their losses if they were to act cooperatively. In fact, in a few groups, the levels of co-operation achieved during the Repeated communication rounds were low, but they were very high for most others. Further, there might be information that players had about their fellow group members carried over from their earlier experience with each other that might have played a role in their decisions during the second stage, once the group had agreed on a strategy for increasing aggregate earnings. This was the case not only for some groups under OSC but also for RC, as shown in the individual sessions' graphs in the Appendix figures on pages 280–82.

Nevertheless, on average, communication helped groups to reduce total extraction and to increase group and individual earnings. The gains were much higher if the communication was repeated. In the OSC treatment, we found no reference in the recorded discussions to detection of intentions by individual players in the discussions of Round 11. The feedback in every round about group outcomes, and the face-to-face interaction, helped players to create a better accounting for the types in the rest of the group.

Given the short duration of the experiment, it is not very likely that some subjects in fact changed their types. What is more likely to have happened is that communication made it common knowledge that there are conditional co-operators in a group. Game-theoretic models show that when there are conditional co-operators, egoists have incentives to co-operate until near the end of a repeated game (Kreps *et al.*, 1982). Communication plays two roles in this context. First, it allows conditional co-operators to express their intentions. Announcing their preferences may be regarded as cheap talk. But there is evidence that humans do a pretty good job of judging whether someone is telling the truth (see Ahn, Janssen, and Ostrom, 2004, for a review of the literature). Brosig (2002) explores how individuals are capable of recognizing the types and how such recognition and signaling of one's own type help, particularly cooperators, to detect the type of their anonymous partner. Further, she finds that pre-play communication induces not only co-operators but also individualistic types to act co-operatively and therefore increase the gains from social exchange.

Once the existence of conditional co-operators becomes common knowledge, there still remains the problem of equilibrium selection among all players. That is, even with a substantive proportion of conditional co-operators, universal defection is still an equilibrium. Players need to be convinced that others would co-operate, and that others also think that a sufficient number of other group members would also co-operate and so on (Chwe, 2001). That is, in addition to the existence of the conditional co-operators, the play of co-operative equilibrium also needs to be established as common knowledge. Repeated face-to-face communication is the most effective means of achieving such common knowledge, as the increased rates of co-operation in repeated communication setting of the reported experiment show.

Further work will follow. Separating type-detection from other confounding effects at work under communication is a natural next step in our research. Also, classifying individual decisions data into the three types we discuss will shine some light on how group co-operation may emerge and be sustained under different rules or institutional environments. Individual-level data in many of these experiments show that players change their strategies over time within the same session. The evidence from these experiments in the field, while not testing preference evolution directly, is consistent with two necessary assumptions of Güth's indirect, evolutionary approach: first, that conditional co-operators exist; and second, that they engage in efforts to detect the presence of other conditional co-operators. Communication is one of the major techniques for group dynamics to emerge, leading to higher levels of co-operation in dilemma and trust games when there are multiple types of players. Repeated face-to-face communication helps to establish the existence of conditional co-operators in a group who can then develop common knowledge that they will play the co-operative equilibrium in a repeated game.

Appendix

Table 12.A1 Payoffs table experimental design

		My level of extraction							Their Average
		1	2	3	4	5	6	7	8	Extraction
	4	758	790	818	840	858	870	878	880	1
	5	738	770	798	820	838	850	858	860	1
	6	718	750	778	800	818	830	838	840	2
	7	698	730	758	780	798	810	818	820	2
	8	678	710	738	760	778	790	798	800	2
	9	658	690	718	740	758	770	778	780	2
	10	638	670	698	720	738	750	758	760	3
	11	618	650	678	700	718	730	738	740	3
	12	598	630	658	680	698	710	718	720	3
	13	578	610	638	660	678	690	698	700	3
	14	558	590	618	640	658	670	678	680	4
Their Extraction Level	15	538	570	598	620	638	650	658	660	4
	16	518	550	578	600	618	630	638	640	4
	17	498	530	558	580	598	610	618	620	4
	18	478	510	538	560	578	590	598	600	5
	19	458	490	518	540	558	570	578	580	5
	20	438	470	498	520	538	550	558	560	5
	21	418	450	478	500	518	530	538	540	5
	22	398	430	458	480	498	510	518	520	6
	23	378	410	438	460	478	490	498	500	6
	24	358	390	418	440	458	470	478	480	6
	25	338	370	398	420	438	450	458	460	6
	26	318	350	378	400	418	430	438	440	7
	27	298	330	358	380	398	410	418	420	7
	28	278	310	338	360	378	390	398	400	7
	29	258	290	318	340	358	370	378	380	7
	30	238	270	298	320	338	350	358	360	8
	31	218	250	278	300	318	330	338	340	8
	32	198	230	258	280	298	310	318	320	8

280

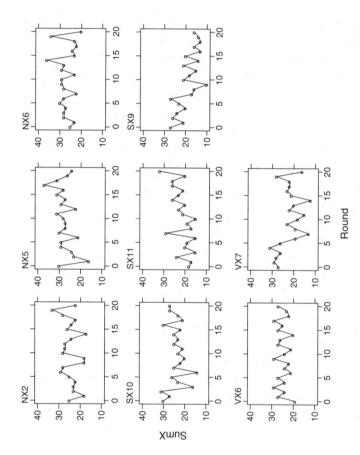

Figure 12.A1   Group extraction (SumX) by session (B, Baseline treatment)

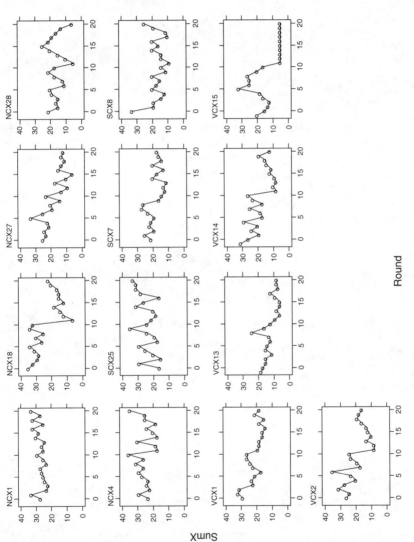

Figure 12.A2 Group extraction (SumX) by session (OSC, one-shot communication)

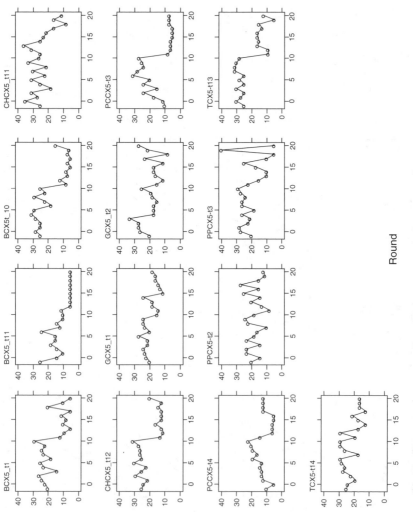

Figure 12.A3 Group extraction (SumX) by session (RC, repeated communication)

Notes

1 See also Carpenter and Cardenas (2003) for a further survey of experimental studies of ultimatum, dictator and trust experiments with non-student subjects across several countries and cultures.

2 For an example on how heterogeneity regarding the actual wealth and occupation of experimental subjects is used strategically in similar experiments in the field, see Cardenas (2003b). There it is shown that, as the distance in wealth across group players increases, the levels of co-operation during face-to-face communication rounds decreases.

3 The table used by players works as follows. Each column corresponds to the individual level of extraction. Once the monitor has announced the total for the group, each player is able to subtract his/her individual extraction from the total to obtain 'their level of extraction', that is, the row in the table with which the player is able to realize earnings in that specific round. The values in the cells are based on a pay-off function in which extraction increases individual earnings at a decreasing rate, and where group extraction reduces individual earnings at a linear rate, producing the typical common-pool resource dilemma as in the model used in the previous section. The values in the cells correspond to Col.$ pesos at an exchange rate, at the time, of about Col.$2700/1$US.

4 The open discussions did not allow them, however, to agree to transfer points or earnings once the experiment was concluded, replicating the work of Ostrom, *et al.* (1994).

5 A multiple regression using the average levels of extraction in Rounds 8 to 10 for each group as the dependent variable, with the independent variables being the average level of extraction in Rounds 18 to 20 and dummy variables for the particular treatment as independent variables, generates a positive and significant coefficient. The adjusted R-square value for the regression is slightly higher than 0.5.

References

Ahn, T. K., Janssen, Marco A. and Ostrom, Elinor (2003) 'Signals, Symbols, and Human Cooperation', in Robert W. Sussman and Audrey R. Chapman (eds) *The Origins and Nature of Sociality*, New York: de Gruyter, 122–39.

Alchian, Armen A. (1950) 'Uncertainty, Evolution, and Economic Theory', *Journal of Political Economy*, 58, 3, 211–21.

Bochet, Oliver, Page, Talbot and Putterman, Louis (2002) 'Communication and Punishment in Voluntary Contribution Experiments', Working paper, Providence, RI, Brown University. See <http://www.econ.brown.edu/2002/wp2002-29.pdf>.

Bohnet, Iris and Frey, Bruno S. (1999) 'The Sound of Silence in Prisoner's Dilemma and Dictator Games', *Journal of Economic Behavior and Organization*, 38, 1, 43–57.

Bolton, Gary E. and Ockenfels, Axel (2000) 'ERC: A Theory of Equity, Reciprocity, and Competition', *American Economic Review*, 90, 166–93.

Bouas, K. S. and Komorita, S. S. (1996) 'Group Discussion and Cooperation in Social Dilemmas', *Personality and Social Psychology Bulletin*, 22, 11, 1144–50.

Brosig, Jeannete (2002) 'Identifying Cooperative Behavior: Some Experimental Results in a Prisoner's Dilemma Game', *Journal of Economic Behavior and Organization*, 47, 275–90.

Camerer, Colin F. (2003) *Behavioral Game Theory: Experiments in Strategic Interaction*, Princeton, NJ, Princeton University Press.

Cardenas, Juan Camilo (2003a) 'Bringing the Lab to the Field: More Than Changing Subjects', Paper presented at the International Meeting of the Economic Science Association, Pittsburgh, Pa., June.

Cardenas, Juan Camilo (2003b) 'Real Wealth and Experimental Cooperation: Evidence from Field Experiments', *Journal of Development Economics*, 70, 263–89.

Carpenter, Jeffrey and Cardenas, Juan-Camilo (2003) 'Experimental Development Economics: A Review of the Literature and Ideas for Future Research', Working paper, Middlebury, Vt., Middlebury College, Department of Economics.

Casari, Marco and Plott, Charles R. (2003) 'Decentralized Management of Common Property Resources: Experiments with a Centuries-old Institution', *Journal of Economic Behavior and Organization*, 51, 217–47.

Chwe, Michael Suk-Young (2001) *Rational Ritual: Culture, Coordination, and Common Knowledge*, Princeton, NJ, Princeton University Press.

Dawes, Robyn M., McTavish, Jeanne and Shaklee, Harriet (1977) 'Behavior, Communication, and Assumptions about Other People's Behavior in a Commons Dilemma Situation', *Journal of Personality and Social Psychology*, 33, 1, 1–11.

Dawes, Robyn M., van de Kragt, Alphons J. C. and Orbell, John M. (1990) 'Cooperation for the Benefit of Us – Not Me, or My Conscience', in Jane Mansbridge (ed.) *Beyond Self-Interest*, Chicago, Ill., University of Chicago Press, 97–110.

Dawkins, Richard (1976) *The Selfish Gene*, Oxford University Press.

Eckel, Catherine C. and Wilson, Rick K. (2003) 'The Human Face of Game Theory: Trust and Reciprocity in Sequential Games', in Elinor Ostrom and James Walker (eds), *Trust and Reciprocity: Interdisciplinary Lessons from Experimental Research*, New York, Russell Sage Foundation, 245–74.

Fischbacher, Urs, Gächter, Simon and Fehr, Ernst (2001) 'Are People Conditionally Cooperative? Evidence from a Public Goods Experiment', *Economics Letters*, 71, 397–404.

Frank, Robert H. (1988) *Passions within Reason: The Strategic Role of the Emotions*, New York, W. W. Norton.

Frohlich, Norman and Oppenheimer, Joe (1998) 'Some Consequences of E-Mail vs. Face-to-Face Communication in Experiments', *Journal of Economic Behavior and Organization*, 35, 3, 389–403.

Güth, Werner (1995a) 'On Ultimatum Bargaining Experiments – A Personal Review', *Journal of Economic Behavior and Organization*, 27, 3, 329–43.

Güth, Werner (1995b) 'An Evolutionary Approach to Explaining Cooperative Behavior by Reciprocal Incentives', *International Journal of Game Theory*, 24, 323–44.

Güth, Werner and Kliemt, Hartmut (1998) 'The Indirect Evolutionary Approach: Bridging the Gap between Rationality and Adaptation', *Rationality and Society*, 10, 3, 377–99.

Güth, Werner and Yaari, Menahem (1992) 'An Evolutionary Approach to Explaining Reciprocal Behavior in a Simple Strategic Game', in Ulrich Witt (ed.), *Explaining Process and Change: Approaches to Evolutionary Economics*, Ann Arbor, Mich., University of Michigan Press, 23–34.

Güth, Werner, Kliemt, Hartmut and Peleg, Bezalel (2000) 'Co-evolution of Preferences and Information in Simple Games of Trust', *German Economic Review*, 1, 1, 83–110.

Güth, Werner, Schmittberger, Rolf and Schwarze, Bernd (1982) 'An Experimental Analysis of Ultimatum Bargaining', *Journal of Economic Behavior and Organization*, 3, 367–88.

Isaac, R. Mark, and Walker, James M. (1988). 'Group Size Effects in Public Goods Provision: The Voluntary Contribution Mechanism', *Quarterly Journal of Economics*, 103, February, 179–200.

Isaac, R. Mark, McCue, Kenneth F. and Plott, Charles R. (1985). 'Public Goods Provision in an Experimental Environment', *Journal of Public Economics*, 26, 51–74.

Kerr, N. L. and Kaufman-Gilliland, C. M. (1994) 'Communication, Commitment, and Cooperation in Social Dilemmas', *Journal of Personality and Social Psychology*, 66, 3, 513–29.

Kollock, Peter (1998) 'Social Dilemmas: The Anatomy of Cooperation', *Annual Review of Sociology*, 24, 183–214.

Kopelman, Shirli, Weber, J. Mark and Messick, David M. (2002) 'Factors Influencing Cooperation in Commons Dilemmas: A Review of Experimental Psychological Research', in Elinor Ostrom, T. Dietz, N. Dolsak, S. Stouick, E. U. Weber (eds), *The Drama of the Commons*, Washington, DC, National Academy Press, 113–57.

Kreps, David, Milgrom, Paul, Roberts, John and Wilson, Robert (1982) 'Rational Cooperation in the Finitely Repeated Prisoner's Dilemma', *Journal of Economic Theory*, 27, 245–52.

Kurzban, Robert and Houser, Daniel (2003) 'An Experimental Investigation of Cooperative Types in Human Groups: A Complement to Evolutionary Theory and Simulations', Working paper, University of Pennsylvania, Philadelphia.

Ledyard, John O. (1995) 'Public Goods: A Survey of Experimental Research', in John Kagel and Alvin Roth (eds), *The Handbook of Experimental Economics*, Princeton, NJ, Princeton University Press, 111–94.

Liebrand, W. B. G. (1984) 'The Effect of Social Motives, Communication and Group Size on Behavior in an *N*-Person Multi-Stage Mixed-Motive Game', *European Journal of Social Psychology*, 14, 239–64.

McCabe, Kevin and Smith, Vernon L. (2003) 'Strategic Analysis in Games: What Information Do Players Use?', in Elinor Ostrom and James Walker (eds), *Trust and Reciprocity: Interdisciplinary Lessons from Experimental Research*, New York, Russell Sage Foundation, 275–301.

McClintock, C. G. and Liebrand, W. B. G. (1988) 'Role of Interdependence Structure, Individual Value Orientation, and Another's Strategy in Social Decision Making: A Transformational Analysis', *Journal of Personality and Social Psychology*, 55, 396–409.

Messick, D. M. and Brewer, M. B. (1983) 'Solving Social Dilemmas: A Review', *Review of Personality and Social Psychology*, 4, 11–33.

Orbell, John M., van de Kragt, Alphons J. C. and Dawes, Robyn M. (1988) 'Explaining Discussion-Induced Cooperation', *Journal of Personality and Social Psychology*, 54, 5, 811–19.

Ostrom, Elinor (1998) 'A Behavioral Approach to the Rational Choice Theory of Collective Action', *American Political Science Review*, 92, 1, March, 1–22.

Ostrom, Elinor and Walker, James (1991) 'Communication in a Commons: Cooperation without External Enforcement', in Thomas R. Palfrey, (ed.), *Laboratory Research in Political Economy*, Ann Arbor, Mich., University of Michigan Press, 287–322.

Ostrom, Elinor, Gardner, Roy and Walker, James (eds) (1994) *Rules, Games, and Common-Pool Resources*, Ann Arbor, Mich., University of Michigan Press.

Ostrom, Elinor, Walker, James and Gardner, Roy (1992) 'Covenants With and Without a Sword: Self-Governance Is Possible', *American Political Science Review*, 86, 2, June, 404–17.

Plott, Charles R. (1986) 'Rational Choice in Experimental Markets', *Journal of Business*, 59, 301–27.

Rocco, Elena (1998) 'Trust Breaks Down in Electronic Contexts but Can Be Repaired by Some Initial Face-to-Face Contact', *Proceedings of CHI 1998*, New York, ACM Press, 496–502.

Rocco, Elena and Warglien, Massimo (1996) 'Computer Mediated Communication and the Emergence of "Electronic Opportunism"', Working paper 1996–01, Trento, Italy, University of Trento.

Sally, David (1995) 'Conversation and Cooperation in Social Dilemmas: A Meta-Analysis of Experiments from 1958 to 1992', *Rationality and Society*, 7, 58–92.

Scharlemann, Jörn P. W., Eckel, Catherine C., Kacelnika, Alex and Wilson, Rick K. (2001) 'The Value of a Smile: Game Theory with a Human Face', *Journal of Economic Psychology*, 22, 5, October, 617–40.

Schmidt, Karen L. and Cohn, Jeffrey F. (2001) 'Human Facial Expressions as Adaptations: Evolutionary Questions in Facial Expression Research', *Yearbook of Physical Anthropology*, 44, 3–24.

Simon, Adam and Gorgura, Heather (2003) 'Say the Magic Word: Effective Communication in Social Dilemmas', Working paper, University of Washington, Department of Political Science, Seattle.

Smith, Vernon (1962) 'An Experimental Study of Competitive Market Behavior', *Journal of Political Economy*, 70, 111–27.

Walker, James M., Gardner, Roy and Ostrom, Elinor (1990) 'Rent Dissipation in a Limited-Access Common-Pool Resource: Experimental Evidence', *Journal of Environmental Economics and Management*, 19, 203–11.

13
Price Competition: The Role of Gender and Education

Martin Dufwenberg, Uri Gneezy and Aldo Rustichini *

Introduction

'Tell me and I forget, teach me and I remember, involve me and I learn.' These famous words of Benjamin Franklin seem to aptly capture the spirit in which Werner Güth, our hero of the occasion here, interacts with others.[1] Consider, for example, his own explanation of how he brings the implied perspective to the classroom:

> I teach experimental economics completely differently compared to other things, I teach it in an 'apprentice' fashion. I use learning-by-doing in small groups from the start. This is a very nice aspect of experimental economics actually.[2]

But Werner involves other people outside the classroom too. Browsing through his publication list, we find more co-authors than we are able to count. We know for a fact that interacting with Werner is a learning experience, and a lot of fun, not least from 1995–7 when we were on the CentER faculty at Tilburg University, which Werner often visited.

The research we report on in this chapter derives from projects we started in the Tilburg days, concerning price competition on the one hand and gender effects on the other. Our research objectives are to examine the impact on price competition of, respectively, gender and education. We shall discuss these two issues in turn.

Gender and competition

Gender gaps are observed in a variety of economic and social environments. Recent research has pointed out various differences between men

* We are grateful to a referee who provided valuable comments. The research was supported by the NSF grant No. 0318378 and GIF (Gneezy) and NSF/SES-0136556 (Rustichini).

and women that may be important in economic interaction. We shall focus on a specific one:[3] when competition is introduced into an environment in which there is no gender gap in performance, the different reaction to competition by men and women might create a gap (Gneezy *etal.*, 2003; Gneezy and Rustichini, 2003). In particular, when men and women compete in the same group of players, men exhibit more competitiveness than women. These results have been derived from contexts where winning is associated with a high payoff: for example, the experiment described in Gneezy *etal.* (2003) used a winner-take-all design in which the participant with the highest performance is the only one paid, and the amount won equals the total payment to subjects in the non-competitive treatment.

This evidence agrees with the general idea found in evolutionary psychology, that gender gaps derive from the different position of the genders in the reproductive process: since males can mate successfully with many different women, at small cost, they face large reproductive gains from competing intensely with other males. Females, on the other hand, can mate successfully only once per period, and thanks to the competition of males they are likely to find at least one mate. Hence their attention is not devoted so much to winning over other females, but rather to selecting the right partner.[4]

It is reasonable to think that the same effect (stronger competition among males) holds in environments where conditions similar to the reproductive 'game' are present: stronger competition produces a larger prize. This qualification (a larger prize for more competitive behaviour) is important, and offers an indirect test of the explanation we have just recalled. If the condition that higher competition provides a higher prize is missing, then the difference in competitive behaviour should vanish. That is, of course, unless one thinks that the preference for competition is a blind force, oblivious to the advantages (for example, in terms of reproductive fitness) that competition might provide.

To test these predictions we need a game where strong competition provides a low payoff to the winner. Such is the case in Bertrand models of price competition, where stronger competition leads to a smaller (in fact, at equilibrium, the smallest possible) prize. We use a simple Bertrand game, similar to that introduced by Dufwenberg and Gneezy (2000), and examine whether men and women make different choices.

Education and competition

Our second research objective concerns the impact of education. There is much interest in experimental economics concerning the impact of experience on behaviour. The most common means by which experience is obtained is by letting subjects perform some decision task many times, so that they gain familiarity with the situation. Our experiment, by contrast, may be seen as an exploration of the impact of another form of experience: experience by education.

Education may matter not only by increasing familiarity with a situation, but also by 'suggesting a solution' to the subjects. In the particular case of Bertrand price competition, the classical solution is precise and stark, eliminating competitors' profits. Perhaps educated subjects, on balance, move in that direction? The experiment comprises one session of each of two treatments, which differ in terms of whether or not the participants had been taught the theory of Bertrand oligopoly. We test whether and how this matters to behaviour.

Outline of the chapter

The chapter proceeds as follows: in the next section we describe the experimental design and specify our hypotheses. In the third section we describe the results of a first experiment, in the fourth section we present the results of a different experimental design, aimed at testing the robustness of the results and the fifth section concludes.

The experimental design

In this section we present (in three subsections) the game that we examine, the design of the experiment and our treatments, and finally the hypotheses we test.

The Bertrand game

The core of the experiment is the following game, similar to that introduced by Dufwenberg and Gneezy (2000): each one of a large (and commonly known) set of players chooses a number from the set $\{0, 1, \ldots, 1000\}$. The player who chooses the lowest number wins as many units (Swedish kronor, in our case) as the number bid, and the others make no money.[5] If there are ties for the winning bid, these are split equally among the winners.

The game may be interpreted as a Bertrand duopoly market with price competition: the players are firms; the chosen numbers are prices; and the payoffs are profits. Admittedly, this account of real-life price competition is stylized. Yet the key feature of such interaction is present in the game: a tension between incentives for high prices that lead to high profits and incentives for low prices that undercut those of competitors.

An attractive feature of the game is its simplicity. It can easily be explained verbally to participants in an experiment without the use of complicated payoff tables.

Assuming that the competitors wish to maximize expected profit, the game has a unique Nash equilibrium in undominated strategies, in which each competitor chooses a price of 1. There are additional Nash equilibria, in which two or more players use the weakly dominated strategy of 0, but from an economic point of view all equilibria are similar: the firms make almost no profit relative to what is available in the market.

Design and treatments

Subjects played the game described only once. The experiment consisted of two separate sessions, each conducted in two different microeconomics classes at Stockholm University. In their course the students were introduced to basic tools of game theory, a fair amount of time was spent doing or discussing experiments, and a few lectures were devoted to topics of industrial organization. The first session took place in the autumn of 2000, at the beginning of the very first lecture and *before* the students had had any exposure to the core topics of the course. The second session (with a different class) took place in the autumn of 2001, in the end of a lecture towards the end of the course and *after* the students had been introduced to the core topics. In particular, they had been taught the theory of Bertrand oligopoly.

This explains how we manipulated the level of education of the subjects. The other key feature of our design concerns the participants' gender, which we needed to record. When the experiment started, each student received an instruction (see the Appendix on page 295) describing the game. The instruction sheet informed students that participation entailed stating a 'code number' consisting of the student's initials plus the last four digits of his/her Swedish social security number. Such numbers have ten digits, specifying year-month-day of birth plus four digits where the penultimate one is *even* for a woman and *odd* for a man. Our design made crucial use of this last feature. It allowed us to separate the data according to the gender of the player.[6]

Our design generates four categories of observation, depending on the gender of a participant, and on the level of his/her education. Table 13.1 summarizes how many observations we have for each category.

Hypotheses

In a recent line of research, Gneezy *et al.* (2003) and Gneezy and Rustichini (2003) found that men react more strongly to competition incentives than do women. In particular, when men and women compete, men exhibit more competitiveness than women.[7]

In the price competition game, the lower the chosen prices, the fiercer is the competition in the market. If men care about winning *per se* more than women, we should observe significantly lower price offers from men. We test the following hypothesis:

H1: *Men and women choose the same prices*

Table 13.1   Number of observations

	Uneducated	Educated
Female	33	31
Male	37	41

Our second hypothesis is related to the influence of education on behaviour. One may take two perspectives on this issue. First, as explained in the beginning of the second section of this chapter, the Bertrand model makes a stark prediction that entails fierce competition and very low prices. One may thus expect that exposure to the microeconomics course moves prices downwards.

Second, education may matter simply because it helps subjects to become more familiar with a situation. In this case there would be no a priori reason to expect the prices to move in a particular direction. The null hypothesis we test is the following:

H2: *The same prices are chosen regardless of the level of education*

Results

The two hypotheses cannot be rejected. Let us look at the details.

The distribution of prices is dispersed, although concentrated on the lower values

The mode and median for the overall sample are both 100, and the mean is 250. There is also a bump in the distribution at high values, with 7.75 per cent of the offers in the 900–1000 range. Such choices are far from equilibrium; the percentage of subjects playing the equilibrium strategy is in the overall sample is 4.93 per cent (seven subjects), with 5.5 per cent (four subjects) in the sample of non-educated subjects and 4.3 per cent (three subjects) in the sample of educated subjects.

There is no significant difference in the behaviour of subjects across gender

This conclusion holds in several different tests. The Mann–Whitney test for the overall set gives a p-value of 0.63. Even if one focuses on special subsets of the prices, the behaviour is indistinguishable: for example, there is no significant difference across genders in the proportion of subjects placing a price lower than 100 (the values are 55.1 and 54.6 for male and female subjects, respectively). The distribution of bids according to gender is presented in Figure 13.1.

The difference across education is weak

The non-parametric tests on the price variable shows no significant difference between the two groups (of educated and non-educated subjects) ($p = 0.187$ for the Mann–Whitney test). But a simple analysis of the histograms seems to indicate that the prices are more concentrated among the low values for the educated group than for the non-educated. A way to study this difference is to study the distribution of a variable separating high prices and low prices (say, equal to one when the price is low, and zero otherwise).

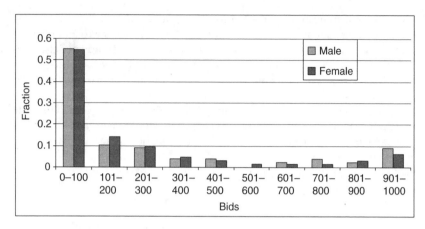

Figure 13.1   Distribution of bids by gender

The difficulty is to find a convincing cutoff price. We observe that 54.93 per cent of the prices for the overall population are below 100, and choose this as the cutoff. One has to note, however, that different cutoffs (50 and 200) do *not* give a significant difference across education levels. Similarly, the upper tail (900 to 1000) does not give significant differences ($p = 0.6130$). If the cutoff price is chosen to be 100 (corresponding to the lowest 10 per cent) then the non-parametric analysis confirms that there is a significant difference between the frequency of low prices in the educated and non-educated population ($p = 0.0301$). It is interesting to note that even conditional on low (less than 100) price offers, the distribution of price across genders is not significantly different ($p = 0.59$ in the Mann–Whitney test).[8] Figure 13.2 presents the bids according to the level of education.

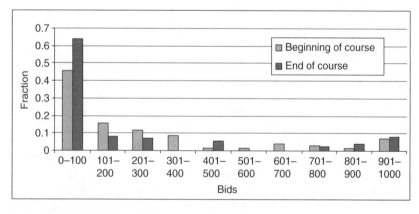

Figure 13.2   Distribution of bids by education

The gender results are robust to alternative designs

The lack of gender effects suggests that some robustness check may be warranted. We therefore ran two additional experiments checking the basic result.

Both experiments were run at the University of Minnesota in autumn 2003. The subjects were undergraduate students in economics, who had just begun to learn the basic concepts. They had no familiarity with the concept of market games, or Bertrand competition, so they would correspond to the non-educated subjects in the previous experiment. Subjects were asked to report several personal characteristics, including gender, *after* they had chosen the price. They did not know in advance that they were going to be asked these questions. No special incentive to answer was provided, but none of the subjects refused to answer, in either of the two experiments.

In the first group of thirty-four subjects (sixteen men and eighteen women) the design was similar to the basic design discussed earlier: subjects made simultaneous offers of prices between 0 and 100 dollars. The lowest offer was paid an amount exactly equal to the offer (that is, one dollar offered was one dollar paid), with ties split.

The distribution of price offers was concentrated at the low end. Over a range from 0 to 100, the mean was 9.14 (10.93 for men; 7.55 for women), the median 4 (5.5 for men; 4 for women) and the mode 1 (1 for men and 1 for women); 55 per cent of the offers were 4 dollars or less. The difference across gender (for example, in the in the average offer) is not statistically significant ($p = 0.49$ in the Mann–Whitney test that the distribution of prices of the two genders are the same).

The fact that prices are concentrated at the low end suggests that both in this experiment and in the previous one a potential gender difference might be hidden under the general pattern of low price offers. An alternative design was developed to address this potential limitation. In this design, subjects could make an offer ranging from 0 to 200 dollars. Then two out of the participants were selected randomly, and the price offers of these two subjects compared: the subject among the two with the lowest offer got a payment equal to the offer.

An experiment according to this design was again run among undergraduate students at the University of Minnesota, in a population similar to the one of the previous experiment. The sample was of 462 students (258 men and 204 women). The results according to gender are presented in Figure 13.3.

The distribution of price offers was indeed less concentrated in the lower end: over a range from 0 to 200, the mean was 76.6 (70.8 for men; 76.6 for women), the median 75 (70.5 for men; 75 for women) and the mode 75 (75 for men and 75 for women). Only 4.7 per cent of the offers are below 8 dollars. This is a price distribution that is not concentrated on low values,

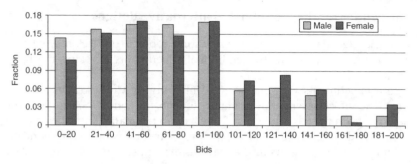

Figure 13.3 Distribution of bids by gender (second experiment)

so there is room for gender differences to appear. But again, the price distribution is not significantly different across genders. This is clear from the data on the mean, median and mode; moreover $p = 0.20$ in the Mann–Whitney test of the hypothesis that the price offer differs across genders.

This game too shares the feature, although to a lesser degree, that winning is associated with low price. We conclude that there is no significant gender difference in the price offers in a simultaneous Bertrand competition game.

Discussion

The primary aim of this chapter was to test the hypothesis that different preferences in competition across genders is not likely to be present if one important condition is missing: that more competitive behaviour is more likely to gain a higher prize for the winner. This condition is clearly missing in the Bertrand games that we studied. The results show that the possibility of winning is not of itself enough to induce more competitive behaviour in male subjects.

The lack of difference in behaviour between genders in this specific experiment does not contradict the explanations that are usually provided for the more competitive behaviour of men, or the results in existing experimental literature on gender differences in competition. Competition there (for example in Gneezy *et al.*, 2003) is understood as the differential performance in a task where effort and talent have to be provided. The outcome in the task might then be interpreted (at least by the subject who is informed of this outcome) as a social signal of skill or talent. In the present experiment, although the interaction between the subjects is 'competitive', the information about personal talent is absent. In addition, the winner in this game gets the lowest possible prize, so competition does not appear to be a way of selecting a winner, but rather a sucker who (although earning slightly more than other participants) is willing to take such small bait.

An important gender issue has been left open by the present study, and should be a subject of future research. Genders may differ with respect to the degree of risk aversion (see, for example, Eckel and Grossman, 2003). A subject in our game is choosing a price facing the price distribution induced by the other players. The results of the experiment justify a belief that this price distribution is not concentrated near the equilibrium (price equal to 1). Let $\mu(p)$ be the probability that the lowest price among the opponents' prices is larger than p, so that μ is decreasing. The choice of a price is the choice among a set of lotteries indexed by p, where the p lottery gives a prize equal to p with probability $\mu(p)$, and a prize of 0 dollars with probability $1 - \mu(p)$. (This avoids the possibility of tied winners, at price $p + 1$.) Consider now two subjects, one man and one woman, who have the same belief about the price distribution. If the woman is less competitive but more risk averse, she might choose the same price as the man, because the effect of preference for competition and risk aversion go in opposite ways. While it seems unlikely that the two effects balance so precisely in the different populations we have tested, to produce behaviour that only looks the same, further analysis of these two characteristics is important.

A second aim of our chapter was to examine the effect of economic education on behaviour. This effect is, as we have seen, more subtle than the effects of gender. Education has *some* effect: low prices (with respect to the cutoff of 100) are significantly more frequent after education than they were before. Subjects who have been told the theory of Bertrand competition, and who have thus been made aware of the idea that there will be cut-throat price competition, choose somewhat lower prices. In this way, education seems to bring subjects closer to the prediction of the Nash equilibrium concept. But this effect is weak, if anything. In addition, this effect is the same across genders.

Against the backdrop of this result, in closing this chapter we connect again to the issue concerning the difference between teaching and involving. We have shown that education may have very little effect on behaviour in a price-competition context. By contrast, it is well known that, if subjects interact over and over again in such games, then behavior tends to change over time.[9] This difference is indicative that being taught and being involved implies different forms of experience, just as Franklin said.

Appendix

Instructions (translated from Swedish)

You are invited to participate in an experiment where you may make money. All students in this room have received instructions with identical wording. Please do not talk to anyone during the duration of the experiment.

The task of each participant in the experiment is to choose a whole number in the interval from 0 to 1000. You must thus choose one of the numbers 0, 1, 2, 3,..., and so on up to the number 1000.

Money will be paid out as follows: One of the participants who has chosen the lowest number will win as many kronor as the number he/she chose. The participant to be paid will be determined by a random draw. Those participants who do not choose the lowest number will not receive any money.

The person who wins money should contact the experimenter in his office before 20 October 2000, and the money will then be paid.

You participate in the experiment by responding to and handing in the following:

I CHOOSE THE FOLLOWING NUMBER: ....

MY INITIALS ARE: ....

THE LAST FOUR DIGITS OF MY SOCIAL SECURITY NUMBER ARE: ....

Notes

1 In this connection, we note that Werner is in good company with at least one other famous experimental economist: Vernon Smith quoted and celebrated Franklin in his toast at the 2002 Nobel Banquet.
2 The quote is from Güth (2002), an interview with Fredrik Andersson and Håkan Holm.
3 Among those gender gaps we do *not* focus on, let us mention that in some situations women are more risk-averse than men (for a survey, see Eckel and Grossman, 2003). For a more popular discussion of differences between men and women, and the evolutionary reasons for these, see Pinker (2002).
4 For a modern exposition of the view of evolutionary psychology on this point, see Daly and Wilson (2001) or Buss (1999).
5 At the time of the experiment, there were approximately ten kronor to the euro, or to the dollar.
6 For example, a person with social security number 440202-5678 must be male. The technique has been used previously by Dufwenberg and Muren (2000). The winner of the game was entitled to private payment by visiting Martin Dufwenberg's office during the following weeks. However, the winner's code number was publicly announced in class. This implies that the degree of anonymity between subjects was relatively low. A code number does not automatically reveal identity, but with some effort one might be able to work this out.
7 While we base our hypothesis on the results of the previous studies, the design there was different in many aspects from the current design. For example, in this chapter, subjects do not make real effort, but rather choose a number. Still there is a cost associated with each choice: a more competitive choice corresponds to a lower payoff.
8 Our finding of a weak effect of education accords with results discussed by Rubinstein (1999, pp. 156, 167–8) (obtained in collaboration with six graduate students) where 'the responses of economics students to daily strategic situations before and after a course in game theory' showed little difference.
9 See Plott (1989) or Holt (1995) for surveys on experimental IO that cover the topic, or Dufwenberg and Gneezy (2000), who show this for game similar to that studied here.

References

Buss, D. M. (1999) *Evolutionary Psychology*, Boston, Mass., Allyn and Bacon.

Daly, M. and Wilson, M. (2001) 'Risk Taking, Intra-sexual Competition, and Homicide', *Nebraska Symposium on Motivation*, 47, 1–36.

Dufwenberg, M. and Gneezy, U. (2000) 'Price Competition and Market Concentration: An Experimental Study', *International Journal of Industrial Organization*, 18, 7–22.

Dufwenberg, M. and Muren, A. (2000) 'Discrimination by Gender and Social Distance', Working paper 2002:2, Department of Economics, Stockholm University.

Eckel, C. C. and Grossman, P. J. (2003) 'Men, Women and Risk Aversion: Experimental Evidence', in C. R. Plott and V. L. Smith (eds), *The Handbook of Experimental Economics Results*, Amsterdam; Elsevier.

Gneezy, U. and Rustichini, A. (2003) 'Gender and Competition at a Young Age', Discussion paper, University of Minnesota.

Gneezy, U., Niederle, M. and Rustichini, A. (2003) 'Preferences in Competitive Environments: Gender Differences', *Quarterly Journal of Economics*, 118, 1049–74.

Güth, W. (2002) 'Werner Güth', Interview by F. Andersson and H. Holm in F. Andersson and H. Holm (eds), *Experimental Economics: Financial Markets, Auctions, and Decision Making*, Dordtecht, Kluwer.

Holt, C. (1995) 'Industrial Organization: A Survey of Laboratory Research', in J. Kagel and A. Roth (eds), *Handbook of Experimental Economics*, Princeton, NJ, Princeton University Press.

Pinker, S. (2002) *The Blank Slate*, Harmondsworth, Penguin.

Plott, C. (1989) 'An Updated Review of Industrial Organization: Applications of Experimental Economics', in R. Schmalensee and R. Willig (eds), *Handbook of Industrial Organization*, vol. II, Amsterdam, North-Holland.

Rubinstein, A. (1999) 'Experience from a Course in Game Theory: Pre- and Post-class Problem Sets as a Didactic Device', *Games and Economic Behavior*, 28, 155–70.

14
Parity, Sympathy and Reciprocity
Werner Güth and Menahem E. Yaari[*]

Introduction

In this chapter, we consider the results of an experiment in which subjects had deviated systematically from the pursuit and maximization of personal gain. We hypothesize that these departures from self-seeking behaviour are caused by one or more of the following factors: (a) *Parity* (also known as *inequality aversion*): in choosing among actions, individuals may be attempting to promote equality of outcomes, even at the cost of some reduction in unilateral personal gain; (b) *Sympathy* (also known as *altruism*): in choosing among actions, individuals may be taking into account not only their own unilateral gains (or losses) but also the gains (or losses) of others; (c) *Reciprocity*: in choosing among actions, individuals may be motivated, to some extent, by a desire to apply *measure for measure* – that is, to reward kindness and unkindness in like manner.

Significant departures of decision-makers from purely self-seeking behaviour have been noted, and studied, for a long time (see Dawes and Thaler, 1988). Experimental investigations of such departures date back at least fifty years (see Sally (1995) for a fairly recent survey).

Parity ('inequality aversion') has been recognized as a possible motivational factor in several experimental settings (see, for example, Bolton, (1991); Fehr and Schmidt, (1999); Bolton and Ockenfels, (2000)). In some cases, a concern for parity, or equity, is assumed to exist only when the decision-maker's position is *inferior*, relative to that of his/her opponent ('one-sided inequality aversion'). In other cases, this concern is assumed to operate symmetrically, regardless of who is holding the advantage – oneself or one's opponent.

---

[*] This chapter is an essay that was written, largely, in the early 1990s, based on experimental evidence gathered in the late 1980s. The second author, whose hesitancy had prevented its publication until now, wishes, by presenting it here, to pay a debt of affection and gratitude to the first author.

Sympathy (the capacity to 'feel with one's neighbour as with oneself') has enjoyed prominence since antiquity. David Hume thought that sympathy was at the basis of the virtues that characterize any civilized society. Experimentally, sympathy (sometimes labelled 'altruism') has often been used to account for co-operative behaviour that cannot be rationalized on the basis of self-interest alone (see, for example, Dawes and Thaler, 1988; Cooper *et al.*, 1996).

Finally, reciprocity ('measure for measure') has also long been identified as a force that motivates behaviour, and has been treated as such in several recent theories of human interaction (see, for example, Rabin, 1993; Fehr *et al.*, 1997; Bolton and Ockenfels, 2000).

Background

Consider two sums of money, say a dollars and ε dollars, and think of a as being quite a bit larger than ε. Figure 14.1 shows a 2-player game-form in which the outcomes are dollar payments to be received by the two players.

If each player's evaluation of outcomes is strictly in accordance with his/her dollar earnings (the higher, the better) then this game-form becomes a simple run-of-the-mill Prisoner's Dilemma. We shall refer to preferences of this type (determined solely, and monotonically, by one's own monetary earnings) as *self-seeking*. Experimental interactions can easily be designed, where subjects play out this game-form, and this has been done many times (for a survey see, for example, Roth, 1995, section IIIA). When this is done, one finds that, as the ratio a/ε increases, subjects tend to opt, systematically and with increasing frequency, for the 'co-operative' strategy pair (C, C). Under self-seeking preferences, C is a strongly dominated strategy, so observing the strategy pair (C, C) would seem to indicate that behaviour is either irrational or non-self-seeking.

	C		D	
C	a	a	$-\varepsilon$	$a+\varepsilon$
D	$a+\varepsilon$	$-\varepsilon$	0	0

Figure 14.1 Two-player game form

Various arguments have been proposed for why people play C in the above game-form. In some notable cases (see Rabin, 1993), these arguments have been 'interactive' in nature – that is, they have to do with the player's expectation of what his/her opponent is going to do. Something like the following: 'I expect my opponent to play C. Given this expectation, for me to play D would be mean. Therefore, I should play C.' Conversely, a player's deliberation might go, say, as follows: 'I expect my opponent to play D.

Given this expectation, the problem with my playing C is not so much the loss of a measly ε (with D, 0 is all I would get) but rather my coming out the sucker, which I abhor. For this reason, I should play D.' In both cases, the underpinning of the final outcome – (C, C) in the first, (D, D) in the second – rests with 'interactive' considerations; that is, with what players expect their opponents to do. In this chapter, we deliberately neutralize these interactive considerations, hoping thereby to isolate the pure effects of parity, sympathy and reciprocity. (In the latter case, we shall consider players moving in sequence, so a minimal interactive element will be retained.) If we take the game-form shown in Figure 14.1 and 'dissect' it into two separate game-forms, we get what is shown in Figure 14.2.

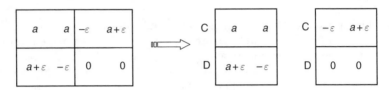

Figure 14.2 Dissecting the game

The two resulting game-forms, on the right, are extremely primitive. Only one of the two players has a move, and that move alone determines both players' monetary payments. Primitive encounters of this type are sometimes referred to as 'dictator games' (see Bolton and Ockenfels, 2000). The question of what one's opponent is expected to do becomes either empty (opponent has no move) or moot (I have no move). Thus, 'interactive' considerations are no longer relevant. Nevertheless, in our experimental setting, the 'cooperative' move C is observed surprisingly often, with increasing frequency as the ratio a/ε increases. How can one account for this evidence, given that repeated play was completely ruled out? We claim that such an account would have to rely on the introduction on non-self-seeking elements – such as parity, sympathy and reciprocity – into the players' preferences; that is, into the ways by which players convert game-forms into outright games.

The experimental setting

The subjects whose responses make up the evidence that will be reported here were 1st-year and 2nd-year students at the Hebrew University of Jerusalem. In a certain week during the academic year, experimenters were allowed to take over the last half-hour of class sessions in two large enrolment courses, namely 'Introduction to Logic' and 'Introduction to Constitutional Law'. The students who were in attendance were invited to spend the remaining class time participating in a simple and 'potentially fairly lucrative' experiment. They were told that, by the end of the period, each one of them would have earned, possibly, as much as five times the University's official hourly student wage, and that average earnings were in fact guaranteed to exceed twice that standard hourly wage. In each of the classes approached in this manner, two or three students decided not to participate, with the rest staying on for the experiment. There were 248 subjects in all, divided about equally between Logic and Constitutional Law. Each subject was given a set of instructions, describing three tasks to be performed. To complete a task, the subject had to select an action from a 'menu' consisting of a small number (usually two) of simple available actions, and then to carry out the selected action. Physically, the actions consisted in positioning a coloured sticker in some designated space.

In each of the three tasks, the subject was in fact a player in a 2-player set-up. The game-forms to be played were fully specified, but presumably only the players (subjects) themselves could convert these game-forms into outright games. In all the game-forms, players were completely anonymous *vis-à-vis* each other – that is, players never knew who their opposing players were. In all cases, the opposing player was described, in the subject's instructions sheet, as 'another person, also participating in this experiment.' And the next sentence was: 'There is no way for you to know the identity of this other person, just as there is no way for this person to know your own identity.' Subjects knew, broadly, that the experiment was being carried out in several large class sessions, over several days. But they did not know which courses, nor how many courses, would be involved. Thus, the only inference a subject could draw was that the person on the opposite side was a student in one of the large-attendance courses of the Hebrew University. Moreover, it was made clear to the subjects that opponents were selected through random matchings and that, in any two tasks, a player would in all likelihood be dealing with two different randomly selected individuals on the opposing side.

In two out of three tasks (to be labelled Task A and Task B) the game-form being played was degenerate, in the sense that only one player, the subject, had a move, with the other player accepting the consequences passively. In the third task (Task C), interaction was genuine, albeit strategically rather primitive.

Tasks A and B: description and results

In the second section, above, we discussed a 'dissected' (or 'sliced') version of the prisoner's dilemma – that is, game-forms obtained when one considers each column of the prisoner's dilemma separately (see Figure 14.2). The first two tasks our subjects were asked to perform were designed to capture the patterns of behaviour when such primitive game-forms are being played out. The subjects' choice of problems in Tasks A and B were as shown in Figure 14.3.

It will be noticed that earnings in these tasks differ somewhat from the relevant payoffs in Figure 14.1 ($-\varepsilon$ has been replaced by 0 in Task A, and $a + \varepsilon$ has been replaced by a in Task B). This was because of our concern not to stretch

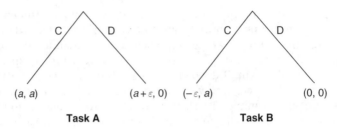

Task A　　　　　　　　　　　　　**Task B**

Figure 14.3　Tasks A and B

credibility by making statements such as 'somebody you don't know will be giving up some money'. Substantively, the change is of no consequence.

The actions labelled here as C and D were identified in the experiment in terms of the colours of two stickers, with one of the stickers to be chosen and placed in some designated space. There was no reason to suppose that the stickers' colours, in themselves, affected the subjects' choices in any way. (In fact, the same colour was used to designate C in some cases and D in others.) It is therefore safe to assume that subjects judged the available actions solely by their consequences. The consequence of any given action was a pair of monetary payments, to be made by the experimenter after the completion of the experiment. All the payments, exactly as described in the subject's instructions sheet, were to be made fully and squarely, without lotteries, auctions or other hanky-panky. The payees in all cases were, first, the subject him/herself and, second, some other unidentified person. There was no way for the subject and this other person to obtain any information about each other, save for the fact that both of them were participants in the experiment. In Figure 14.3, the first and second components of the outcome pair are the amount to be earned by the subject and by 'the other person', respectively. All payments were denominated in new Israeli shekels (NIS).[1]

The results obtained for Tasks A and B are given in Table 14.1.

Table 14.1　Data for Tasks A and B

Task A					Task B					
		C		D				C		D	
a (NIS)	ε (NIS)	N	%	N	%	a (NIS)	ε (NIS)	N	%	N	%
5	0.10	46	85	8	15	10	0.10	37	71	15	29
5	0.25	40	78	11	22	10	0.25	26	74	9	26
5	0.50	71	81	17	19	10	0.50	58	62	36	38
5	0.75	19	68	9	32	10	0.75	26	59	29	41
5	1.00	16	59	11	41	10	1.00	9	39	14	61
All parameter values		192	77	56	23	All parameter values		156	63	92	37

Tasks A and B: analysis

The evidence contained in Table 14.1 is obviously incompatible with subjects being strictly self-seeking. Moreover, the design of the experiment makes it virtually impossible to attribute this departure from self-seeking behaviour to considerations of repeated interaction or long-term reputation. It is therefore to be concluded that, in choosing among actions, subjects had taken into account not only payments due to be made to themselves but also payments due to be made to other individuals, about whom they knew very little. Let (x, y) be the outcome of an action being contemplated by the subject, with x and y being the monetary payment levels for the subject and for the other individual, respectively. Our hypothesis was that subjects would evaluate such an outcome by calculating the quantity:

$$U(x, y) = (1 - \sigma)x + \sigma y - \eta|x - y|$$

Where σ and η ($0 \leqslant \sigma \leqslant 1$ and $\eta \geqslant 0$) are, respectively, the subject's 'coefficient of *sympathy*' and 'coefficient of *parity*'. The coefficient of sympathy measures the subject's sensitivity to 'what's happening to the other fellow', and the coefficient of parity measures the subject's sensitivity to inequality. The condition $\sigma = \eta = 0$ characterizes a purely self-seeking subject.

Notice that in Task A, both sympathy and parity tend to push the subject towards the co-operative action, C. That is, in Task A, as the values of σ and η become higher, choosing C becomes more likely. In Task B, on the other hand, sympathy still works in the direction of reinforcing C, while parity now works in the opposite direction – that is, higher values of η tend to reinforce self-seeking behaviour. This fact makes it possible to use our data to separate out the effects of sympathy and parity. Our hypothesis concerning the evaluation of outcomes by the subjects may in fact be rewritten as follows:

$$U(x, y) = \begin{cases} (1 - \alpha)x + \alpha y & \text{if } x \geqslant y \\ (1 - \beta)x + \beta y & \text{if } x \leqslant y \end{cases}$$

where $\alpha = \sigma + \eta$ and $\beta = \sigma - \eta$. Clearly, Task A involves α while Task B involves β. Consider a subject whose sympathy and parity coefficients are given, accordingly, by $\sigma = (\alpha + \beta)/2$ and $\eta = (\alpha - \beta)/2$. In task A, this subject would choose the action C if $\alpha \geqslant \varepsilon/(a + \varepsilon)$ and, similarly, in Task B, s/he would choose C if $\beta \geqslant \varepsilon/(a + \varepsilon)$. Table 14.1 contains evidence on the frequency of subjects choosing C or D in both tasks. It follows, therefore, that Table 14.1 contains the frequencies with which the inequalities $\alpha \geqslant t$ and $\beta \geqslant t$ are satisfied, for various values of t. In short, Table 14.1 contains information about the *distributions* of α and β in the population from which our subjects were drawn. We can, in fact, use Table 14.1 to obtain 5 points

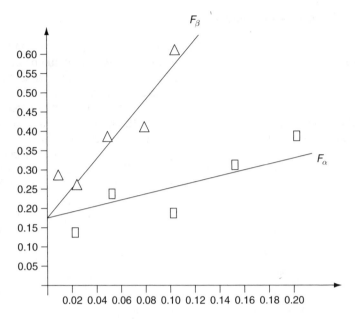

Figure 14.4   Distribution of α and β

on the distribution of α and 5 points on the distribution of β. These points are drawn in Figure 14.4.

The points in Figure 14.4 can be used to estimate two distribution functions, one for α and one for β, using linear regression. The estimated distribution functions, obtained in this way, are also drawn in Figure 14.4.

Let the distribution functions of α and β be denoted F_α and F_β, respectively. The regression equations that have been estimated were:

$$F_\alpha(t) = c + at$$

$$F_\beta(t) = c + bt$$

The two equations were constrained to have a common intercept, c. Removal of this constraint – that is, allowing the two equations to have separate intercepts, yields an insignificant difference between the two estimated intercepts. Theoretically, equality of the two intercepts is a consequence of our *comonotonicity* assumption (see below).

The results of estimating the regression coefficients in the above equations were as follows:

$c = 0.164$ (*t*-value 5.07)

$a = 1.039$ (*t*-value 3.48)

$b = 4.065$ (*t*-value 6.84)

All the coefficients are significant at the 0.01 level, and the overall fit is quite good ($R^2 = 0.87$, adjusted $R^2 = 0.83$). The estimated distribution functions for α and β (with the coefficients rounded off slightly) come out as:

$$F_\alpha(t) = \min\{0.16 + t, 1\}$$

$$F_\beta(t) = \min\{0.16 + 4t, 1\}$$

We now proceed to use these equations in order to retrieve the distributions of the preference characteristics σ and η, recalling that $\sigma = (\alpha + \beta)/2$ and $\eta = (\alpha - \beta)/2$. In order to do this, we must say something about the degree of *dependence* among the variables being studied. Postulating independence is clearly inappropriate, because we expect sensitivity to the position of the other to go hand-in-hand (at least to some extent) with sensitivity to inequality. In the absence of an estimate for the strength of this coherence of tendencies, we postulate complete dependence. More precisely, we suppose the two tendency variables, σ and η, to be (strictly) *co-monotone*. By this we mean that the inequality:

$$(\sigma_i - \sigma_j)(\eta_i - \eta_j) > 0$$

holds whenever $(\sigma_i - \sigma_j) \neq 0$, where σ_i and η_i are, respectively, the σ-value and η-value for some individual i, and similarly σ_j and η_j for some individual j. In other words, ordering the individuals according to their σ-values leads to the same ranking as ordering them according to their η-values. Another way of expressing this is to say that there exists a strictly increasing function, say f, such that $\eta = f(\sigma)$.

Given the co-monotonicity of σ and η, we turn now to the other two ('synthetic') variables, α and β. Clearly, α and σ are also co-monotone, as are α and η. This follows immediately from the definition, $\alpha = \sigma + \eta$. As for the other variable, β, its definition is given by $\beta = \sigma - \eta$, which neither implies nor contradicts the co-monotonicity of β with the other variables. We do know, however, that there exists a real function, say g, such that $\beta = g(\alpha)$. While g is not, a priori, an increasing function, all we need to do is to exhibit an increasing function g such that the equation $\beta = g(\alpha)$ agrees with the distributions of α and β, as estimated above. In other words, we wish to find g such that $F_\alpha(t) = F_\beta(g(t))$ holds for all t. This leads to $g(t) = t/4$, which is strictly increasing. We are now in a position to write the equation $\beta = \alpha/4$, which, together with the relationships $\sigma = (\alpha + \beta)/2$ and $\eta = (\alpha - \beta)/2$, can now be used to retrieve the distribution functions of σ and η, to be denoted F_σ and F_η. The result is:

$$F_\sigma(t) = \min\{0.16 + 1.6t, 1\}$$

$$F_\eta(t) = \min\{0.16 + 2.7t, 1\}$$

for $0 \leqslant t \leqslant 1$. The population from which our subjects were drawn may thus be characterized in the following way. There is a hard core (an atom) of pure self-seekers (that is, individuals satisfying $\sigma = \eta = 0$) comprising one-sixth of the population. The remaining five-sixths exhibit the properties of sympathy and parity – that is, they have positive σs and ηs, with the σ-values and η-values of this remaining five-sixths of the population being uniformly distributed. By our co-monotonicity assumption, sympathy and parity go hand-in-hand – that is, the two variables are perfectly correlated. The highest possible value of σ (that is, the upper bound of the support of σ), as estimated from our raw (*unrounded*) data, is given by 0.505, which is remarkably close to the theoretical maximum, $\sigma = 1/2$ ($\sigma = 1/2$ is the case of an agent who regards a dollar paid to oneself and a dollar paid to the other person as equally desirable – 'Love Thy Neighbour as Thyself'.) As for the parity coefficient, η, the highest possible value allowed by our estimated distribution is given, approximately, by $\eta = 1/3$. In other words, the *highest* degree of inequality-aversion occurs when an individual is willing to pay 1 dollar to correct a 3-dollar gap in incomes. The *average utility function* implied by our data that is, the utility U, as defined above, using the estimated *means* of σ and η, comes out to be:

$$U^{av}(x, y) = 0.78x + 0.22y - 0.14|x - y|$$

We know of no a priori discussion of parity and sympathy that would tell us whether this representation of an average individual is 'reasonable' or not. It should be remembered also that the subjects in our experiment were informed that 'the person on the opposing side' was also someone participating in the same experiment. All our estimates and results are, of course, specific to this context.

Task C

In Tasks A and B, subjects were dealing with degenerate game-forms, where strategic interaction was altogether absent. The game-form of Task C, in contrast, involved genuine interaction, albeit very limited in scope. The tree diagram in Figure 14.5 describes the simple setting in which the subjects found themselves:

Each subject was assigned the role of either Opener (Player I) or Responder (Player II). Once again, subjects were informed that the person playing the other role was also a participant in the experiment, whose identity was to remain unknown, and for whom the subject's own identity would equally remain unknown. Subjects in the role of Player I were asked to perform their selected action by placing a sticker in a designated space, using the colour of the sticker to indicate the action being taken (L or R in Figure 14.5). Subjects in the role of Player II were asked to submit *strategies*, by placing two coloured

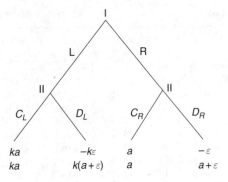

Figure 14.5 Task C

stickers in designated spaces, each sticker indicating what the action should be, given the colour that Player I will have selected. It was natural for the subjects to think in terms of strategies, in view of the information that the person on the other side was someone who was going to be matched to you (that is, to the subject) at random, *after* the entire experiment had been run. It was possible, therefore, that the of the opponent's opening move had not even been played at the time of the subject's own action. Thus, specifying the selected action conditionally, for every possible opening move of the opponent, was clearly seen to be the only way.

Note that Player II in Task C is in fact facing a situation that is the same as that of the decision-maker in Task A. Indeed, Player II's choice is between a 'co-operative' move, which would result in both him/herself and the person on the other side receiving equal positive payments, and a 'defecting' move, which yields a somewhat greater payment to oneself but inflicts a loss on the person on the other side. This is true at both nodes where Player II has a move. The difference is that at the left-hand node the stakes are uniformly higher than at the right-hand node, by a factor $k > 1$. Thus, all that Player I's move does is to determine the size of the stakes. Moving left, by Player I, can be thought of as 'being trusting' and moving right as 'being cautious'. As before, all payments were denominated in new Israeli shekels (NIS), and specifically with values $a = 5$, 10 and $\varepsilon = 0.50$, 1. The values of the scale factor were set at $k = 2, 3$. Taking all the combinations of these parameter values yielded a $2 \times 2 \times 2$ design, so subjects were divided at random into eight groups of roughly equal size, with each group filling one of the cells in the $2 \times 2 \times 2$ design. On average, about thirty subjects were assigned to each cell, divided about equally between those in the role of Player I (Opener) and Player II (Responder).[2] Subjects who are uniformly self-seeking would always play R ('cautious') in the role of Player I and (D_R, D_L) in the role of Player II – that is, they would play out the unique subgame perfect equilibrium of the game in which individually received money payments act as payoffs. As we

Table 14.2 Data, Task C

	Parameters			Player I (Opener)		Player II (Responder)			
	a	*ε*	*k*	R	L	$C_R C_L$	$C_R D_L$	$D_R C_L$	$D_R D_L$
1	5	0.5	2	3 (20)	12 (80)	8 (62)	1 (8)	2 (15)	2 (15)
2	5	1	2	2 (14)	12 (86)	7 (58)	1 (8)	1 (8)	3 (25)
3	5	0.5	3	4 (29)	10 (71)	9 (60)	2 (13)	2 (13)	2 (13)
4	5	1	3	4 (24)	13 (76)	7 (44)	2 (13)	2 (13)	5 (31)
5	10	0.5	2	1 (7)	14 (93)	11 (65)	1 (6)	0 (0)	5 (29)
6	10	1	2	4 (24)	13 (76)	12 (67)	0 (0)	3 (17)	3 (17)
7	10	0.5	3	2 (11)	16 (89)	14 (93)	0 (0)	0 (0)	1 (7)
8	10	1	3	2 (14)	12 (86)	8 (53)	3 (20)	0 (0)	4 (27)
Overall				22 (18)	102 (82)	76 (63)	10 (8)	10 (8)	25 (21)

already know, from Tasks A and B, subjects were not uniformly self-seeking, and this observation was reinforced by the results for Task C, which are summarized in Table 14.2 (numbers in parentheses are percentages).

In our analysis of Tasks A and B, we had assumed that an individual evaluates an action that yields x shekels to him/herself and y shekels to the person on the other side by calculating the quantity:

$$U(x, y) = (1 - \sigma)x + \sigma y - \eta|x - y|$$

where the pair (σ, η) designates the individual's type. Adopting this framework also for Task C, we find that an individual of type (σ, η) in the role of Player II (Responder) would play (C_R, C_L) if $(1 - 2\sigma)/(\sigma + \eta) \leqslant a/\varepsilon$, and (D_R, D_L) otherwise. (Note that this condition is independent of the scale factor k.) Given the distributions of σ and η, as estimated above, and given our co-monotonicity assumption, the predicted behaviour pattern for Player II (the Responder) would be as specified in Table 14.3. In order to facilitate comparison with the observed data, each cell in Table 14.3 contains both the predicted frequency (at the lower left) and the observed frequency (at the upper right). Observed frequencies have been calculated, in each case, by pooling together the data for the two values of the scale parameter k.

The hypothesis underlying the predicted frequencies in Table 14.3 is *not* the hypothesis that we had formulated at the outset for task C. We did not expect the framework developed for Tasks A and B to be adequate for explaining behaviour in Task C. Rather, we had expected that, in Task C, subjects in the role of Player II would display a tendency to *reciprocate*, a tendency that cannot come into play in the interaction-free settings of Tasks A and B but can (and would, we thought) come into play in Task C.

Table 14.3 Player II's strategies

Parameter values		Frequencies of Player II's strategies (in %)			
a	ε	$C_R C_L$	$C_R D_L$	$D_R C_L$	$D_R D_L$
5	0.50	61	11	14	14
		75	0	0	25
5	1	51	10	10	29
		67	0	0	33
10	0.50	79	3	0	18
		79	0	0	21
10	1	60	10	8	22
		75	0	0	25

An agent is said to engage in reciprocity if s/he tends to be kind to an opponent who is perceived to have been kind and co-operative, and to be unkind – even vindictive – to an opponent who is perceived to have been unkind and disregarding. Consider an agent in the role of Player II in Task C. We expected that, in this situation, at least some subjects would regard the opening move L of Player I as kind and trusting, and the opening move R of Player I as unkind and distrusting. These subjects (we thought) would then tend to reciprocate, by rewarding a trusting opponent and snubbing a distrusting one. In our setting, reciprocity would be exhibited through Player II's use of the strategy (D_R, C_L), which one might refer to as the *reciprocating strategy*. Our hypothesis was that the frequency of appearance of this response strategy (D_R, C_L), would be positive and significantly greater than the frequency of appearance of its mirror image (C_R, D_L), which we had expected to be negligible. Under reciprocity, the frequencies of the two 'straight' strategies (C_R, C_L) and (D_R, D_L), would both be lower than the values predicted by the model of Tasks A and B, with both reductions together making up the frequency of the reciprocating strategy (D_R, C_L). In the event (see Table 14.3), subjects in the role of Player II did select the reciprocating strategy in about 8 percent of the cases, on average, but a similar number of subjects opted for the mirror-image strategy (C_R, D_L), thereby displaying a pattern that seems to run contrary to reciprocity. The case for reciprocity in our data, if there is one, is very weak at best. Possibly this is because the opening move R is not being perceived by Responders as necessarily being unkind.

Now let us consider the behaviour of Player I (the Opener). The choice as to whether to play R or L clearly hinges on Player I's assessment of who it is out there, at the opposite side of the interaction. The first hypothesis to be checked out in this respect is *rational expectations*. Under rational expectations, one assumes that the agent in the role of Player I takes the distribution of types in the population as given and picks an action optimally, given this distribution and given his/her type. One then proceeds to check whether the distribution of agents implicit in the observed actions of subjects in the role of Player I coincides with the underlying distribution, which Player I had taken as given. In the present context, rational expectations will have been confirmed if the distribution of types that was derived above for Tasks A and B were a fixed point of this kind. Checking the data, we find that this is not the case. Indeed, if the subjects in the role of Player I had known that the distribution of types was as described in the previous section, then they would all play L ('trusting') regardless of their own type – that is, even a strictly self-seeking Player I would play L. While our data do show the rates of playing L to be quite high, they fall far short of 100 per cent L in all cases.

Given the failure of rational expectations, we are led to consider the following alternative hypothesis: when contemplating the question of who it could be, sitting out there at the opposite end of the interaction, the subject's answer might be 'it's probably someone more or less like myself'. This leads us to the hypothesis that a subject of type (σ, η) in the role of Player I will act on the assumption that the opponent (Player II) is also of type (σ, η). We refer to this naïve mode of forming expectations regarding the opponent's type as the *egomorphic expectations* hypothesis.

It is straightforward to calculate what an agent of type (σ, η) would do, in the role of Player I, if s/he assumed the opponent to be of the same type. Using the distribution of types derived in the previous section, one can determine, for each parameter configuration, the predicted frequencies with which the two actions, L and R, will be selected. This calculation comes out as follows: for parameter values (a, ε, k), a fraction $(a + 8\varepsilon)/(6a + 12\varepsilon)$ of the population will play R ('cautious'), independently of the value of the scale parameter k. A comparison of these predicted frequencies with observed frequencies is displayed in Table 14.4. (Observed frequencies reflect a pooling of responses over the two values of the parameter k.)

We see that the move R ('cautious') is observed systematically less frequently than would be predicted under egomorphic expectations. However, when we consider the four parameter configurations separately, we find that in each case the difference between the observed frequency and the predicted frequency is not statistically significant. In this sense we are entitled to conclude that the egomorphic expectations hypothesis has been confirmed.

Table 14.4 Predicted and observed behaviour by Player I

Parameter values		Frequency (in %) of Move R ('Cautious') by Player I	
a	*ε*	Predicted	Observed
5	0.50	25	24
5	1	31	26
10	0.50	21	18
10	1	25	19

Relevance

Even after its recent triumphs, experimental work in economics is still viewed with suspicion, in some quarters, on the grounds of relevance. The argument runs roughly as follows: evidence regarding the behaviour of subjects in experimental settings may be of considerable interest to psychologists studying individual behaviour *per se*. Impeccable and robust as their findings might be, these findings can only be taken as evidence of how subjects tend to behave in the given experimental setting. In a market setting, the *very same* subjects may well behave differently, if only because in the marketplace a person's very livelihood is at stake, and this can never be the case in an experiment. Thus, experimental evidence becomes 'economically irrelevant' almost by definition.

Similar scepticism can presumably be voiced in the present context: it is true that we find subjects veering off systematically from self-seeking behaviour, but in a market setting 'things are *serious*', so these very subjects may well revert to being strict self-seekers. Presumably, this would render our findings economically irrelevant. Quite surprisingly – and inadvertently – we find ourselves in possession of evidence on this issue.

Let us recall some of the design features of our experiment. Subjects who had participated in the experiment were entitled to receive certain monetary payments. The exact amount due to any given subject was determined not only by this subject's own action, but also by the actions of others, who would be matched to this subject later. Thus, payments could only be made after a delay, a day or two after the conclusion of all the experimental sessions. This was clearly the only way to guarantee that everyone received his/her due. Subjects were therefore instructed to write their names on their completed task sheets, and their instructions told them that they could collect their individually calculated earnings at some designated office on campus at any time during working hours in the week following the running of the experiments.

What happened was this: at the very start of the payment week, a queue of people anxious to collect their rightful earnings formed, outside the door of the designated payment office. After the people in the queue had been paid, individuals continued turning up steadily throughout the working hours, with the pace of arrivals slackening off gradually. This process continued throughout the week. At the week's end, it became apparent that only about 70 per cent of the subjects had in fact turned up to collect their earnings. It therefore became necessary to declare an extension of the payment deadline, so another week was allowed for subjects to come and collect their money. This decision was announced clearly and conspicuously on all the appropriate bulletin boards. Most of the remaining subjects did show up within the extension period, but a small residual, consisting of about 5 per cent of all subjects, never did turn up to collect their rightful earnings. (It was decided not to pursue these remaining subjects individually.)

This course of events made it possible for us to explore the relationship, if any, between 'taking money seriously' and a person's tendency to veer away from strictly self-seeking behaviour. Consider the view that non-self-seeking behaviour is a kind of frivolity that will tend to disappear 'when things get serious'. If this view is correct, then the more seriously the subject regards monetary earnings, the less likely this subject would be to depart from strictly self-serving actions. Fortunately, our observations made it possible for us to test this hypothesis in a straightforward manner. We defined two integer-valued variables, t and s, such that t_i is subject i's time of arrival to collect his/her earnings, and s_i is the number of i's non-self-seeking moves ('moves to the left') in Tasks A, B, and C. (The exact definitions of t and s are simple and unimportant.) On calculating the correlation between these two variables, we found exactly nothing: no statistical relationship whatsoever was found to exist between acting selfishly and taking money seriously. People who were very eager to meet their money were as likely to depart from self-seeking behaviour as were people who were extremely relaxed about it.

Conclusion

The fact that people do not always act in a purely self-interested manner has long been recognized. But a detailed analysis of such behaviour, with a view to identifying the systematic elements in it (if any) is only recently being attempted. Our intent, in the present chapter, has been to try to contribute to this newly emerging research effort.

Notes

1   At the time, the official rate of exchange was approximately 1.00 NIS = 0.50 US$.
2   In the course of running the experiment, discrepancies developed in some cases between numbers of Openers and numbers of matching Responders. In these cases, final payments were determined by randomly forming a few 2-to-1 matchings.

References

Bolton, G. E. (1991) 'A Comparative Model of Bargaining: Theory and Evidence', *American Economic Review*, 81, 5, 1096–136.

Bolton, G. E. and Ockenfels, A. (2000) 'ERC: A Theory of Equity, Reciprocity, and Competition', *American Economic Review*, 90, 1, 166–93.

Cooper, R., DeJong, D. V. Forsythe, R. and Ross, T. W. (1996) 'Cooperation without Reputation: Experimental Evidence from Prisoner's Dilemma Games', *Games and Economic Behavior*, 12, 187–218.

Dawes, R. M. and Thaler, R. H. (1988) 'Anomalies: Cooperation', *Journal of Economic Perspectives*, 2, 3, 187–97.

Fehr, E. and Schmidt, K. (1999) 'A Theory of Fairness, Competition, and Cooperation', *Quarterly Journal of Economics*, 114, 3, 817–68.

Fehr, E., Gächter, S. and Kirchsteiger, G. (1997) 'Reciprocity as a Contract Enforcement Device: Experimental Evidence', *Econometrica*, 65, 4, 833–60.

Rabin, M. (1993) 'Incorporating Fairness into Game Theory and Economics', *American Economic Review*, 83, 5, 1281–302.

Roth, A. E. (1995) 'Introduction to Experimental Economics', in J. H. Kagel and A. E. Roth (eds), *The Handbook of Experimental Economics*, Princeton, NJ: Princeton University Press.

Sally, D. (1995) 'Conversation and Cooperation in Social Dilemmas', *Rationality and Society*, 7, 1, 58–92.

15
Fairness in Stackelberg Games

Steffen Huck, Manfred Königstein and Wieland Müller

Introduction

In experiments, it is often observed that subjects do not play according to the subgame perfect equilibrium (SPE). There is overwhelming evidence for this claim coming from sequential bargaining games. In its simplest form, the ultimatum game, introduced by Güth *et al.* (1982), the Proposer is predicted to claim (almost) the entire pie for him/herself while the Responder should accept all positive offers. Contrary to this prediction, one usually observes that the modal offer made by Proposers is a 50–50 split, and that Responders reject substantial positive offers.

Further evidence for the failure of the SPE to predict behaviour in sequential games comes from Stackelberg duopoly markets. Huck *et al.* (2001) report on an experiment designed to compare Stackelberg and Cournot duopoly markets with quantity competition. A main finding of this study is that Stackelberg markets yield, as predicted, higher outputs than do Cournot markets. However, leaders' output is much lower than predicted, and followers' empirical reaction function is far from the rational reaction function. In fact, the observed reaction function is sometimes even upward-sloping.

One reason for subjects' reluctance to play according to the SPE might be that payoffs in the SPE are unequal. In fact, in the ultimatum game, the Proposer is predicted to receive almost the entire pie, leaving (virtually) nothing for the Responder. In the linear Stackelberg game implemented in Huck *et al.* (2001) the Stackelberg leader is supposed to produce twice as much as the Stackelberg follower. This implies a payoff for the leader that is twice as high as that of the follower.

It has been argued that social preferences might interfere with subgame perfect behaviour. For example, Binmore *et al.* (2002) ask whether, given payoff-interdependent preferences, players respect backward induction. For this purpose, they split backward induction into its component parts: subgame consistency and truncation consistency. They analyse both conditions by comparing the outcomes of two-stage bargaining games with one-stage games with varying rejection payoffs. They find systematic violations of both subgame and truncation consistency. Johnson *et al.* (2002) analyse whether it is the subjects' limited computation ability or their social preferences that are responsible for the failure of backward induction. They run experiments on a two-person, three-period, alternating-offer bargaining game and show that both social preferences and the fact that many subjects do not instinctively apply backward induction are responsible for the failure of the SPE prediction to emerge.

In this chapter we add another piece to the mosaic by asking the following question: what happens in a Stackelberg duopoly in which payoffs in the SPE are equal? We report on experiments testing this hypothesis. We find that equalizing the payoffs in the SPE outcome drives behaviour closer to the SPE prediction. However, the effect is moderate and the behaviour of both, leaders and followers, is still distinctively different from the SPE.

This deviation may be explained by the players' fairness motives. Assuming preferences described by Fehr and Schmidt (1999) we find an aversion to disadvantageous inequality. Interestingly, however, some participants seem to enjoy advantageous inequality, a motivation that is usually discarded in the literature as being irrelevant.

Markets, treatments and procedures

Consider the following duopoly market. Two firms, the Stackelberg leader L and the Stackelberg follower F, face linear inverse demand:

$$p(Q) = \max\{30 - Q, 0\} \qquad Q = q^L + q^F$$

while linear costs are given by:

$$C(q^i) = 6q^i \qquad i = L, F$$

First, the Stackelberg leader decides on its quantity q^L, then – knowing q^L – the Stackelberg follower decides on its quantity q^F. The SPE solution is given by $q^L = 12$ and the follower's best-reply function $q^F(q^L) = 12 - \frac{1}{2}q^L$, yielding $q^F = 6$ in equilibrium. This implies payoffs $\pi^L = 72$ and $\pi^F = 36$. Joint-profit maximization implies an aggregate output of $Q^J = 12$, and symmetric joint-profit maximization implies $q^J = 6$ ($\pi^J = 72$).[1]

Huck *et al.* (2001) implemented the above market in the laboratory. We refer to these data as treatment SYM, since the payoff functions of the players are symmetric. In the experiment, the participants got a payoff table (see Huck *et al.*, 2001) in which all possible combinations of quantity choices and the corresponding profits were shown. The numbers given in the payoff table were measured in a fictitious currency. Each firm could choose a quantity from the set $\{3, 4, \ldots, 15\}$. The payoff table was generated according to the demand and cost functions given above.[2]

In a second treatment (treatment ASYM) we implement a duopoly market as described above, but with asymmetric cost functions:

$$C^L(q^L) = 6q^L - 18 \quad \text{and} \quad C^F(q^F) = 6q^F + 18$$

Introducing fixed costs does not change the theoretical predictions as worked out above. Specifically, SPE quantities are still $q^L = 12$ and $q^F = 6$. However, it does influence the distribution of payoffs across players. SPE payoffs are now $q^L = q^F = 54$; that is, our manipulation of the payoff functions induces equal SPE payoffs.

In the data analysis below we investigate whether behaviour varies between treatments. If participants were rational, money-maximizing agents, we should observe no difference, since the players should choose SPE quantities in both treatments. But we predict that in treatment ASYM the leader's quantity is higher and the follower's quantity lower than in treatment SYM. We interpret this as evidence for fairness motives, since this shift in behaviour partially compensates the cost asymmetry induced in treatment ASYM.

Both treatments employed a fixed matching scheme. The experiments for treatment SYM were conducted at Humboldt University in Berlin in June and July 1998. They were run in three sessions with 18, 14, and 16 participants, respectively. Thus, in total, 48 subjects (24 pairs) participated in these experiments. The experiments for treatment ASYM were conducted at University College London in February and March 2002 when a total of 40 subjects (20 pairs) participated in three sessions (12 + 10 + 18). At both locations, subjects were either recruited randomly from a pool of potential participants or invited to participate via leaflets distributed around the University campus.

All experiments were run with pen and paper. Subjects were seated such that communication was prevented between players. The player positions of a leader or follower were randomly assigned to subjects and remained constant throughout the entire experiment. Each session consisted of ten rounds, with individual feedback between rounds. Sessions lasted between 60 and 75 minutes. Subjects' average earnings in treatment SYM were DM 15.67 (including a flat payment of DM 5) (about us $9 at the time of the experiment). Subjects' average earnings in treatment ASYM were £9 (including a flat payment of £4) which was about us $14 at the time of the experiment.

After reading the instructions, participants were allowed to ask the experimenters questions privately. In the instructions (see Huck *et al.*, 2001) subjects were told that they were to act as a firm which, together with another firm, produces an identical product and that, in each round, both have to decide what quantity to produce. Before the first round started, subjects were asked to answer a control question (which was checked) in order to make sure that everybody fully understood the payoff table.

The firms were labelled A (Stackelberg leader) and B (Stackelberg follower). In each of the ten rounds the Stackelberg leaders received a decision sheet on which they had to note their code number and their decision by entering one of the possible quantities in a box. These sheets were then passed on to the subjects acting as followers. Subjects were not able to observe how the Stackelberg leaders' decision sheets were allocated to the followers. After collecting the sheets from the Stackelberg leaders, one experimenter left the room to collate the sheets into their final order.

Followers also had to enter their code number, and then made their decision on the same sheet. In doing so, they immediately had complete information about what happened in the course of the actual round. After the round, the sheets were collected and passed back to the Stackelberg leaders who now also had information about this round's play. Again, one of the experimenters left the room with the decision sheets. After the collection of the sheets, the next round started.

Results

Table 15.1 provides essential summary statistics at an aggregate level for both treatments. Recall that the outcome in the SPE is $(q^L, q^F) = (12, 6)$. The first column in Table 15.1 reveals that average individual quantities in treatment SYM are $(9.13, 7.92)$ whereas they are $(10.12, 7.31)$ in treatment ASYM. Thus, in treatment ASYM both players' quantities are closer to SPE quantities than in treatment SYM. At the same time, these shifts (increase in leader quantity and decrease in follower quantity) compensate partially for the leader's cost disadvantage that has been introduced by treatment ASYM. Total quantity is lower and total profits higher in treatment SYM.

Total output

Recall that industry output in the SPE outcome is 18, whereas in the Cournot outcome it is 16. Moreover, all individual quantity combinations summing up to 12 maximize joint profit. Figure 15.1 shows the distribution of total outputs in the two treatments. There is one key observation to be made in this figure. The most frequent industry output in treatment SYM is $Q^J = 12$ (15 per cent of all case). In contrast to that, in treatment ASYM there is only

Table 15.1 Aggregate data (averages)

Treatment	Individual quantity	Total quantity	Total profits
SYM	9.13/7.92	17.05	118.49
	(2.67/2.00)	(3.67)	(45.99)
ASYM	10.12/7.31	17.43	114.50
	((3.03/2.44))	(3.79)	(39.89)

Note: Standard deviations in parentheses.

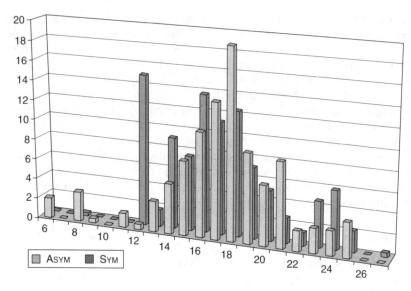

Figure 15.1   Distribution of total output (all rounds)

one such outcome (or 0.5 per cent of all cases) and the most frequent output coincides with the SPE output of 18 (19 per cent of all cases). Thus, whereas joint-profit maximization plays a significant role in treatment Sm, it is not an issue in treatment ASYM.

This finding is natural if the players entertain fairness motives. Namely, in treatment SYM joint-profit maximization can be achieved by equal-quantity choices of 6 by both players, which induces equal payoffs of 72. In treatment ASYM, quantity choices of 6 by both players also maximize joint profit but induce unequal payoffs ($\pi^L = 56, \pi^F = 90$). Furthermore, all quantity combinations that sum to 12 induce unequal payoffs. We conclude that a joint-profit maximizing and fair outcome is harder to achieve in treatment ASYM than in treatment SYM. This compares to Mason *et al.* (1992) who report on a series of (simultaneous choice) Cournot duopolies with either symmetric or asymmetric costs and find that subjects in asymmetric experiments are significantly less co-operative than subjects in symmetric experiments. The Stackelberg duopoly is *per se* an asymmetric game because of the sequentiality of choices. None the less, co-operation (joint-profit maximization) may be achieved more easily with symmetric cost (treatment SYM). As the comparison of our two treatments shows, joint-profit maximization is possible in Stackelberg markets as long as it is feasible with equal payoffs. We summarize this in:

OBSERVATION 1   There is substantially more joint-profit maximization in treatment SYM where this is feasible with equal payoffs for both players.

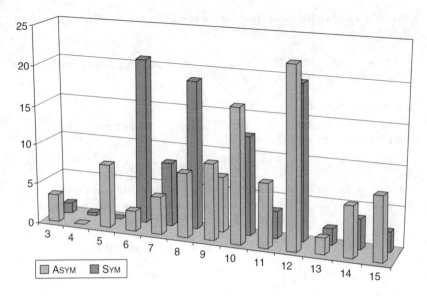

Figure 15.2 Distribution of leaders' quantities (all rounds)

Leaders' behaviour

Table 15.1 revealed that leaders in treatment ASYM choose on average one unit more than leaders in treatment SYM (10.12 versus 9.13). More evidence is provided in Figure 15.2 showing histograms of the leaders' choices over all ten rounds. In treatment SYM the most frequent choice of leaders is 6 (with more peaks at quantities 12 and 8). However, in treatment ASYM the mode of leaders' choices is 12.

To investigate the behaviour of leaders in more detail, we estimate the panel regression model:

$$q_{it}^L = \alpha_0 + \alpha_1 TREAT + \nu_i + \varepsilon_{it}$$

where q_{it}^L is the individual quantity set by leader firm i in period t, ν_i is the subject-specific random error component and ε_{it} is the overall error component. We include *TREAT*, which is a dummy variable indicating the treatment. That is, $TREAT = 0$ if the decision stems from treatment SYM and $TREAT = 1$ for decisions coming from treatment ASYM. The result is as follows (*t*-statistics in parentheses)

$$
\begin{array}{ccc}
q_{it}^L = & \alpha_0 & + \alpha_1 TREAT + \nu_i + \varepsilon_{it} \\
& 9.133 & 0.982 \\
& (27.18) & (1.97)
\end{array}
$$

Since the coefficient α_1 is positive and significant, we have statistical support[3] for:

> OBSERVATION 2   Leaders in treatment ASYM produce significantly more than leaders in treatment SYM.

Followers' behaviour

To assess followers' behaviour more thoroughly, we estimate the following panel regression model:

$$q_{it}^F = \beta_0 + \beta_1 q_{it}^L + \beta_2 TREAT + \beta_3 TREAT \times q_{it}^L + v_i + \varepsilon_{it}$$

where q_{it}^F is the individual quantity set by follower firm i in period t, q_{it}^L is the quantity of the leader in period t with whom follower i was matched. All other variables are as above. The result is as follows:

$$
\begin{array}{cccccc}
q_{it}^F = & \beta_0 & + \beta_1 q_{it}^L & + \beta_2 TREAT & + \beta_3 TREAT \times q_{it}^L & +v_i + \varepsilon_{it} \\
& 6.349 & 0.172 & 1.428 & -0.218 & \\
& (11.10) & (3.14) & (1.76) & (-2.99) &
\end{array}
$$

The estimated reaction function in treatment SYM is $q^F(q^L) = 6.349 + 0.172q^L$. It has a much lower intercept than the rational reaction function and is *upward* sloping. In contrast, in treatment ASYM the reaction function is $q^F(q^L) = 7.777 - 0.046q^L$. It has a higher intercept and is more or less flat.[4] On average, it seems, followers simply choose roughly the Cournot quantity $q^C = 8$. Both reaction functions are shown in Figure 15.3.

> OBSERVATION 3   The empirical reaction function in treatment ASYM has a higher intercept and a lower slope than the reaction function in treatment SYM. In particular, whereas the reaction function in treatment SYM is upward sloping, it is essentially flat in treatment ASYM.

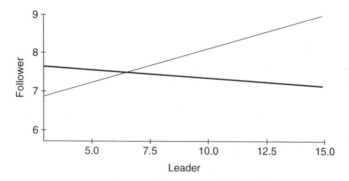

Figure 15.3 Observed reaction functions in treatments SYM (thin line) and ASYM (thick line)

Discussion and concluding remarks

In both treatments, SYM and ASYM, individual quantities deviate substantially from the predicted SPE levels. The leader's quantity is lower and the follower's quantity is higher than in SPE. Moreover, the behaviour of both players differs between treatments, which is inconsistent with SPE. A possible explanation for our findings are fairness motives of the experimental participants. Fair allocations in distribution conflicts (for example, in sequential bargaining games) may be characterized as allocations in which a player's share of a joint outcome is proportional to his/her contribution to the outcome. This is known as equity theory (see, for example, Homans, 1961; Güth, 1994). The Stackelberg game is not a pure distribution conflict, since production of the joint outcome cannot be separated from its distribution. Furthermore, equity theory does not account for the fact that the leader has a strategic advantage compared to the follower.[5] Nevertheless, the participants may entertain perceptions of fairness. In treatment SYM equal payoffs should be perceived as the fair outcome of the game, since equal quantity choices (contributions to joint outcome) induce equal payoffs. Empirically, the leader earns somewhat more than the follower, which resembles the partial first-mover exploitation that has been observed repeatedly in ultimatum games. In treatment ASYM, the follower is favoured and the leader is disfavoured compared to treatment SYM. Taking a fairness perspective we predicted and found a decrease of the follower's quantity and an increase in the leader's quantity.

Fairness motives have been combined with rational choice theory by appropriate assumptions about the utility functions of players. An example is the model by Fehr and Schmidt (1999) which allows for advantageous and disadvantageous inequality aversion. Huck *et al.* (2001) analysed the implications of the Fehr – Schmidt model for the behaviour in the Stackelberg market with symmetric cost (that is, treatment SYM). Accordingly, follower behaviour can be explained by aversion to disadvantageous inequality – that is, a positive value of the parameter α in the terminology of Fehr and Schmidt (1999). Followers are willing to give up some amount of money in order to reduce the gap between their own (low) payoffs and the leaders' (higher) payoffs. Interestingly, however, Huck *et al.* (2001) also report that some players seem to enjoy advantageous inequality – that is, they exhibit a *negative* value for the parameter β – a motivation that has been ruled out as being irrelevant by Fehr and Schmidt. To some extent, we reproduce these findings in treatment ASYM. There are leaders who produce more than they should if they were money-maximizers and if they anticipated follower behaviour correctly. This is evidence of disadvantageous inequality aversion. But there are also followers who choose higher quantities than their money-maximizing quantity, which increases the gap between their own *high* payoffs and the leaders' lower payoffs. These participants seem to enjoy advantageous inequality. This evidence is intriguing, but identification of individual preference characteristics would require further analysis that accounts, for example, for possible repeated game dynamics. These issues are beyond the scope of this chapter and are left for further studies.

Appendix: translated instructions

Instructions used in Huck et al., 2001.
Welcome to our experiment! Please read these instructions carefully! Do not talk to your neighbours and be quiet during the entire experiment. If you have a question, let us know. Queries will be answered privately.

In our experiment you can earn different amounts of money, depending on your behaviour and that of other participants who are matched with you.

You play the role of a firm which produces the same product as another firm in the market. Both firms always have to make a single decision, namely what quantities they want to produce. In the attached table [not printed here], you can see the resulting profits of both firms for all possible quantity combinations.

The table is read as follows: the head of the row represents one firm's quantity (Firm A) and the head of the column represents the quantity of the other firm (Firm B). Inside the little box where row and column intersect, Firm A profit matching this combination of quantities is up to the left and Firm B profit matching these quantities is down to the right. The profit is denoted in a fictitious unit of money we call Taler.

So far, so simple. But how do you make your decision? Take a look at your code number: if it starts with an A, you are an A firm, if it starts with a B, you are a B firm. The procedure is that the A firm always starts. This means that the A firm chooses its quantity (selects a line in the table) and the B firm is informed about the A firm's choice. Knowing the quantity produced by the A firm, the B firm decides on its quantity (selects a column in the table). The B firm then, of course, already knows its own profit. The A firm will be informed about it (or rather about B's choice). The decisions are marked on a separate decision-sheet, which we shall hand out soon to all participants with role A.

During the entire experiment, anonymity among participants and instructors will be maintained, since your decisions will only be identified by your code number. You therefore have to keep your code card carefully. Only when you show the code card later will you receive your payment.

Concerning the payment, note the following: at the end of the experiment two of the ten rounds will be randomly chosen to count for payment. The sum of your profits in Taler of these two rounds above determines your payment in DM. For each ten Taler you will be paid 1 DM. In addition to this, you will receive 5 DM independent of your profits in the ten rounds.

Notes

1 In an equivalent (simultaneous) Cournot market, Nash equilibrium play implies $q^C = 8$.

2 Because of the discreteness of the strategy space, such a payoff table typically induces multiple equilibria. To avoid this, the bi-matrix representing the payoff

table was manipulated slightly. By subtracting one payoff unit in 14 of the 169 entries we ensured the uniqueness of both the Cournot–Nash equilibrium and the subgame perfect Stackelberg equilibrium.

3 That leaders choose higher quantities in treatment ASYM than in treatment SYM can also be validated by a robust statistical test using a player's average quantity choice across ten periods as unit of analysis. A one-tailed Mann–Whitney U-test rejects the null-hypothesis at the 5 per cent level ($N = 44, p = 0.028$).

4 The slope coefficient differs significantly between treatments, but is not significantly different from zero in treatment ASYM.

5 That the equity concept may be ambiguous and may be operationalized differently is shown in, for example, Königstein (2000).

References

Binmore, K., McCarthy, J., Ponti, G., Samuelson, L. and Shaked, A. (2002) 'A Backward Induction Experiment', *Journal of Economic Theory*, 104, 48–88.

Fehr, E. and Schmidt, K. (1999) 'A Theory of Fairness, Competition, and Cooperation', *Quarterly Journal of Economics*, 114, 817–68.

Güth, W. (1994) 'Distributive Justice: A Behavioral Theory and Empirical Evidence', in H. Brandstätter and W. Güth (eds), *Essays on Economic Psychology*, Berlin, Springer-Verlag, 153–76.

Güth, W., Schmittberger, R. and Schwarze, B. (1982) 'An Experimental Analysis of Ultimatum Bargaining', *Journal of Economic Behavior and Organization*, 3, 376–88.

Huck, S., Müller, W. and Normann, H.-T. (2001) 'Stackelberg Beats Cournot: On Collusion and Efficiency in Experimental Markets', *Economic Journal*, 111, 749–65.

Homans, G. C. (1961) *Social Behavior: Its Elementary Forms*, New York: Harcourt.

Johnson, E. J., Camerer, C., Sen, S. and Rymon, T. (2002) 'Detecting Failures of Backward Induction: Monitoring Information Search in Sequential Bargaining', *Journal of Economic Theory*, 104, 16–47.

Königstein, M. (2000) *Equity, Efficiency and Evolutionary Stability in Bargaining Games with Joint Production*, Lecture Notes in Economics and Mathematical Systems series, vol. 483, Berlin, Springer-Verlag.

Mason C. F., Phillips, O. R. and Nowell, C. (1992) 'Duopoly Behavior in Asymmetric Markets: An Experimental Evaluation', *Review of Economics and Statistics*, 74, 662–70.

16
Learning from (and in) the Ultimatum Game: An Interview with Alvin E. Roth

Steffen Huck

SH: *Al, do you remember when you heard or read about the ultimatum game for the first time? And can you describe what you first thoughts about the results were back then?*

AR: I don't recall when I first heard about the ultimatum game, which is a kind of testimony to how important a role it plays in contemporary economic thought. Like the prisoner's dilemma, one can't recall not knowing of it. But I certainly already knew of it shortly after the publication of Güth, Schmittberger and Schwarz (1982), since Werner came to visit us in Pittsburgh around that time. (I recall that he and I went running in the park by my house, and he was in better shape than I was . . .)

And were you surprised? Some people's immediate reaction was after all that the results had to be wrong . . .

Well, no, I don't think I was surprised (except that he ran faster ☺) Remember, I had been doing bargaining experiments too, and seeing that fairness mattered – for example, in my 1979 paper with Mike Malouf, or my 1982 paper with Keith Murnighan (or the 1981 paper with both of them). We also saw a lot of what appeared to be fairness-related disagreements, particularly in Roth and Murnighan (1982). But I certainly recall that some people's immediate, and not so immediate, reaction was as you describe. Years later, at a famous university that I won't name, when I was presenting the results of the (1991) four-country study of ultimatum games and markets, I was surprised to find that almost my whole talk was given over to answering sceptical questions about the basic ultimatum game results.

What do you think is the most important impact the ultimatum game had on the profession? And do you believe we would think differently today about any topic had it never been conducted?

It's hard to imagine that economists would think as they do about fairness if not for the ultimatum game. It affected our ideas about the robustness and importance of people's ideas about fairness. Ultimatum game results have also entered some formal theories of fairness as parameter values (I'm thinking here in particular of the work of Bolton and Ockenfels, 2000). And, of course, they helped to focus our ideas about perfect equilibria and simple monetary preferences, and helped to get us thinking about other kinds of models to explain the ultimatum game results and unify them with other things we noticed that didn't quite fit standard models.

And outside the game theory and experimental econ community? In many branches of the economics literature people still seem quite happy to stick to the orthodox model . . .

Well, I'm not sure that, in many branches of economics, they *shouldn't* be happy with the orthodox model. I was once asked to respond at a conference in Torino to Amos Tversky's question 'Why are economists so reluctant to abandon the rational model?' My reply was called 'Individual Rationality as a Useful Approximation . . . ', and the idea was that we like some models because they are simple, even if they are not exactly correct – that is, we like some models because they are useful approximations. And then we have to ask, for what tasks are they useful, despite not being exactly correct, and for which tasks does the fact that they aren't correct make them not so useful. For example, the model of the earth as perfectly round is obviously false, and useless for planning your next mountain-climbing vacation. But it's a very useful approximation for computing the earth's orbit; much more useful, in fact, then the more realistic and complex model of the earth as a rough-surfaced oblate spheroid. So, for some purposes we should use the round earth model, and for some we should abandon it. In the same way, if I'm interested in why the interest rate on bonds is sensitive to how far in the future they mature, the orthodox economic model might be the right one. So, I think that, as experimenters, our job isn't over when we show that some prediction from received theory is false, or even when we propose another model that better accounts for some of our results. Part of our job is to help think about when our results imply that standard models won't be good approximations, not only in the lab but in the world, and to offer evidence that our new models are sufficiently robust so that they can be useful approximations.

In your first contribution to the ultimatum literature with Jack Ochs you discovered a substantial amount of disadvantageous counter-offers – a similar phenomenon to rejections in the one-stage ultimatum game. Did this come as a surprise to you or was it expected?

It did come as a surprise. Some of the earlier papers reporting experiments motivated by the ultimatum game had suggested that the simplicity of the ultimatum game made it somehow 'special', and that, in multi-period bargaining games, we might see behaviour that looked more like perfect equilibrium, and less like fairness or other distributional concerns playing a really big role. So we were quite surprised to see bargainers rejecting unequal splits, only to come back the next period to propose a division of the discounted pie that, if accepted, would give them even less than they had declined. And once we found this in our data, we went back and looked at the data from those earlier experiments, and found it there too. So the ultimatum game turned out not to be so special after all, just especially simple, and a good tool with which to investigate hypotheses about bargaining.

Around that time you also stressed that a model where some players have other elements in their utility function besides absolute payoffs might be a good way to account for the data, and that was pretty much what Gary Bolton came up with in his 1991 paper. Can you describe to what extent your views were captured by Gary's model at the time?

Gary Bolton was one of the most independent students I've ever had, so I don't want to claim to have influenced him too much, although *my* views have certainly been influenced by *his* work. He had been our research assistant for the experiments reported in Ochs and Roth (1989) and so he was certainly familiar with those results and our views on them. Regarding a theory of fairness, I don't think Jack Ochs and I intended to do more than note that, if one thought that people had preferences that were revealed by their choices, then the disadvantageous counter-offers meant that people preferred more equal distributions even if that meant receiving smaller personal payoffs. Gary's 1991 paper reported how he had tried to formalize a theory of fairness in his dissertation, and he and Axel Ockenfels proposed a much more comprehensively articulated theory in their 2000 paper.

Werner was, of course, very much opposed to any of this tinkering with utility functions. Yet though his work on the so-called indirect evolutionary approach he made it possible to study why agents might have such non-standard preferences. Basically, one might say that his later work enable evolutionary justifications of Gary's model. Do you see any irony in this?

No, no irony, just some convergence. I think that's to be expected with debates that are about how to *interpret* a set of facts. I recall that my early discussions about these things with Werner were a lot like the joke about economists and sociologists, which says that economists study how people make choices, and sociologists study why people don't have any choices. But lots of the same facts can be organized within a framework of fixed preferences, or a framework of social norms, or of boundedly rational learning, or of evolution. So debates over frameworks aren't strictly speaking empirical debates, but rather debates about which ways of looking at things will unify the broadest classes of phenomena most usefully, and provide the most interesting and insightful predictions.

There are now several papers taking particular models, such as Bolton and Ockenfels or Fehr and Schmidt (1999), or variants of both, to applied problems – for example, in mechanism design and contract theory. Do you find this risky – in particular, if these models are not supported by new additional data but simply take, say, inequality aversion as a fact?

Sure it's risky, everything new is risky. But it's also risky not to explore new things. Experimenters certainly know that we have to test predictions that aren't yet supported by data. So as long as we don't claim more than we're entitled to, I think it's inevitable, and good, that the implications of new theories will be explored in applications. And evaluating the success of these efforts will help us figure out which of these new theories are useful approximations, and for what . . .

Let me come back to the early 1990s, when there was also the debate about the best-shot game and Werner argued that the inefficiency of equal splits was key to understanding the difference between the two games. Today Gary Charness and Matt Rabin (2002) advocate again the role of efficiency in their approach to modelling social preferences. On the other hand, you and Vesna Prasnikar argued that the difference between the two games was mainly due to different off-equilibrium properties of the game. Have you ever reassessed your views on that?

Often. I think that observation was the beginning of my interest in models of learning. The ultimatum game, the best-shot game, and the multi-player market game we studied (also in our four-country experiment) all had similar perfect equilbria, at which one player got essentially everything. If you think of players as having to learn to play a game from experience, then the whole landscape of the payoff surface starts to matter, and not just the equilibria. If players don't start playing a game by playing equilibria, then what they learn, and how fast, depends on the feedback they get away from equilibrium.

So the mid-1990s saw you putting more and more emphasis on the role of learning, in particular in your work with Ido Erev (1998) and also later in your Econometrica *article on high-stake ultimatum games with Bob Slonim (1998). A key feature in the latter is that, with high stakes, rejection rates decline over time, which is picked up by proposers who get less generous. While this is in line with simple learning models one might alternatively suspect that responders get increasingly frustrated. If you put somebody in an adverse environment, say a repressive country, they might rebel initially and get tired later – which I would rather model as a change in preferences. But I reckon you won't agree on this.*

How I would feel about you modelling a change in behaviour as a change in preferences would probably depend on how you modelled it, and whether your model allowed me to successfully predict other changes of behaviour in different environments. That is what I like about the learning models: they allow us to predict a broad range of behaviour with a single, simple model. So, for the ultimatum game, the learning models predict that responders will learn to reject small offers more slowly than proposers will learn not to make them. The reason is that whether a responder accepts or rejects a small offer doesn't make much difference in his payoff – since the offer is small. So it doesn't change his behaviour quickly, and if he has some initial propensity to reject small offers, this can persist for a long time. But a proposer who has a small offer rejected earns zero, while if he has a more moderate offer accepted he earns a lot, and so if a more moderate offer has a higher chance of acceptance, that can have a big effect on his payoff, and hence on his subsequent behaviour. We tried a direct test of this in Cooper, Feltovich, Roth and Zwick (2003), in which we reported an experiment in which Responders played the ultimatum game twice as often as Proposers. (That is, in each session there were twice as many Proposers as Responders, and half the Proposers played in each round.) So, when a Proposer was making his tenth offer, a Responder was responding to his twentieth. This essentially had the effect of letting the Responders learn at twice their usual speed compared to Proposers. And, as predicted by simple learning models, this increased the speed at which Responders learned to accept lower offers, relative to the speed at which Proposers learned not to make them.

But Responders' behaviour does not only depend on the feedback they get about their own payoff but also on what they know about the Proposers' payoffs. And one way of reading experiments such as Gary Bolton's work with Klaus Abbink, Karim Sadrieh and Fang Fang Tang (2001) is that they refute simple adaptive learning models by showing that behaviour is sensitive to changes in the informational structure that are irrelevant for those models.

Certainly, one can refute any simple model if the proposition being tested is that it predicts all observed behaviour perfectly. But if you think of a model as a useful approximation, the fact that it can be refuted doesn't necessarily change your beliefs: approximations are not exactly correct, so sometimes their predictions are wrong: they leave things out. In this respect, simple models are sort of like dancing bears: the wonder of it is not that the bear dances as well as a person, but that it dances at all. So, there are many things that can't be accounted for by a simple reinforcement model of learning. Thank goodness you and I and the rest of humanity are a bit more complex than that. The attraction of simple learning models is how much they can explain. And the attraction of the simplest, reinforcement learning models, which use only a player's own experience, and not his other information, is that they show just how much you can predict even without using that extra information. That's especially important in view of how much information we normally assume players have in game theoretic models, even, of course, in games of incomplete information. And it makes reinforcement learning models potentially applicable to a broad range of economic environments, since players always get some feedback from their own actions, but they may or may not know other player's actions, options, payoffs, etc. (Think about trying to apply a theory of fairness to a situation in which you don't know others' payoffs, for example . . .) Of course, being *potentially* applicable to a broad range of environments is only a virtue if the predictions you get when you ignore other information when players have it aren't too bad. That's why it has been encouraging to look at reinforcement learning models in high information situations as well as the low information environments they might seem to be designed for.

So, we shouldn't spend too much time on 'horse races' comparing the predictive power of different models in a given set of situations, but rather explore the boundaries of the domains on which specific models do a good job?

I don't think I'm prepared to say that one of those is a good thing and the other is not. But I definitely like simple models that are robust, and can be used for prediction, rather than models with lots of parameters or special cases that can be fit[ted] to many circumstances, but that have little predictive value.

Let me get back to your work with Cooper, Feltovich and Zwick (2003). That was your most recent piece on ultimatum games. Is it also going to be your last one?

The future is the hardest thing to predict, isn't it? The ultimatum game has proved to be so useful that I don't dare predict that I'll never study it again. Right now, I don't have any purely ultimatum game experiments planned. But one thing we've learned from Werner and the ultimatum game is that a simple, well-formulated game can be an exceptionally useful research tool. So, with Uri Gneezy and Ernan Haruvy (2003), and Brit Grosskopf (2003), I've written two papers that deal with a related game we call the *reverse ultimatum game*. In the two-person version, one Proposer plays with one Responder. The Proposer proposes a division of (in our experiments) 25 tokens to the Responder. If the Responder accepts, then the game ends with this division as the outcome. If the Responder rejects the offer, the Proposer is then allowed to make another offer, as long as that offer is *strictly higher* by a minimum increment (1 token), and as long as both players' proposed shares remain strictly positive. In addition, the Proposer may end the bargaining at any point, in which case both players receive 0. That is, the game ends either when the responder accepts a proposal, or when, following a rejection, the proposer declines to make a better offer.[1] We call this a 'reverse' ultimatum game because the Responder gets to issue incremental 'reverse' ultimata to the Proposer, of the form 'increase your offer or we'll both get zero'. And it's easy to verify that the unique subgame perfect equilibrium of this game gives all but one token to the *Responder*, so it is the reverse of the ultimatum game in that respect also. Not surprisingly, if the Responder rejects all the 'fair' offers, he may find that the Proposer declines to continue bargaining, and so, as in the ultimatum game, the observed outcomes are much nearer equal divisions than the perfect equilibrium. But this game, and variants with more than one Responder, prove to be easier to manipulate than the ultimatum game, with treatments that reverse the perfect equilibrium prediction. So we have found it to be a useful tool for studying issues related to deadlines (if there's a deadline, the Proposer can convert a reverse ultimatum game into an ultimatum game), and some issues relating to the design of 'right of first refusal' contracts.

Al, one final question: is there any specific message you want to convey to Werner?

You mean, aside from 'Happy birthday'? I hope he keeps running fast for a long time to come.

Note

1 The game also ends if the responder rejects an offer of 24 tokens, since the Proposer cannot make another offer without decreasing his/her own share to zero, and the rules require that all shares be positive.

References

Abbink, K., Bolton, G., Sadrieh, K. and Tang, F.-F. (2001) 'Adaptive Learning versus Punishment in Ultimatum Bargaining', *Games and Economic Behavior*, 37, 1–25.

Bolton, G. (1991) 'A Comparative Model of Bargaining: Theory and Evidence', *American Economic Review*, 81, 1096–136.

Bolton, G. and Ockenfels, A. (2000) 'ERC: A Theory of Equity, Reciprocity and Competition', *American Economic Review*, 90, 166–93.

Charness, G. and Rabin, M. (2002) 'Understanding Social Preferences with Simple Tests', *Quarterly Journal of Economics*, 117, 817–69.

Cooper, David J., Feltovich, Nick, Roth, Alvin E. and Zwick, Rami (2003) 'Relative versus Absolute Speed of Adjustment in Strategic Environments: Responder Behavior in Ultimatum Games', *Experimental Economics*, 6, 2, 181–207.

Erev, Ido and Roth, A. E. (1998) 'Predicting How People Play Games: Reinforcement Learning in Experimental Games with Unique, Mixed Strategy Equilibria', *American Economic Review*, 88, 4, September, 848–81.

Fehr, E. and Schmidt, K. (1999) 'A Theory of Fairness, Competition, and Co-operation, *Quarterly Journal of Economics* 114, 817–68.

Gneezy, Uri, Haruvy, Ernan and Roth, Alvin E. (2003) 'Bargaining Under a Deadline: Evidence from the Reverse Ultimatum Game', *Games and Economic Behavior*, Special Issue in Honor of Robert W. Rosenthal, 45, 2, November, 347–68.

Grosskopf, Brit and Roth, Alvin E. (2003) 'If You Are Offered the Right of First Refusal, Should You Accept? An Investigation of Contract Design', Mineo, Harvard University July.

Güth, W. and Yaari, M. (1992) 'Explaining Reciprocal Behavior in Simple Strategic Games: An Evolutionary Approach', in U. Witt (ed.), Ann Arbor, Mich., University of Michigan Press, 23–34.

Güth, W., Schmittberger, R. and Schwarze, B. (1982) 'An Experimental Analysis of Ultimatum Bargaining', *Journal of Economic Behavior and Organization*, 3, 367–88.

Ochs, J. and Roth, A. E. (1989) 'An Experimental Study of Sequential Bargaining', *American Economic Review*, 79, 355–84.

Prasnikar, V. and Roth, A. E. (1992) 'Considerations of Fairness and Strategy: Experimental Data From Sequential Games', *Quarterly Journal of Economics*, August, 865–88.

Roth, A. E. (1996) 'Individual Rationality as a Useful Approximation: Comments on Tversky's "Rational Theory and Constructive Choice", in K. Arrow, E. Colombatto, M. Perlman and C. Schmidt (eds), *The Rational Foundations of Economic Behavior: Proceedings of the IEA Conference held in Turin, Italy*, London, Macmillan, 198–202.

Roth, A. E. and Erev, I. (1995) 'Learning in Extensive-Form Games: Experimental Data and Simple Dynamic Models in the Intermediate Term', *Games and Economic Behavior*, Special Issue: Nobel Symposium, 8, January, 164–212.

Roth, A. E. and Malouf, M. K. (1979) 'Game-Theoretic Models and the Role of Information in Bargaining', *Psychological Review*, 86, 574–94.

Roth, A. E. and Murnighan, J. K. (1982) 'The Role of Information in Bargaining: An Experimental Study', *Econometrica*, 50, 1123–42.

Roth, A. E., Malouf, M. and Murnighan, J. K. (1981) 'Sociological versus Strategic Factors in Bargaining', *Journal of Economic Behavior and Organization*, 2, 153–77.

Roth, A. E., Prasnikar, V., Okuno-Fujiwara, M. and Zamir, S. (1991) 'Bargaining and Market Behavior in Jerusalem, Ljubljana, Pittsburgh, and Tokyo: An Experimental Study', *American Economic Review*, 81, December, 1068–95.

Slonim, R. and Roth, A. E. (1998) 'Learning in High Stakes Ultimatum Games: An Experiment in the Slovak Republic', *Econometrica*, 66, 3, May, 569–96.

Index